OPTIMAL
ESTIMATION

OPTIMAL ESTIMATION

WITH AN INTRODUCTION TO STOCHASTIC CONTROL THEORY

FRANK L. LEWIS

School of Electrical Engineering
Georgia Institute of Technology
Atlanta, Georgia

A WILEY-INTERSCIENCE PUBLICATION

JOHN WILEY & SONS

New York • Chichester • Brisbane • Toronto • Singapore

Library of Congress Cataloging in Publication Data:

Lewis, Frank L.
  Optimal estimation.

  "A Wiley-Interscience publication."
  Bibliography: p.
  Includes index.
  1. Control theory.   2. Mathematical optimization.
3. Stochastic processes.   I. Title.
QA402.3.L4875   1986   629.8′312   85-26554
ISBN 0-471-83741-5

Printed in the United States of America

10 9 8 7 6 5 4 3 2 1

To my parents

to whom one always owes more
than is said

PREFACE

This book is intended for use in a second graduate course in modern control theory. A background in both probability theory and the state variable representation of systems is assumed. One advantage of modern control theory is that it employs matrix algebra, which results in a simplification in notation and mathematical manipulations when dealing with multivariable systems. Appendix A provides a short review of matrices for those who need it.

The book is also intended as a reference. Equations of recurring usefulness are displayed in tabular form for easy access.

Many examples are used to illustrate the concepts and to impart intuition, and the reader is shown how to write simple software implementations of estimators for use on a computer.

Optimal control and optimal estimation are the dual theories that provide the foundation for the modern study of systems. Optimal control can be studied in a purely deterministic context where the unrealistic assumption is made that perfect information about nature is available. The solutions to control problems are discovered by a backward progression through time.

Optimal estimation, on the other hand, is the problem of casting into a useful form information from an inherently noisy and substantially uncooperative environment; it is, therefore, intimately concerned with probabilistic notions. Solutions to estimation problems are discovered by a forward progression through time.

Stochastic control theory uses information reconstructed from noisy measurements to control a system so that it has a desired behavior; hence, it represents a marriage of optimal estimation and deterministic optimal control.

In Part 1, we discuss the estimation problem, covering in detail a system which extracts information from measured data and which is known as the Kalman filter. Kalman filters are used, for example, in communication networks and in the navigation of ships, aircraft, and space vehicles.

Classical estimation theory, including the Wiener filter, is reviewed in Chapter 1, and throughout the book we point out relationships between the Kalman filter and this older body of knowledge.

In Part 2, we introduce stochastic control theory, treating both state variable systems and polynomial systems.

For a complete foundation in modern system theory and more insight on stochastic control, a course in deterministic optimal control is required.

This book is dedicated to my teachers, J. B. Pearson, E. W. Kamen, and C. Y. Chong, whose knowledge and wisdom I have imperfectly transcribed into this written volume wherein the insight and the better ideas are completely theirs. It is also dedicated to N. B. Nichols, whose achievements have been a source of inspiration and friendship a source of strength.

No thanks are enough for Peggy Knight, Pam Majors, and Mary Ann Tripp, who deciphered my reams of illegible handwritten pages and transformed them into the final typed version of the manuscript, complete with all its figures.

This book has developed continuously since it was begun, and I have never reread it without making some changes and corrections. I dread the day when the printed text is before me and I realize that no more changes are possible, although so many things could obviously be presented so much more clearly. Nevertheless, I hope that with all its flaws and omissions it will be a useful addition to the many excellent books already written in this field.

FRANK L. LEWIS

Atlanta, Georgia
February 1986

CONTENTS

PART 1
OPTIMAL ESTIMATION

CHAPTER 1. CLASSICAL ESTIMATION THEORY   3
    1.1. Mean-Square Estimation   3
    1.2. Maximum-Likelihood Estimation   19
    1.3. The Cramér–Rao Bound   24
    1.4. Recursive Estimation   28
    1.5. Wiener Filtering   33
        Problems   48

CHAPTER 2. DISCRETE-TIME KALMAN FILTER   56
    2.1. Deterministic State Observer   56
    2.2. Linear Stochastic Systems   61
    2.3. The Discrete-Time Kalman Filter   67
    2.4. Discrete Measurements of
        Continuous-Time Systems   80
    2.5. Error Dynamics and Statistical Steady
        State   97
    2.6. Frequency Domain Results   107
    2.7. Correlated Noise and Shaping Filters 117
    2.8. Optimal Smoothing   127
        Problems   134

CHAPTER 3.   CONTINUOUS-TIME KALMAN FILTER   144
             3.1.   Derivation from Discrete Kalman Filter   144
             3.2.   Some Examples   151
             3.3.   Derivation from Wiener–Hopf Equation   159
             3.4.   Error Dynamics and Statistical Steady
                    State   170
             3.5.   Frequency Domain Results   173
             3.6.   Correlated Noise and Shaping Filters   180
             3.7.   Discrete Measurements of
                    Continuous-Time Systems   185
             3.8.   Optimal Smoothing   191
                    Problems   196

CHAPTER 4.   KALMAN FILTER DESIGN AND
             IMPLEMENTATION   205
             4.1.   Modeling Errors, Divergence, and
                    Exponential Data Weighting   205
             4.2.   Reduced-Order Filters and Decoupling   227
             4.3.   Using Suboptimal Gains   239
             4.4.   Scalar Measurement Updating   243
                    Problems   244

CHAPTER 5.   ESTIMATION FOR NONLINEAR
             SYSTEMS   249
             5.1.   Update of the Hyperstate   249
             5.2.   General Update of Mean and Covariance   256
             5.3.   Extended Kalman Filter   260
                    Problems   273

PART 2
OPTIMAL STOCHASTIC CONTROL

CHAPTER 6.   STOCHASTIC CONTROL FOR STATE
             VARIABLE SYSTEMS   283
             6.1.   Dynamic Programming Approach   283
             6.2.   Continuous-Time Linear Quadratic
                    Gaussian Problem   305
             6.3.   Discrete-Time Linear Quadratic
                    Gaussian Problem   313
                    Problems   317

CHAPTER 7. STOCHASTIC CONTROL FOR
POLYNOMIAL SYSTEMS 322
    7.1. Polynomial Representation of Stochastic
Systems 322
    7.2. Optimal Prediction 324
    7.3. Minimum Variance Control 328
    7.4. Polynomial Linear Quadratic Gaussian
Regulator 331
    Problems 339

APPENDICES

APPENDIX A. REVIEW OF MATRIX ALGEBRA 345
    A.1. Basic Definitions and Facts 345
    A.2. Partitioned Matrices 346
    A.3. Quadratic Forms and Definiteness 348
    A.4. Matrix Calculus 350

APPENDIX B. COMPUTER SOFTWARE 353
    B.1. Plotting Routines 354
    B.2. Time Response of Systems 354
    B.3. Simulation of Optimal Controllers 357
    B.4. Kalman Filter Simulation 361

REFERENCES 365

INDEX 369

REFERENCE TABLES
AND FIGURES

Table 2.3-1	Discrete-Time Kalman Filter	69
Table 2.3-2	Alternative Measurement Update Equations	70
Table 2.3-3	Kalman Filter A Priori Recursive Formulation	70
Figure 2.7-2	Some Useful Continuous Spectrum-Shaping Filters	120
Figure 2.7-3	Some Useful Discrete Spectrum-Shaping Filters	121
Table 2.8-1	Discrete Optimal Smoothing Scheme	132
Table 3.1-1	Continuous-Time Kalman Filter	147
Table 3.7-1	Continuous-Discrete Kalman Filter	186
Table 3.8-1	Continuous Optimal Smoothing Scheme	194
Table 5.3-1	Continuous-Discrete Extended Kalman Filter	263
Table 5.3-2	Continuous-Continuous Extended Kalman Filter	265

OPTIMAL ESTIMATION

PART 1

OPTIMAL ESTIMATION

1
CLASSICAL ESTIMATION THEORY

There are many ways to estimate an unknown quantity from available data. This chapter basically covers mean-square estimation and the Wiener filter. A discussion of recursive estimation is included, since it provides a good background for the Kalman filter. The techniques presented here can be described as *classical techniques* in that they represent the state of the art prior to 1960, when Kalman and Bucy developed their approach (Kalman 1960, 1963; Kalman and Bucy 1961).

1.1 MEAN-SQUARE ESTIMATION

Suppose there is an unknown vector X and available data Z which is somehow related to X, so that knowing Z we have some information about X.

Mean-square estimation implies a setting in which there is prior information about both the unknown $X \in R^n$ and the measurement $Z \in R^p$, so that both X and Z are stochastic. That is, we know the joint statistics of X and Z.

Given Z, we would like to make an *estimate* $\hat{X}(Z)$ of the value of X so that the *estimation error*

$$\tilde{X} = X - \hat{X} \qquad (1.1\text{-}1)$$

is small. "Small" can be defined in several ways, leading to different methods of estimation.

Define the *mean-square error* as the expected value of the Euclidean

3

norm squared of $\tilde{X}$,

$$J = \overline{\tilde{X}^T\tilde{X}} = \overline{(X - \hat{X})^T(X - \hat{X})}. \qquad (1.1\text{-}2)$$

Since $\tilde{X}$ is a random variable the expected value (overbar) is required as shown. (We use both overbars and $E(\cdot)$ for expected value.)

The *mean-square estimation criterion* can be stated as follows. Using all available information select an estimate $\hat{X}$ to minimize (1.1-2).

It is worth saying a few words about the "available information." In the next subsection is considered the case where there is no information other than the statistics of X. Next, we consider the case where a random variable Z, the measurement, is given which provides additional information on X. There will be no assumption about how Z derives from X, so this latter result will be completely general. Finally, in the third subsection we will make a useful restricting assumption on the form of the estimate. This presentation is modified from the one in Papoulis (1965).

Mean-square estimation is a special case of *Bayesian estimation*. The general Bayesian estimation problem is to find the estimate that minimizes the *Bayes risk*

$$J = E[C(\tilde{X})] = \int_{-\infty}^{\infty} C(x - \hat{X})f_{XZ}(x, z)\, dx\, dz, \qquad (1.1\text{-}3)$$

where $C(\cdot)$ is a *cost function*. Selecting $C(\tilde{X}) = \tilde{X}^T\tilde{X}$ results in (1.1-2). Another special case is *maximum a posteriori estimation* $[C(\tilde{X}) = 0$ if all components of $\tilde{X}$ have magnitude less than a given ε, $C(\tilde{X}) = 1$ otherwise]. It can be shown that as long as the cost is symmetric and convex, and the conditional probability density function (PDF) $f_{X/Z}(x/z)$ is symmetric about its mean, then all Bayes estimates yield the same optimal estimate (Sorenson 1970). As we shall discover, this is the conditional mean of the unknown given the data, $\overline{X/Z}$.

Mean-Square Estimation of a Random Variable X by a Constant

Suppose that there is not yet given a measurement Z, but that we know only the *a priori statistics* of the unknown X. That is, the PDF $f_X(x)$ is known, or equivalently all moments of X are given. The problem is to choose a constant that will yield the best *a priori estimate* $\hat{X}$ for X in the mean-square sense.

To solve this problem work with the mean-square error (1.1-2):

$$J = \overline{(X - \hat{X})^T(X - \hat{X})} = \overline{X^TX} - \hat{X}^T\bar{X} - \bar{X}^T\hat{X} + \hat{X}^T\hat{X}$$
$$= \overline{X^TX} - 2\hat{X}^T\bar{X} + \hat{X}^T\hat{X},$$

where we have used the fact that, since it is a constant, $\hat{X}$ does not depend on X and so can be taken out from under the expectation operator. Now, differentiating this it is seen that

$$\frac{\partial J}{\partial \hat{X}} = -2\bar{X} + 2\hat{X} = 0,$$

or

$$\hat{X}_{MS} = \bar{X}. \tag{1.1-4}$$

We have thus arrived at the possibly obvious and certainly intuitive conclusion that if only the the statistics of X are known then the best constant to estimate X is the *mean* of X.

To obtain a measure of confidence for estimate (1.1-4), let us find the *error covariance*. Notice first that

$$E(\tilde{X}) = E(X - \bar{X}) = \bar{X} - \bar{X} = 0,$$

so that the expected estimation error is equal to zero. This property is of sufficient importance to give it a name. Accordingly, define an estimate $\hat{X}$ to be *unbiased* if

$$E(\tilde{X}) = 0. \tag{1.1-5}$$

This is equivalent to

$$E(\hat{X}) = \bar{X}, \tag{1.1-6}$$

which says that the expected value of the estimate is equal to the mean of the unknown. (In the absence of data, the optimal mean-square estimate *is* the mean of the unknown, as we have just discovered.)

We can now write for the error covariance

$$P_{\tilde{X}} = \overline{(\tilde{X} - \bar{\tilde{X}})(\tilde{X} - \bar{\tilde{X}})^T} = \overline{\tilde{X}\tilde{X}^T}$$

$$= \overline{(X - \hat{X})(X - \hat{X})^T} = \overline{(X - \bar{X})(X - \bar{X})^T},$$

but this is just equal to the covariance of X. The a priori error covariance is therefore just

$$P_{\tilde{X}} = P_X. \tag{1.1-7}$$

If the error covariance is small, then the associated estimate is a good one. We call the inverse of the error covariance, $P_{\tilde{X}}^{-1}$, the *information matrix*.

Mean-Square Estimation of a Random Variable X Given a Random Variable Z: General Case

Now there is prescribed additional information. Suppose that in addition to the statistics of X, the value of another related random variable Z is known (presumably by some measurement process). We shall make no assumptions on how Z derives from X. It is assumed only that the joint distribution $f_{XZ}(x, z)$ of X and Z is given.

The optimal mean-square estimate depends on the (known) measurement, so write $\hat{X}(Z)$. We call this estimate after taking the measurement the *a posteriori estimate* for X. As the value of Z changes so does $\hat{X}$, but if Z has a fixed value of $Z = z$, then $\hat{X}(Z)$ has the fixed value $\hat{X}(z)$. This fact will be required later.

To find $\hat{X}(Z)$ write (1.1-2) as, by definition,

$$J = \int\int_{-\infty}^{\infty} [x - \hat{X}(z)]^T [x - \hat{X}(z)] f_{XZ}(x, z) \, dx \, dz.$$

Or, using Bayes' rule,

$$J = \int_{-\infty}^{\infty} f_Z(z) \int_{-\infty}^{\infty} [x - \hat{X}(z)]^T [x - \hat{X}(z)] f_{X/Z}(x/z) \, dx \, dz.$$

Since $f_Z(z)$ and the inner integral are nonnegative, J can be minimized by fixing $Z = z$ and minimizing for every such z the value of the *conditional mean-square error*

$$J' = \int_{-\infty}^{\infty} [x - \hat{X}(z)]^T [x - \hat{X}(z)] f_{X/Z}(x/z) \, dx. \qquad (1.1\text{-}8)$$

Now repeat the derivation of (1.1-4) using conditional means. Then

$$J' = \overline{[X - \hat{X}(Z)]^T [X - \hat{X}(Z)]/Z}$$
$$= \overline{X^T X/Z} - 2\hat{X}(Z)^T \overline{X/Z} + \hat{X}(Z)^T \hat{X}(Z), \qquad (1.1\text{-}9)$$

where we have used our previous realization that $\hat{X}(Z)$ is fixed if Z is fixed, so that $\hat{X}(Z)$ comes out from under the conditional expectation operator. Differentiate to find that

$$\frac{\partial J'}{\partial \hat{X}(Z)} = -2\overline{X/Z} + 2\hat{X}(Z) = 0,$$

or finally,

$$\hat{X}_{MS}(Z) = \overline{X/Z}. \qquad (1.1\text{-}10)$$

This is the not so obvious but equally intuitive counterpart to (1.1-4). If a measurement $Z = z$ is given, then we should take as the a posteriori estimate for X the conditional mean of X given $Z = z$. If X and Z are independent then $f_{XZ} = f_X f_Z$ so that

$$\overline{X/Z} = \int x f_{X/Z}(x/z) \, dx = \int x \frac{f_{XZ}(x, z)}{f_Z(z)} \, dx = \bar{X}, \qquad (1.1\text{-}11)$$

and the a priori and a posteriori estimates are the same. Then knowledge of Z does not provide any information about X.

Equation (1.1-10) appears very simple, but the simplicity of its appearance belies the difficulty of its use. In fact, the computation of the conditional mean $\overline{X/Z}$ may be an intractable problem. Worse than this, to find $\overline{X/Z}$ we require f_{XZ}, which may not be known. In some cases, we may know only its first and second moments.

Equation (1.1-10) is the mean-square estimate in the most general case. It is often called the *conditional mean estimate of X*, and under the general conditions we have mentioned, all Bayes estimation schemes lead to this same optimal estimate.

It is sometimes convenient to express the conditional mean $\overline{X/Z}$ in terms of the PDF $f_{Z/X}(z/x)$. To do this, note that by Bayes' rule

$$\overline{X/Z} = \int_{-\infty}^{\infty} x f_{X/Z}(x/z) \, dx$$

$$\overline{X/Z} = \int_{-\infty}^{\infty} x \frac{f_{XZ}(x, z)}{f_Z(z)} \, dx$$

$$= \frac{\int_{-\infty}^{\infty} x f_{XZ}(x, z) \, dx}{\int_{-\infty}^{\infty} f_{XZ}(x, z) \, dx}, \tag{1.1-12}$$

$$\overline{X/Z} = \frac{\int_{-\infty}^{\infty} x f_{Z/X}(z/x) f_X(x) \, dx}{\int_{-\infty}^{\infty} f_{Z/X}(z/x) f_X(x) \, dx}, \tag{1.1-13}$$

which is the desired result.

The conditional mean estimate is unbiased since [recall $\bar{X} = E(\overline{X/Z})$]

$$E\{\tilde{X}\} = E\{\overline{\tilde{X}/Z}\} = E\{\overline{(X - \hat{X})/Z}\} = E\{(X - \overline{X/Z})/Z\}$$

$$= E\{\overline{X/Z} - \overline{\overline{X/Z}/Z}\} = E\{\overline{X/Z} - \overline{X/Z}\} = 0.$$

(Note that taking the same expected value twice is the same as taking it once.)

A measure of confidence in an estimate is given by the error covariance. To find the error covariance associated with the conditional mean estimate write, since $\bar{\tilde{X}} = 0$,

$$P_{\tilde{X}} = \overline{\tilde{X}\tilde{X}^T} = \overline{(X - \hat{X})(X - \hat{X})^T}$$

$$= E\{(X - \overline{X/Z})(X - \overline{X/Z})^T/Z\}$$

$$= E\{P_{X/Z}\},$$

or

$$P_{\tilde{X}} = \overline{P_{X/Z}} \tag{1.1-14}$$

Under some conditions $P_{\tilde{X}} = P_{X/Z}$, so that the unconditional error covariance is the same as the conditional error covariance given a particular measurement Z. In general, however, (1.1-14) must be used since $P_{X/Z}$ may depend on Z.

It is worth noting that J as given by (1.1-2) is equal to trace $(P_{\hat{X}})$ so that $\hat{X}_{MS}$ minimizes the trace of the error covariance matrix.

The next example shows how to compute mean-square estimates. We shall use $(\bar{X}, P_X)$ to denote a random variable with mean $\bar{X}$ and covariance P_X. If X is normal we write $N(\bar{X}, P_X)$.

Example 1.1-1:   Nonlinear Mean-Square Estimation (Melsa and Cohn 1978)

Let unknown X be distributed uniformly according to

$$f_X(x) = \begin{cases} 1, & 0 \le x \le 1 \\ 0, & \text{otherwise.} \end{cases} \tag{1}$$

a.   In the absence of other information, choose a constant to estimate X in the optimal mean-square sense (i.e., find the a priori estimate).

According to (1.1-4), in this case the best estimate is

$$\hat{X}_{MS} = \bar{X} = \int_{-\infty}^{\infty} x f_X(x)\, dx$$

$$= \int_0^1 x\, dx = \frac{1}{2}. \tag{2}$$

b.   Now suppose a Z is measured which is related to X by

$$Z = \ln\left(\frac{1}{X}\right) + V, \tag{3}$$

where V is noise with exponential distribution

$$f_V(v) = \begin{cases} e^{-v}, & v \ge 0 \\ 0, & v < 0, \end{cases} \tag{4}$$

and X and V are independent. Find the conditional mean estimate.

If X and V are independent then (function of a random variable, Papoulis 1965)

$$f_{Z/X}(z/x) = f_V\left[z - \ln\left(\frac{1}{x}\right)\right]$$

$$= \begin{cases} e^{-(z - \ln(1/x))}, & z - \ln\frac{1}{x} \ge 0 \\ 0, & z - \ln\frac{1}{x} < 0 \end{cases}$$

$$= \begin{cases} \dfrac{1}{x} e^{-z}, & x \ge e^{-z} \\ 0, & x < e^{-z}. \end{cases} \tag{5}$$

Now using (1.1-13) we get

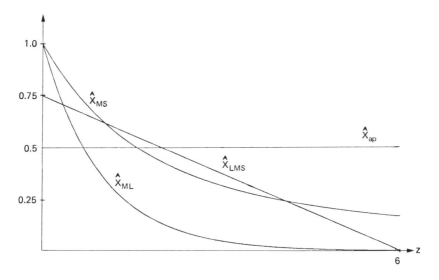

FIGURE 1.1-1   Nonlinear mean-square estimate.

$$\hat{X}_{MS}(z) = \frac{\int_{e^{-z}}^{1} e^{-z}\, dx}{\int_{e^{-z}}^{1} \frac{1}{x}\, e^{-z}\, dx}$$

$$= \frac{xe^{-z}\big|_{e^{-z}}^{1}}{e^{-z}\ln x\big|_{e^{-z}}^{1}}$$

$$= \frac{(1-e^{-z})e^{-z}}{ze^{-z}},$$

or

$$\hat{X}_{MS} = \frac{1-e^{z}}{Z}. \tag{6}$$

This nonlinear estimate is sketched in Fig. 1.1-1, along with the a priori estimate X_{ap} and some other estimates for the same problem which we shall presently discuss. ∎

Example 1.1-2:   *Gaussian Conditional Mean and Covariance*

This is an extremely important example which we shall constantly use in subsequent work.

Let $Y \in {}^{\bullet}R^{n+p}$ be *normal*, symbolized by $N(\bar{Y}, P_Y)$, so that its PDF is

$$f_Y(y) = \frac{1}{\sqrt{(2\pi)^{n+p}|P_Y|}}\, e^{-(1/2)(y-\bar{Y})^T P_Y^{-1}(y-\bar{Y})}. \tag{1.1-15}$$

Now partition Y as $Y = [X^T \quad Z^T]^T$ where $X \in R^n$ and $Z \in R^p$ are themselves both random vectors (RV). Suppose that we perform an experiment where the value z of Z is measured, and that it is then desired to get some information about the unmeasured portion X of Y.

To find an expression for the conditional PDF note first that X and Z are both normal, that is $X \sim N(\bar{X}, P_X)$, $Z \sim N(\bar{Z}, P_Z)$. The cross-covariance is $P_{XZ} = E(X - \bar{X})(Z - \bar{Z})^T$. Then

$$y = \begin{bmatrix} x \\ z \end{bmatrix}, \qquad \bar{Y} = \begin{bmatrix} \bar{X} \\ \bar{Z} \end{bmatrix}, \qquad P_Y = \begin{bmatrix} P_X & P_{XZ} \\ P_{ZX} & P_Z \end{bmatrix}, \tag{1}$$

and so (Appendix A)

$$P_Y^{-1} = \begin{bmatrix} D^{-1} & -D^{-1}P_{XZ}P_Z^{-1} \\ -P_Z^{-1}P_{ZX}D^{-1} & P_Z^{-1} + P_Z^{-1}P_{ZX}D^{-1}P_{XZ}P_Z^{-1} \end{bmatrix} \tag{2}$$

where the Schur complement of P_X is $D = P_X - P_{XZ}P_Z^{-1}P_{ZX}$.

Noting that $[f_Y(y)$ and $f_{XZ}(x, z)$ are the same PDF]

$$f_{X/Z}(x/z) = \frac{f_{XZ}(x, z)}{f_Z(z)} = \frac{f_Y(y)}{f_Z(z)}$$

we can write

$$f_{X/Z}(x/z) = \frac{\sqrt{(2\pi)^p |P_Z|}}{\sqrt{(2\pi)^{n+p} |P_Y|}} \exp\left\{ -\frac{1}{2}[(y - \bar{Y})^T P_Y^{-1}(y - \bar{Y}) - (z - \bar{Z})^T P_Z^{-1}(z - \bar{Z})] \right\}. \tag{3}$$

[To write (1.1-15) as $f_{XZ}(x, z)$ we can substitute from (1).]

By using (1) and (2), write the exponent as

$$\begin{bmatrix} x - \bar{X} \\ z - \bar{Z} \end{bmatrix}^T P_Y^{-1} \begin{bmatrix} x - \bar{X} \\ z - \bar{Z} \end{bmatrix} - (z - \bar{Z})^T P_Z^{-1}(z - \bar{Z}) = (x - \bar{X})^T D^{-1}(x - \bar{X})$$

$$- (x - \bar{X})^T D^{-1} P_{XZ} P_Z^{-1}(z - \bar{Z})$$

$$- (z - \bar{Z})^T P_Z^{-1} P_{ZX} D^{-1}(x - \bar{X})$$

$$+ (z - \bar{Z})^T (P_Z^{-1} + P_Z^{-1} P_{ZX} D^{-1} P_{XZ} P_Z^{-1})(z - \bar{Z})$$

$$- (z - \bar{Z})^T P_Z^{-1}(z - \bar{Z}). \tag{4}$$

Define the *conditional mean* as

$$\overline{X/Z} = \bar{X} + P_{XZ} P_Z^{-1}(Z - \bar{Z}) \tag{1.1-16}$$

and the *conditional covariance* as

$$P_{X/Z} = P_X - P_{XZ} P_Z^{-1} P_{ZX}, \tag{1.1-17}$$

the Schur complement of P_X in P_Y. Now it takes only patience to show that (4) can be written as

$$(x - \overline{X/Z = z})^T P_{X/Z}^{-1}(x - \overline{X/Z = z}). \tag{5}$$

Furthermore, it can be demonstrated (Appendix A) that

$$|P_Y|/|P_Z| = |P_{X/Z}|. \tag{6}$$

We have thus proven that X/Z is a normal RV with PDF

$$f_{X/Z}(x/z) = \frac{1}{\sqrt{(2\pi)^n |P_{X/Z}|}} \exp\left[-\frac{1}{2}(x - \overline{X/Z = z})^T P_{X/Z}^{-1}(x - \overline{X/Z = z}) \right]. \quad (1.1\text{-}18)$$

The conditional mean estimate in the Gaussian case is therefore given by (1.1-16). Note that in the Gaussian case, $P_{X/Z}$ is independent of Z so that the error covariance

$$P_{\tilde{X}} = P_{X/Z} \quad (1.1\text{-}19)$$

is given by (1.1-17).

From (1.1-17) it is clear that $P_{X/Z} \leq P_X$. If in a particular case $P_{X/Z} < P_X$, then after measuring Z we are more certain of the value of X than we were before (i.e., the a posteriori RV X/Z is distributed more closely about its mean $\overline{X/Z}$ than is the a priori RV X about its mean $\bar{X}$). ∎

Example 1.1-3: *Conditional Mean for Linear Gaussian Measurements*

Here we find the Gaussian conditional mean estimate and error covariance in the case where X and Z have a particular relation.

Suppose the unknown X and the data Z are related by the *linear Gaussian measurement model*

$$Z = HX + V, \quad (1.1\text{-}20)$$

where $X \sim N(\bar{X}, P_X)$, $V \sim N(0, R)$ is zero mean *measurement noise*, and measurement matrix H is known and deterministic (i.e., constant). Suppose $R > 0$ and X and V are orthogonal so that $\overline{XV^T} = 0$. Let $X \in R^n$, $V \in R^p$, so that H is a $p \times n$ matrix and $Z \in R^p$.

The measurement Z is then a normal RV, and it is not difficult to determine its mean $\bar{Z}$ and variance P_Z in terms of the known quantities $\bar{X}$, P_X, R, and H. Thus

$$\bar{Z} = \overline{HX + V} = H\bar{X} + \bar{V}$$

by linearity, or

$$\bar{Z} = H\bar{X}. \quad (1.1\text{-}21)$$

Furthermore

$$P_Z = \overline{(Z - \bar{Z})(Z - \bar{Z})^T} = \overline{(HX + V - H\bar{X})(HX + V - H\bar{X})^T}$$
$$= \overline{H(X - \bar{X})(X - \bar{X})^T H^T} + \overline{H(X - \bar{X})V^T} + \overline{V(X - \bar{X})^T H^T} + \overline{VV^T}$$
$$= HP_X H^T + \overline{HXV^T} - \overline{H\bar{X}\bar{V}^T} + \overline{VX^T H^T} - \overline{V\bar{X}^T H^T} + R,$$

or

$$P_Z = HP_X H^T + R. \quad (1.1\text{-}22)$$

Therefore Z is $N(H\bar{X}, HP_X H^T + R)$.

To find the cross-covariance of X and Z write

$$P_{XZ} = \overline{(X - \bar{X})(Z - \bar{Z})^T} = \overline{(X - \bar{X})(HX + V - H\bar{X})^T}$$
$$= \overline{(X - \bar{X})(X - \bar{X})^T H^T} + \overline{XV^T} - \overline{\bar{X}\bar{V}^T}, \quad (1.1\text{-}23)$$

or

$$P_{ZX} = P_{XZ}^T = HP_X, \tag{1.1-24}$$

since P_X is symmetric.

The error covariance is obtained by using the above expressions for P_Z, P_{XZ}, and P_{ZX} in (1.1-17), which yields

$$P_{\tilde{X}} = P_{X/Z} = P_X - P_{XZ}P_Z^{-1}P_{ZX}$$

$$= P_X - P_X H^T (HP_X H^T + R)^{-1} HP_X. \tag{1.1-25}$$

By using the matrix inversion lemma (Appendix A) this can be written as

$$P_{\tilde{X}} = P_{X/Z} = (P_X^{-1} + H^T R^{-1} H)^{-1} \tag{1.1-26}$$

if $|P_X| \neq 0$. Notice that $P_{X/Z} \leq P_X$. If strict inequality holds, measuring RV Z decreases our uncertainty about the unknown RV X.

The conditional mean estimate is found by substituting for P_{XZ}, P_Z, and $\bar{Z}$ in (1.1-16)

$$\hat{X} = \overline{X/Z} = \bar{X} + P_{XZ}P_Z^{-1}(Z - \bar{Z})$$

$$= \bar{X} + P_X H^T (HP_X H^T + R)^{-1}(Z - H\bar{X}). \tag{1.1-27}$$

To find a neater expression for $\overline{X/Z}$, note that

$$P_X H^T (HP_X H^T + R)^{-1} = P_X H^T (HP_X H^T + R)^{-1}[(HP_X H^T + R) - HP_X H^T]R^{-1}$$

$$= P_X H^T [I - (HP_X H^T + R)^{-1} HP_X H^T]R^{-1}$$

$$= [P_X - P_X H^T (HP_X H^T + R)^{-1} HP_X]H^T R^{-1}$$

$$= P_{X/Z} H^T R^{-1}.$$

Hence

$$\hat{X} = \overline{X/Z} = \bar{X} + P_{X/Z} H^T R^{-1}(Z - H\bar{X}). \tag{1.1-28}$$

The PDF $f_{X/Z}(x/z)$ is normal with conditional mean $\overline{X/Z}$ and conditional covariance $P_{X/Z}$.

It is important to note that $\overline{X/Z}$ depends on the value of Z, but that covariance $P_{X/Z}$ is independent of the value of Z. Thus, $P_{X/Z}$ can be computed *before* we take any measurements.

It is not difficult to redo this example for the case of nonorthogonal X and V. See the problems. ∎

Exercise 1.1-4: Kalman Gain

In the case of linear Gaussian measurements, define the *Kalman gain* by

$$K = P_X H^T (HP_X H^T + R)^{-1}. \tag{1.1-29}$$

a. Express $P_{X/Z}$ and $\overline{X/Z}$ in terms of K.
b. Prove that

$$P_{X/Z} = (I - KH)P_X(I - KH)^T + KRK^T = (I - KH)P_X \tag{1.1-30}$$

∎

The Orthogonality Principle

The following *orthogonality principle* is basic to probabilistic estimation theory. Let RVs X and Z be jointly distributed. Then for any function $g(\cdot)$,

$$\overline{g(Z)(X - \overline{X/Z})^T} = 0. \qquad (1.1\text{-}31)$$

That is, any function of Z is orthogonal to X once the conditional mean has been subtracted out. To show this, write

$$\overline{g(Z)(X - \overline{X/Z})^T} = \overline{g(Z)X^T} - \overline{g(Z)\overline{X^T/Z}}$$
$$= \overline{g(Z)X^T} - \overline{\overline{g(Z)X^T/Z}} = \overline{g(Z)X^T} - \overline{g(Z)X^T} = 0.$$

We used the fact that

$$g(Z) \cdot \overline{X^T/Z} = \overline{g(Z)X^T/Z} \qquad (1.1\text{-}32)$$

for any function $g(\cdot)$, since $g(Z)$ is deterministic if Z is fixed.

The idea behind this can be illustrated as in Fig. 1.1-2, where orthogonal RVs are represented as being at right angles. All functions of Z are represented as being in the direction of Z. (In the figure, interpret $\hat{X}_{\text{LMS}}$ as $\overline{X/Z}$.) The orthogonality principle says that the RV

$$\tilde{X} = X - \overline{X/Z}, \qquad (1.1\text{-}33)$$

which is the estimation error, is orthogonal to all other RVs $g(Z)$. It is emphasized that this figure is extremely heuristic, since if nonlinear functions $g(\cdot)$ are allowed, then RVs of the form $g(Z)$ do not form a subspace. The figure should only be considered as a mnemonic device.

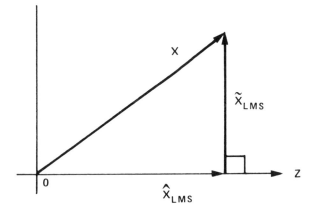

FIGURE 1.1-2   The orthogonality principle.

We can quickly show that the orthogonality principle has an immediate application to estimation theory. In fact, if $g(\cdot)$ is any function, then

$$E\|X - \overline{X/Z}\| \le E\|X - g(Z)\|, \qquad (1.1\text{-}34)$$

where "$\|\cdot\|$" denotes the Euclidean norm. Thus no other function of Z provides a "closer approximation to X" in a probabilistic sense than does $\overline{X/Z}$.

This is a partial statement of the *stochastic projection theorem*. To show it, a probabilistic formulation of Pythagoras' theorem is used: if Y and Z are orthogonal, then

$$E\|Y + Z\|^2 = E\|Y\|^2 + E\|Z\|^2. \qquad (1.1\text{-}35)$$

Now use (1.1-31) to write for any $g(\cdot)$

$$E\|X - g(Z)\|^2 = E\|X - \overline{X/Z} + \overline{X/Z} - g(Z)\|^2$$

$$= E\|X - \overline{X/Z}\|^2 + E\|\overline{X/Z} - g(Z)\|^2 \ge E\|X - \overline{X/Z}\|^2$$

with equality if and only if $g(Z) = \overline{X/Z}$.

Linear Mean-Square Estimation of a Random Variable X Given a Random Variable Z

No assumptions are made here on the relation between Z and X; the results to be presented are valid for any possible nonlinear relation between measurement and unknown. In the realization that (1.1-10) is in general too much trouble to compute, except for the Gaussian case, let us restrict the admissible estimates $X(Z)$ to those that depend linearly on Z, that is,

$$\hat{X}(Z) = AZ + b, \qquad (1.1\text{-}36)$$

for some constant matrix $A \in R^{n \times p}$, and vector $b \in R^n$. This *linear estimate restriction* should not be confused with the linear measurement model (1.1-20).

An additional motive for restricting the admissible estimates to those of the form (1.1-36) is that this form allows processing of the measurements Z by a linear system to obtain $\hat{X}$.

Since the class of admissible estimates is restricted, the *linear mean-square estimate* will not, in general, be as good in the sense of minimizing (1.1-2) as the general mean-square estimate (1.1-10).

The constants A and b depend on the measurement process and on the a priori information available before the measurement is made. They must be chosen to minimize the mean-square error (1.1-2). Recall that $\text{tr}(MN) = \text{tr}(NM)$ for compatible matrices M and N and write

$$J = \overline{(X - \hat{X})^T(X - \hat{X})} = \text{tr}\,\overline{(X - \hat{X})(X - \hat{X})^T}$$

$$= \text{tr}\,\overline{(X - AZ - b)(X - AZ - b)^T}$$

$$= \text{tr}\,\overline{[(X - \bar{X}) - (AZ + b - \bar{X})][(X - \bar{X}) - (AZ + b - \bar{X})]^T}.$$

After some straightforward work we arrive at

$$J = \mathrm{tr}[P_X + A(P_Z + \bar{Z}\bar{Z}^T)A^T + (b - \bar{X})(b - \bar{X})^T$$
$$+ 2A\bar{Z}(b - \bar{X})^T - 2AP_{ZX}]. \tag{1.1-37}$$

From Appendix A, $(d/dA)\,\mathrm{tr}(ABA^T) = 2AB$ and $(d/dA)\,\mathrm{tr}(BAC) = B^T C^T$; so for a minimum

$$\frac{\partial J}{\partial b} = 2(b - \bar{X}) + 2A\bar{Z} = 0 \tag{1.1-38a}$$

and

$$\frac{\partial J}{\partial A} = 2A(P_Z + \bar{Z}\bar{Z}^T) - 2P_{XZ} + 2(b - \bar{X})\bar{Z}^T = 0. \tag{1.1-38b}$$

(Recall $P_{ZX}^T = P_{XZ}$.) From (1.1-38a),

$$b = \bar{X} - A\bar{Z},$$

and substituting this into (1.1-38b)

$$AP_Z - P_{XZ} = 0$$

or

$$A = P_{XZ}P_Z^{-1}.$$

Using this A and b in (1.1-36) the optimal linear mean-square estimate is determined as

$$\hat{X}_{\mathrm{LMS}} = \bar{X} + P_{XZ}P_Z^{-1}(Z - \bar{Z}). \tag{1.1-39}$$

This equation is typical of the form of many estimators. We can consider $\bar{Z}$ as an estimate for the measurement Z. We also define the *residual* as

$$\tilde{Z} = Z - \bar{Z}. \tag{1.1-40}$$

The mean $\bar{X}$ is an a priori estimate for X, and the second term is a correction based on the residual. If we happen to measure exactly what we expect, so that $Z = \bar{Z}$, then the a priori estimate $\bar{X}$ was correct and it does not need to be modified; in that case $\hat{X} = \bar{X}$. The weighting $P_{XZ}P_Z^{-1}$ by which the measurement error $(Z - \bar{Z})$ is incorporated into the estimate $\hat{X}$ is dependent on the second-order joint statistics. Thus, if P_{XZ} is smaller then X and Z depend on each other to a lesser degree and measurement of Z yields less information about X. In fact, if X and Z are independent, then $P_{XZ} = 0$ and Z yields no information about X. In this case $\hat{X} = \bar{X}$ for any value of Z, as given by (1.1-4). Or again, if P_Z is very large, then we have little confidence in Z and so $(Z - \bar{Z})$ is weighted less heavily into $\hat{X}$. If we have great confidence in Z then P_Z is small, and the residual has a greater role in determining $\hat{X}$.

Compare (1.1-39) to the general mean-square estimate $\hat{X}_{\mathrm{MS}}$ given by

(1.1-10). To find $\hat{X}_{MS}$ all the joint statistics of X and Z are required since the PDF $f_{Z/X}$ (or equivalently f_{XZ} and f_Z) is needed. To find the best linear estimate $\hat{X}_{LMS}$, on the other hand, all that is required is the *first-order statistics* (i.e., means) and *second-order statistics* (i.e., covariances).

To find the error covariance associated with $\hat{X}_{LMS}$, first verify that $\hat{X}_{LMS}$ is unbiased by writing

$$\overline{\hat{X}_{LMS}} = \bar{X} + P_{XZ}P_Z^{-1}\overline{(Z - \bar{Z})}$$
$$= \bar{X} + P_{XZ}P_Z^{-1}(\bar{Z} - \bar{Z})$$
$$= \bar{X}.$$

Therefore, $\bar{\tilde{X}} = 0$ and

$$P_{\tilde{X}} = \overline{(\tilde{X} - \bar{\tilde{X}})(\tilde{X} - \bar{\tilde{X}})} = \overline{\tilde{X}\tilde{X}^T}$$
$$= \overline{(X - \hat{X})(X - \hat{X})^T}$$
$$= \overline{[(X - \bar{X}) - P_{XZ}P_Z^{-1}(Z - \bar{Z})][(X - \bar{X}) - P_{XZ}P_Z^{-1}(Z - \bar{Z})]^T}$$
$$= P_X - P_{XZ}P_Z^{-1}P_{ZX} - P_{XZ}P_Z^{-1}P_{ZX} + P_{XZ}P_Z^{-1}P_ZP_Z^{-1}P_{ZX},$$

or

$$P_{\tilde{X}} = P_X - P_{XZ}P_Z^{-1}P_{ZX}. \tag{1.1-41}$$

In this expression, P_X represents the a priori error covariance and the second term represents the reduction in uncertainty because of the measurement. We see again that if X and Z are independent then $P_{XZ} = 0$ and $P_{\tilde{X}} = P_X$, so measuring Z provides no increase in our knowledge of X.

A very important point will now be made. Equations (1.1-39) and (1.1-41) are identical on the right-hand sides to (1.1-16) and (1.1-17). This means that:

> In the Gaussian case, the *optimal* mean-square estimate is *linear*, since $\hat{X}_{MS} = \overline{X/Z}$ and $\hat{X}_{LMS}$ are the same.   (1.1-42)

The similarity of equations (1.1-39) and (1.1-41) to (1.1-16) and (1.1-17) implies another very important result which we shall need later: the expressions derived in Example 1.1-3 are the optimal *linear* mean-square estimate and error covariance for the case of linear measurements even if the measurement noise is not Gaussian.

Exercise 1.1-5:   The Orthogonality Principle with Linear Estimates

If we restrict our attention to linear estimates then we have a particularly nice geometric formulation of the orthogonality principle, for we can talk in terms of vector spaces.

a.   The linear mean-square estimate does not satisfy $\overline{g(Z)(X - \hat{X}_{LMS})^T} = 0$ for all functions $g(\cdot)$. Show that it does satisfy the orthogonality condition

$$\overline{Z(X - \hat{X}_{LMS})^T} = 0. \tag{1.1-43}$$

Thus $(X - \hat{X}_{\text{LMS}})$ is orthogonal to all *linear* functions of Z of the form $g(Z) = AZ$ for constant A.

Since $\hat{X}_{\text{LMS}}$ is unbiased, $(X - \hat{X}_{\text{LMS}})$ is orthogonal to all linear functions $g(Z) = AZ + b$ for constants A and b. Show this.

b. Show that the set of all linear functions $g(Z) = AZ$ is a vector space if the inner product of two elements is defined as $\overline{Z_1^T Z_2} = \text{trace } \overline{Z_2 Z_1^T}$.

c. In terms of these constructions we can now use Fig. 1.1-2, where the axis labeled "Z" represents the subspace of linear functions of Z. This should be contrasted to the situation in the previous subsection where the orthogonality principle was first discussed. There, we did not restrict $g(\cdot)$ to be linear, so we could not speak in terms of subspaces.

d. Show that if $g(\cdot)$ is any linear function, then

$$E\|X - \hat{X}_{\text{LMS}}\| \le E\|X - g(Z)\|, \tag{1.1-44}$$

with equality only if $g(Z) = \hat{X}_{\text{LMS}}$. Thus $\hat{X}_{\text{LMS}}$ provides a "closer approximation" in a probabilistic sense to the unknown X than any other *linear* function $g(Z)$. Does (1.1-44) hold in general for nonlinear $g(\cdot)$?

e. The orthogonality principle can be used to derive $\hat{X}_{\text{LMS}}$. Suppose we want to derive the optimal estimate of the form $\hat{X} = AZ + b$.

First, assume that A is given. Then the best value for b is the optimal estimate for the RV $(X - AZ)$ by a constant. Write it down.

Now to find A show that, assuming $\hat{X}$ is unbiased, the optimal A satisfies

$$\overline{[(X - \bar{X}) - A(Z - \bar{Z})](Z - \bar{Z})^T} = 0. \tag{1}$$

Hence find A. Is $\hat{X}$ unbiased? ■

Example 1.1-6: *Linear Mean-Square Estimation*

Find the optimal linear mean-square estimate for the experiment of Example 1.1-1.

To apply (1.1-39) we need to compute first and second moments of X and Z. Some of the intermediate steps will be left out.

$$\bar{X} = \int_{-\infty}^{\infty} x f_X(x)\, dx = \frac{1}{2},$$

$$f_{XZ}(x, z) = f_{Z/X}(z/x) f_X(x)$$

$$= \begin{cases} \dfrac{1}{x} e^{-z}; & e^{-z} \le x \le 1; \quad 0 \le z \\ 0, & \text{otherwise,} \end{cases}$$

$$f_Z(z) = \int_{-\infty}^{\infty} f_{XZ}(x, z)\, dx$$

$$= \begin{cases} z e^{-z}; & z \ge 0 \\ 0; & z < 0 \end{cases},$$

$$\bar{Z} = \int_{-\infty}^{\infty} z f_Z(z) = 2,$$

$$\overline{XZ} = \int_{-\infty}^{\infty} \int_{-\infty}^{\infty} xz f_{XZ}(x, z)\, dx\, dz$$

$$= \int_{0}^{\infty} \int_{e^{-z}}^{1} ze^{-z}\, dz = \frac{3}{4},$$

$$P_{XZ} = \overline{XZ} - \bar{X}\bar{Z} = \tfrac{3}{4} - 1 = -\tfrac{1}{4},$$

$$\overline{Z^2} = \int_{-\infty}^{\infty} z^2 f_Z(z)\, dz = 6,$$

$$P_Z = \overline{Z^2} - \bar{Z}^2 = 6 - 4 = 2,$$

and finally, using (1.1-39),

$$\hat{X}_{\text{LMS}} = \bar{X} + P_{XZ} P_Z^{-1}(Z - \bar{Z})$$

$$= \tfrac{1}{2} - \tfrac{1}{4}(\tfrac{1}{2})(Z - 2)$$

$$= \tfrac{3}{4} - \tfrac{1}{8} Z.$$

Note that $\hat{X}_{\text{LMS}}$ is unbiased since $\bar{\hat{X}}_{\text{LMS}} = \tfrac{3}{4} - \tfrac{1}{8}\bar{Z} = \tfrac{1}{2}$.

To compare this to $\hat{X}_{\text{MS}}$ found previously examine Fig. 1.1-1. Note that $\hat{X}_{\text{LMS}}$ cannot generally be obtained simply by linearizing $\hat{X}_{\text{MS}}$. ■

This example has not presented linear mean-square estimation under the most flattering of circumstances. The computation of the moments is tedious, and finding $\hat{X}_{\text{LMS}}$ is more work in this case than is finding $\hat{X}_{\text{MS}}$. However, it is not always possible to find the conditional mean, but given first and second moments, $\hat{X}_{\text{LMS}}$ can always be found. Besides this, the means and variances can often be determined from the measurement process without knowing the joint PDF f_{XZ} itself. Linear mean-square estimation comes into its own as we proceed in our development.

Example 1.1-7: Quadratic Mean-Square Estimation

It is desired to find the optimal estimate $\hat{X}$ which depends on Z quadratically, that is,

$$\hat{X}(Z) = CZ^2 + AZ + b, \tag{1}$$

where for simplicity it is supposed that X and Z are scalars. The mean-square criterion becomes

$$J = \overline{(X - \hat{X})^T(X - \hat{X})} = \overline{(X - CZ^2 - AZ - b)^2},$$

and to find the optimal C, A, b write

$$\frac{\partial J}{\partial b} = -2\overline{(X - CZ^2 - AZ - b)} = 0$$

$$\frac{\partial J}{\partial A} = -2\overline{Z(X - CZ^2 - AZ - b)} = 0$$

$$\frac{\partial J}{\partial C} = -2\overline{Z^2(X - CZ^2 - AZ - b)} = 0,$$

which can be collected into the "normal equations,"

$$\begin{bmatrix} 1 & \bar{Z} & \overline{Z^2} \\ \bar{Z} & \overline{Z^2} & \overline{Z^3} \\ \overline{Z^2} & \overline{Z^3} & \overline{Z^4} \end{bmatrix} \begin{bmatrix} b \\ A \\ C \end{bmatrix} = \begin{bmatrix} \bar{X} \\ \overline{XZ} \\ \overline{XZ^2} \end{bmatrix}. \qquad (2)$$

Several things are interesting about these equations. First, clearly if $C = 0$ they provide us with an alternative derivation of (1.1-39). Second, we have seen that the optimal linear mean-square estimate depends on the joint statistics of X and Z through order 2. It is now apparent that the quadratic mean-square estimate depends on joint statistics through order 4. ∎

1.2 MAXIMUM-LIKELIHOOD ESTIMATION

Nonlinear Maximum-Likelihood Estimation

The connotation of "maximum likelihood" is a setting in which nothing is known a priori about the unknown quantity, but there is prior information on the measurement process itself. Thus, X is deterministic and Z is stochastic. The conditional PDF of Z given the unknown X, $f_{Z/X}(z/X)$, contains information about X, and if it can be computed then X may be estimated according to the *maximum-likelihood estimation criterion*, which can be stated as follows. Given a measurement Z, the maximum-likelihood estimate $\hat{X}_{\mathrm{ML}}$ is the value of X which maximizes $f_{Z/X}$, the likelihood that X resulted in the observed Z.

The PDF $f_{Z/X}$ is a *likelihood function*. Since $\ln(\cdot)$ is a monotonically increasing function, the estimate $\hat{X}$ could equally well be chosen to maximize $\ln(f_{Z/X})$, which is often more convenient. Thus $\ln(f_{Z/X})$ is also a likelihood function.

If the first derivatives are continuous and $f_{Z/X}(z/X)$ has its maximum interior to its domain as a function of X, then the maximum-likelihood criterion can be expressed as

$$\left. \frac{\partial f_{Z/X}(z/X)}{\partial X} \right|_{X = \hat{X}} = 0 \qquad (1.2\text{-}1a)$$

or

$$\left. \frac{\partial \ln f_{Z/X}(z/X)}{\partial X} \right|_{X = \hat{X}} = 0. \qquad (1.2\text{-}1b)$$

This is called the *likelihood equation*.

Note that since X is deterministic it makes no sense here to talk about $f_{X/Z}$.

Example 1.2-1: Nonlinear Maximum-Likelihood Estimation

Let measurement Z and unknown X be related as in Example 1.1-1b, where X is now deterministic (i.e., no a priori statistics are given on X).

The likelihood function is given by (5) in Example 1.1-1 (where lowercase x's should be replaced by uppercase since X is deterministic). To maximize the likelihood function, select X as small as possible: so the maximum-likelihood estimate is

$$\hat{X}_{ML} = e^{-Z}.$$

See Melsa and Cohn (1978).

Note that the estimate is given by $\hat{X}_{ML} = e^{-Z}$, while the *value* of the estimate given the data $Z = z$ is $\hat{X}_{ML} = e^{-z}$. See Fig. 1.1-1. ∎

Linear Gaussian Measurements

For our applications, assume now that Z depends linearly on X according to

$$Z = HX + V. \tag{1.2-2}$$

Assume the measurement noise is normal so that

$$f_V(v) = \frac{1}{\sqrt{(2\pi)^p |R|}} e^{-(1/2)v^T R^{-1} v}. \tag{1.2-3}$$

Then the likelihood function is given by a shifted version of $f_V(v)$,

$$f_{Z/X}(z/X) = f_V(z - HX)$$
$$= \frac{1}{\sqrt{(2\pi)^p |R|}} e^{-(1/2)(z-HX)^T R^{-1}(z-HX)}. \tag{1.2-4}$$

To maximize $f_{Z/X}$, we can equivalently maximize $\ln(f_{Z/X})$, or minimize

$$J = \tfrac{1}{2}(z - HX)^T R^{-1}(z - HX). \tag{1.2-5}$$

Differentiating,

$$\frac{\partial J}{\partial X} = H^T R^{-1}(z - HX) = 0$$

so that the optimal estimate is given by

$$\hat{X}_{ML} = (H^T R^{-1} H)^{-1} H^T R^{-1} Z. \tag{1.2-6}$$

Evidently, to obtain the maximum-likelihood estimate the measurement matrix H must have full rank.

Note that the mean-square performance index (1.1-2) depends on the *estimation error*, while (1.2-5) depends on the *residual*, since $\hat{Z} = H\hat{X}$ is an estimate for Z so that the residual is

$$\tilde{Z} = Z - H\hat{X}. \tag{1.2-7}$$

To find the error covariance, first note that the estimation error is

$$\tilde{X} = X - \hat{X} = X - (H^T R^{-1} H)^{-1} H^T R^{-1} (HX + V)$$
$$= -(H^T R^{-1} H)^{-1} H^T R^{-1} V. \qquad (1.2\text{-}8)$$

(Of course we cannot actually compute $\tilde{X}$ by this equation since noise V is unknown!) Now, taking expected values

$$\bar{\tilde{X}} = -(H^T R^{-1} H)^{-1} H^T R^{-1} \bar{V} = 0, \qquad (1.2\text{-}9)$$

so that the maximum-likelihood estimate for the linear Gaussian case is unbiased.

Now the error covariance becomes

$$P_{\tilde{X}} = \overline{(\tilde{X} - \bar{\tilde{X}})(\tilde{X} - \bar{\tilde{X}})^T} = \overline{\tilde{X}\tilde{X}^T}$$
$$= (H^T R^{-1} H)^{-1} H^T R^{-1} \overline{VV^T} R^{-1} H (H^T R^{-1} H)^{-1},$$

or

$$P_{\tilde{X}} = (H^T R^{-1} H)^{-1}. \qquad (1.2\text{-}10)$$

In terms of $P_{\tilde{X}}$ the optimal estimate can be expressed as

$$\hat{X}_{\text{ML}} = P_{\tilde{X}} H^T R^{-1} Z. \qquad (1.2\text{-}11)$$

Note that if $H = I$ then the error covariance $P_{\tilde{X}}$ is equal to R, the noise covariance, which makes sense intuitively.

An *estimator* will be taken to mean an estimate plus its error covariance. The estimator described by (1.2-10) and (1.2-11) is called the *Gauss–Markov* estimator.

The absence of a priori information on X can be expressed as

$$\bar{X} = 0, \qquad P_X \to \infty, \qquad (1.2\text{-}12)$$

which models complete ignorance. Using these values in (1.1-26), (1.1-28) we obtain (1.2-10), (1.2-11). The maximum-likelihood estimate is, therefore, in the linear Gaussian case, a limiting case of the mean-square estimate.

The error covariance $P_{\tilde{X}}$ is independent of the measurement Z, so that once we have designed an experiment described by $Z = HX + V$, $P_{\tilde{X}}$ can be computed off-line to see if the accuracy of the experiment is acceptable *before* we build the equipment and take any measurements. This is an important result.

Example 1.2-2: Experiment Design

To estimate a deterministic voltage X there are given two alternatives for designing an experiment. Two expensive meters can be used, each of which add $N(0, 2)$ Gaussian noise to X, or four inexpensive meters can be used, each of which add $N(0, 3.5)$ Gaussian noise to X. Which design would result in the more reliable estimate for X?

Design number one is described by

$$Z = \begin{bmatrix} 1 \\ 1 \end{bmatrix} X + V,$$

where

$$V \sim N\left(0, \begin{bmatrix} 2 & 0 \\ 0 & 2 \end{bmatrix}\right).$$

This gives an error covariance of

$$P_{\tilde{X}} = (H^T R^{-1} H)^{-1} = \left([1 \quad 1] \begin{bmatrix} \frac{1}{2} & 0 \\ 0 & \frac{1}{2} \end{bmatrix} \begin{bmatrix} 1 \\ 1 \end{bmatrix}\right)^{-1} = 1.$$

Design number two is described by

$$Z = \begin{bmatrix} 1 \\ 1 \\ 1 \\ 1 \end{bmatrix} X + V, \qquad V \sim N(0, 3.5I),$$

where I is the identity matrix, so

$$P_{\tilde{X}} = (H^T R^{-1} H)^{-1} = \tfrac{7}{8}.$$

Design two using four inexpensive meters is therefore a more reliable scheme since it has a lower error covariance. ■

The noise covariance matrix R must be determined before maximum-likelihood methods can be used in the linear Gaussian case. In the problems the effects of inaccurately known noise covariance are explored. The accurate determination of the noise parameters is a crucial part of the estimation process.

The next examples provide further insight into maximum-likelihood estimation.

Example 1.2-3:   Independent Measurements of a Scalar Unknown

If the measurements are independent then R is diagonal. Suppose an experiment is described by three scalar measurements (one vector measurement)

$$Z = \begin{bmatrix} 1 \\ 1 \\ 1 \end{bmatrix} X + V, \qquad V \sim N\left(0, \begin{bmatrix} \sigma_1^2 & & 0 \\ & \sigma_2^2 & \\ 0 & & \sigma_3^2 \end{bmatrix}\right),$$

with X deterministic. Then the error covariance is given by

$$P_{\tilde{X}} = (H^T R^{-1} H)^{-1} = \left(\frac{1}{\sigma_1^2} + \frac{1}{\sigma_2^2} + \frac{1}{\sigma_3^2}\right)^{-1}, \qquad (1.2\text{-}13)$$

and the estimate is

$$\hat{X}_{\mathrm{ML}} = P_{\bar{x}} H^T R^{-1} Z$$

$$= \frac{(Z_1/\sigma_1^2 + Z_2/\sigma_2^2 + Z_3/\sigma_3^2)}{(1/\sigma_1^2 + 1/\sigma_2^2 + 1/\sigma_3^2)} \tag{1.2-14}$$

where $Z = [Z_1 \quad Z_2 \quad Z_3]^T$ is the measurement vector, with Z_i being scalar random variables. This is a very natural and appealing result, as can be seen by noticing a few points.

First, suppose that $\sigma_3^2 \to \infty$, indicating that we have no confidence in Z_3. Then $1/\sigma_3^2 \to 0$ and the effects of Z_3 simply drop out, leaving a two measurement experiment. Now, suppose that $\sigma_3^2 \to 0$, indicating that we have great confidence in Z_3. Then $P_{\bar{x}} \to \sigma_3^2$ and $\hat{X}_{\mathrm{ML}} \to Z_3$, so that the maximum-likelihood procedure would say to disregard Z_1, Z_2 in this situation. What we have here, then, is an estimator that weights Z_i into $\hat{X}_{\mathrm{ML}}$, depending on our measure of confidence σ_i^2 in Z_i.

The generalization to an arbitrary number of measurements Z_i is immediate. If all σ_i are equal to σ, then (1.2-14) becomes

$$\hat{X}_{\mathrm{ML}} = \frac{Z_1 + Z_2 + Z_3}{3},$$

the *sample mean*, and (1.2-13) becomes

$$P_{\bar{x}} = \frac{\sigma^2}{3}. \qquad \blacksquare$$

Example 1.2-4:   An Electric Circuit Analog

Examine Fig. 1.2-1. Let voltages V_i be applied between one end of resistances R_i and ground. The other ends of the resistances are connected to a common point at potential V. Find the voltage V if $R_{\mathrm{in}} = \infty$.

This problem can be solved by using Kirchhoff's current law $I_1 + I_2 + I_3 = I$,

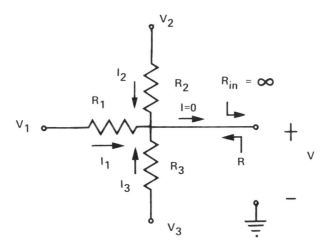

FIGURE 1.2-1   An electric circuit analog to maximum-likelihood estimation.

where $R_{in} = \infty$ means that $I = 0$. Now apply Ohm's law to each I_i to obtain

$$I_1 + I_2 + I_3 = \frac{V_1 - V}{R_1} + \frac{V_2 - V}{R_2} + \frac{V_3 - V}{R_3} = 0,$$

or

$$\frac{V_1}{R_1} + \frac{V_2}{R_2} + \frac{V_3}{R_3} = V\left(\frac{1}{R_1} + \frac{1}{R_2} + \frac{1}{R_3}\right).$$

The solution is

$$V = \left(\frac{1}{R_1} + \frac{1}{R_2} + \frac{1}{R_3}\right)^{-1}\left(\frac{V_1}{R_1} + \frac{V_2}{R_2} + \frac{V_3}{R_3}\right), \tag{1}$$

which corresponds exactly to (1.2-14)!

The "effective resistance seen by V" is

$$R = \left(\frac{1}{R_1} + \frac{1}{R_2} + \frac{1}{R_3}\right)^{-1}. \tag{2}$$

This corresponds to the error covariance $P_{\tilde{X}}$ in (1.2-13). In electric circuit theory R is written as the *parallel combination* of the R_i, that is, $R = R_1 // R_2 // R_3$, which is just a shorthand notation for "the reciprocal of the sums of the reciprocals." With this notation (1.2-13) can be written as

$$P_{\tilde{X}} = \sigma_1^2 // \sigma_2^2 // \sigma_3^2 \tag{1.2-15}$$

so that the error covariance is just the parallel combination of the individual noise variances. As such, it is smaller than any individual σ_i^2, so our confidence in $\hat{X}$ after the experiment is greater than our confidence in any one measurement.

Equation (1) is called *Millman's theorem*, and it is an exact analog to maximum-likelihood estimation of a scalar unknown with independent measurements. ∎

1.3   THE CRAMÉR–RAO BOUND

To examine the performance of the maximum-likelihood estimate we can use the following lower bound on the variance of any unbiased estimate.

If $\hat{X}$ is any unbiased estimate of a *deterministic* variable X based on measurement Z, then the covariance of the estimation error $\tilde{X} = X - \hat{X}$ is bounded by

$$P_{\tilde{X}} \geq J_F^{-1}, \tag{1.3-1}$$

where the *Fisher information matrix* is given by

$$J_F = E\left\{\left[\frac{\partial \ln f_{Z/X}(z/X)}{\partial X}\right]\left[\frac{\partial \ln f_{Z/X}(z/X)}{\partial X}\right]^T\right\} \tag{1.3-2}$$

$$= -E\left[\frac{\partial^2 \ln f_{Z/X}(z/X)}{\partial X^2}\right]. \tag{1.3-3}$$

Equality holds in (1.3-1) if and only if

$$\frac{\partial \ln f_{Z/X}(z/X)}{\partial X} = k(X)(X - \hat{X}). \tag{1.3-4}$$

It is assumed that

$$\frac{\partial f_{Z/X}}{\partial X} \quad \text{and} \quad \frac{\partial^2 f_{Z/X}}{\partial X^2}$$

exist and are absolutely integrable.

Equation (1.3-4) is a limitation on the form of the derivative: it must be proportional to $\hat{X}$ with constant of proportionality at most a function of the unknown X. An estimate $\hat{X}$ is *efficient* if it satisfies the *Cramér–Rao bound* (1.3-1) with equality, that is, if (1.3-4) holds.

The proof of the Cramér–Rao bound can be found in Van Trees (1968) or Sorenson (1970a, 1970b).

In general $\hat{X}_{\text{ML}}$ is not necessarily efficient. However, it is not hard to show that if (1.3-4) is ever satisfied so that there exists efficient estimate, then this estimate is given by $\hat{X}_{\text{ML}}$. The Cramér–Rao bound thus demonstrates the importance of the maximum-likelihood estimate.

To gain some insight into the Fisher information matrix, consider the linear Gaussian measurement model. From (1.2-4) there results

$$\frac{\partial \ln f_{Z/X}(z/X)}{\partial X} = H^T R^{-1}(z - HX) \tag{1.3-5}$$

and differentiating again we find

$$J_F = H^T R^{-1} H = P_{\bar{X}}^{-1}, \tag{1.3-6}$$

the inverse of the error covariance matrix.

According to (1.2-13), we have for Example 1.2-3 that $J_F = P_{\bar{X}}^{-1} = 1/\sigma_1^2 + 1/\sigma_2^2 + 1/\sigma_3^2$, which can be phrased in the intuitively appealing statement that in the scalar case of independent measurements, the information J_F at the end of an experiment is the *sum* of the information provided by each measurement. An electrical analog is provided by the conductance $1/R$ which adds according to Example 1.2-4 equation (2) when resistances are put in parallel.

The maximum-likelihood estimate has several nice properties as the number of independent measurements N goes to infinity. To discuss them we need the concept of *stochastic convergence*. The sequence $Y_1, Y_2, \ldots$ of RVs *converges in probability* to RV Y if for all $\varepsilon > 0$

$$P(\|Y_N - Y\| > \varepsilon) \to 0 \quad \text{for} \quad N \to \infty. \tag{1.3-7}$$

Under fairly general conditions the maximum-likelihood estimate $\hat{X}_{\text{ML}}$ has the following limiting properties.

1. $\hat{X}_{\text{ML}}$ converges in probability to the correct value of X as $N \to \infty$. (Any estimate with this property is said to be *consistent*.)

2. $\hat{X}_{ML}$ becomes efficient as $N \to \infty$.

3. As $N \to \infty$ $\hat{X}_{ML}$ becomes Gaussian $N(X, P_{\tilde{X}})$.

For a *stochastic* unknown X, a result similar to the Cramér–Rao bound can be stated as follows.

If $\hat{X}$ is any estimate of a stochastic variable X based on measurement Z, then the covariance of the estimation error $\tilde{X} = X - \hat{X}$ is bounded by

$$P_{\tilde{X}} \geq L^{-1}, \tag{1.3-8}$$

where the *information matrix* L is given by

$$L = E\left\{ \left[\frac{\partial \ln f_{XZ}(x, z)}{\partial x} \right] \left[\frac{\partial \ln f_{XZ}(x, z)}{\partial x} \right]^{T} \right\} \tag{1.3-9}$$

$$= -E\left[\frac{\partial^2 \ln f_{XZ}(x, z)}{\partial x^2} \right]. \tag{1.3-10}$$

Equality holds in (1.3-8) if and only if

$$\frac{\partial \ln f_{XZ}(x, z)}{\partial x} = k(x - \hat{X}), \tag{1.3-11}$$

where k is a constant. It is assumed that

$$\frac{\partial f_{XZ}}{\partial x} \quad \text{and} \quad \frac{\partial^2 f_{XZ}}{\partial x^2}$$

exist and are absolutely integrable with respect to both variables, and that

$$\lim_{x \to \infty} B(x) f_X(x) = 0 \tag{1.3-12a}$$

and

$$\lim_{x \to -\infty} B(x) f_X(x) = 0 \tag{1.3-12b}$$

where

$$B(x) \triangleq \int_{-\infty}^{\infty} (x - \hat{X}) f_{Z/X}(z/x) \, dz. \tag{1.3-13}$$

An estimate $\hat{X}$ for which the bound is satisfied with equality [i.e., condition (1.3-11) on the form of f_{XZ} holds] is said to be *efficient*.

The proof is similar to that of the Cramér–Rao bound, but expectations are now taken with respect to Z *and* X (Van Trees 1968, Sorenson 1970a, 1970b).

Several points should be noted.

1. This bound depends on the joint PDF f_{XZ} while the Cramér–Rao bound (for deterministic unknown X) depends on the likelihood function $f_{Z/X}$.
2. The Cramér–Rao bound deals with unbiased estimates $\hat{X}$; this bound substitutes for unbiasedness the requirements (1.3-12).
3. The efficiency condition (1.3-11) in terms of the *constant* k should be contrasted with the efficiency condition of the Cramér–Rao bound in terms of the function $k(X)$ of X.
4. Differentiating (1.3-11) yields the equivalent condition for efficiency

$$\frac{\partial^2 \ln f_{XZ}(x, z)}{\partial x^2} = k, \tag{1.3-14}$$

or [since $f_Z(z)$ is independent of X]

$$\frac{\partial^2 \ln f_{X/Z}(x/z)}{\partial x^2} = k. \tag{1.3-15}$$

Integrating twice and taking the antilog yields

$$f_{X/Z}(x/z) = e^{-(k/2)x^T x + k\hat{X}^T x + c}; \tag{1.3-16}$$

so that the a posteriori PDF must be Gaussian for an efficient estimate to exist.

It is interesting to relate information matrix L to the Fisher information matrix J_F. Since $f_{XZ} = f_{Z/X}f_X$ it follows that

$$L = J_F - E\left[\frac{\partial^2 \ln f_X(x)}{\partial x^2}\right], \tag{1.3-17}$$

so that the stochastic description of X enters independently of the noise statistics.

We can gain a little more insight by considering the linear Gaussian measurement model. Thus let measurement Z and unknown X be related by (1.2-2) with $V = N(0, R)$. Then by (1.3-6) we know that $J_F = H^T R^{-1} H$. To find L, note that

$$f_X(x) = \frac{1}{\sqrt{(2\pi)^n |P_X|}} e^{-(1/2)(x-\bar{X})^T P_X^{-1}(x-\bar{X})} \tag{1.3-18}$$

so that

$$\frac{\partial \ln f_X(x)}{\partial x} = -P_X^{-1}(x - \bar{X}). \tag{1.3-19}$$

Then by (1.3-17)

$$L = H^T R^{-1} H + P_X^{-1}. \tag{1.3-20}$$

The total information available has been *increased* by a priori information on X. See equation (1.1-26).

1.4   RECURSIVE ESTIMATION

Sequential Processing of Measurements

Up to this point we have considered *batch processing* of the measurements in which all information is incorporated in one single step into the estimate. This leads to the necessity for inverting $p \times p$ matrices, where p is the number of measurements. The next examples demonstrate a more sensible approach when p is large.

Example 1.4-1:   Recursive Computation of Sample Mean and Variance

Given k measurements Z_i the sample mean is

$$\hat{X}_k = \frac{1}{k} \sum_{i=1}^{k} Z_i. \tag{1}$$

Suppose that $\hat{X}_k$ has been computed based on measurements Z_i for $i = 1, \ldots, k$. Now one more measurement Z_{k+1} is made. The new sample mean could be computed as

$$\hat{X}_{k+1} = \frac{1}{k+1} \sum_{i=1}^{k+1} Z_i, \tag{2}$$

but this clearly involves duplication of some of the work we have already done.

To avoid this, $\hat{X}_{k+1}$ can be computed in terms of $\hat{X}_k$ and Z_{k+1} by proceeding as follows.

$$\hat{X}_{k+1} = \frac{k}{k+1} \left(\frac{1}{k} \sum_{i=1}^{k} Z_i \right) + \frac{1}{k+1} Z_{k+1}$$

$$= \frac{k}{k+1} \hat{X}_k + \frac{1}{k+1} Z_{k+1}.$$

To make this look like some of our previous results, write it as

$$\hat{X}_{k+1} = \hat{X}_k + \frac{1}{k+1} (Z_{k+1} - \hat{X}_k). \tag{1.4-1}$$

This recursive equation can be represented as a time-varying linear system as in Fig. 1.4-1. The quantity $\tilde{Z}_{k+1} = Z_{k+1} - \hat{X}_k$ can be interpreted as a residual. If the Z_i are measurements of a constant unknown X, then $\hat{X}_k$ is the estimate of X based on the first k measurements and (1.4-1) is the *estimate update equation* which adds in the effects of the next measurement Z_{k+1}. Compare this equation to (1.1-28). Conceptually, the two are identical; they both express the current estimate as a linear combination of the previous estimate and the residual. Note that as k increases, the measurements Z_{k+1} have less and less effect as more weighting is given to $\hat{X}_k$.

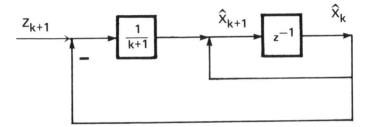

FIGURE 1.4-1 Linear system for recursive generation of sample mean.

A recursive method can also be found to compute the sample variance. To wit, after k measurement we have

$$\sigma_k^2 = \frac{1}{k} \sum_{i=1}^{k} (Z_i - \hat{X}_k)^2. \tag{3}$$

Introducing one more measurement Z_{k+1} there results

$$\sigma_{k+1}^2 = \frac{1}{k+1} \sum_{i=1}^{k+1} (Z_i - \hat{X}_{k+1})^2$$

$$= \frac{1}{k+1} \sum_{i=1}^{k+1} [(Z_i - \hat{X}_k) + (\hat{X}_k - \hat{X}_{k+1})]^2.$$

After some manipulation we achieve

$$\sigma_{k+1}^2 = \frac{k}{k+1} \sigma_k^2 + \frac{1}{k+1} (Z_{k+1} - \hat{X}_k)^2 - (\hat{X}_{k+1} - \hat{X}_k)^2. \tag{4}$$

Now, using (1.4-1) write

$$(\hat{X}_{k+1} - \hat{X}_k)^2 = \frac{1}{(k+1)^2} (Z_{k+1} - \hat{X}_k)^2$$

so that

$$\sigma_{k+1}^2 = \frac{k}{k+1} \sigma_k^2 + \frac{k}{(k+1)^2} (Z_{k+1} - \hat{X}_k)^2,$$

$$\sigma_{k+1}^2 = \sigma_k^2 + \frac{1}{k+1} \left[\frac{k}{k+1} (Z_{k+1} - \hat{X}_k)^2 - \sigma_k^2 \right]. \tag{1.4-2}$$

■

Schemes such as the above in which the current estimate depends on the previous estimate and the current measurement are called *recursive* or *sequential estimation* schemes. Such schemes rely on *sequential processing* of data, where the measurements are processed in stages.

Example 1.4-2

Suppose there are four measurements of a scalar unknown X given by $Z_1 = 10$, $Z_2 = 12$, $Z_3 = 11.5$, $Z_4 = 11$. Find the sample mean and variance recursively.

Using (1.4-1) several times with $\hat{X}_0 = 0$,

$$\hat{X}_1 = \hat{X}_0 + (Z_1 - \hat{X}_0) = Z_1 = 10$$
$$\hat{X}_2 = 10 + \tfrac{1}{2}(12 - 10) = 11$$
$$\hat{X}_3 = 11 + \tfrac{1}{3}(11.5 - 11) = 11\tfrac{1}{6}$$
$$\hat{X}_4 = 11\tfrac{1}{6} + \tfrac{1}{4}(11 - 11\tfrac{1}{6}) = 11\tfrac{1}{8}.$$

Thus, after such each measurement Z_k, $\hat{X}_k$ provides an up to date estimate of X. These estimates are up to date in the maximum-likelihood sense, since there is no a priori information about X.

To find σ_k^2 recursively, let $\sigma_0^2 = 0$. Then, applying (1.4-2)

$$\sigma_1^2 = 0$$
$$\sigma_2^2 = \tfrac{1}{4}(Z_2 - \hat{X}_1)^2 = 1$$
$$\sigma_3^2 = 1 + \tfrac{1}{3}[\tfrac{2}{3}(11.5 - 11)^2 - 1] = 0.722$$
$$\sigma_4^2 = 0.722 + \tfrac{1}{4}[\tfrac{3}{4}(11 - 11\tfrac{1}{6})^2 - 0.722] = 0.547$$

The sample variance σ_k^2 yields a measure of confidence in $\hat{X}_k$ after each measurement Z_k. Sequence σ_k^2 is decreasing, indicating that each measurement increases the reliability of $\hat{X}_k$. ∎

Sequential Maximum-Likelihood Estimation

Let us generalize our results by deriving a sequential maximum-likelihood estimator. Let there be given a linear Gaussian measurement Z of an unknown X

$$Z = HX + V, \tag{1.4-3}$$

where $X \in R^n$, $Z \in R^p$, $V \in R^p$, $V \sim N(0, R)$, $|R| \neq 0$. Recall that "maximum likelihood" implies ignorance of any statistics of X.

Write (1.4-3) explicitly in terms of the components of vectors as

$$\begin{bmatrix} z_1 \\ z_2 \\ \vdots \\ z_p \end{bmatrix} = \begin{bmatrix} h_1^T \\ h_2^T \\ \vdots \\ h_p^T \end{bmatrix} X + \begin{bmatrix} v_1 \\ v_2 \\ \vdots \\ v_p \end{bmatrix},$$

where h_i^T is the ith row of H, and suppose that $R = \text{diag}\{\sigma_i^2\}$. This assumption says that the measurements are independent and is sufficient to guarantee that a recursive estimation procedure exists. (Note that in this subsection lowercase letters are used to define components of random vectors which are themselves random variables.) Define

$$
Z_k = \begin{bmatrix} z_1 \\ z_2 \\ \cdot \\ \cdot \\ \cdot \\ z_k \end{bmatrix}, \qquad H_k = \begin{bmatrix} h_1^T \\ h_2^T \\ \cdot \\ \cdot \\ \cdot \\ h_k^T \end{bmatrix}, \qquad V_k = \begin{bmatrix} v_1 \\ v_2 \\ \cdot \\ \cdot \\ \cdot \\ v_k \end{bmatrix},
$$

as the first k components of Z, H, and V, respectively. Define $R_k = \mathrm{diag}\{\sigma_1, \sigma_2, \ldots, \sigma_k\}$.

Considering the first k components of Z we have

$$
Z_k = H_k X + V_k, \qquad V_k \sim N(0, R_k).
$$

Based on this information the error covariance P_k is (1.2-10), or

$$
P_k = (H_k^T R_k^{-1} H_k)^{-1} \tag{1.4-4}
$$

and the estimate $\hat{X}_k$ is (1.2-11), or

$$
\hat{X}_k = P_k H_k^T R_k^{-1} Z_k. \tag{1.4-5}
$$

Now it is necessary to include the effects of measurement component z_{k+1} to update $\hat{X}_k$ to $\hat{X}_{k+1}$. To do this, write

$$
Z_{k+1} = H_{k+1} X + V_{k+1}, \qquad V_{k+1} \sim N(0, R_{k+1}),
$$

or

$$
Z_{k+1} = \begin{bmatrix} H_k \\ h_{k+1}^T \end{bmatrix} X + V_{k+1}, \qquad V_{k+1} \sim N\left(0, \begin{bmatrix} R_k & 0 \\ 0 & \sigma_{k+1}^2 \end{bmatrix}\right).
$$

The error covariance P_{k+1} based on the first $k+1$ measurement components is

$$
\begin{aligned}
P_{k+1} &= (H_{k+1}^T R_{k+1}^{-1} H_{k+1})^{-1} \\
&= \left(\begin{bmatrix} H_k \\ h_{k+1}^T \end{bmatrix}^T \begin{bmatrix} R_k & 0 \\ 0 & \sigma_{k+1}^2 \end{bmatrix}^{-1} \begin{bmatrix} H_k \\ h_{k+1}^T \end{bmatrix}\right)^{-1} \\
&= \left(H_k^T R_k^{-1} H_k + \frac{h_{k+1} h_{k+1}^T}{\sigma_{k+1}^2}\right)^{-1},
\end{aligned}
$$

or

$$
P_{k+1} = \left(P_k^{-1} + \frac{h_{k+1} h_{k+1}^T}{\sigma_{k+1}^2}\right)^{-1}. \tag{1.4-6}
$$

This is the *error covariance update* equation. To find an alternative form for (1.4-6) use the matrix-inversion lemma to write

$$
P_{k+1} = P_k - P_k h_{k+1}(h_{k+1}^T P_k h_{k+1} + \sigma_{k+1}^2)^{-1} h_{k+1}^T P_k. \tag{1.4-7}
$$

To find the estimate update equation, we have

$$\hat{X}_{k+1} = P_{k+1} H_{k+1}^T R_{k+1}^{-1} Z_{k+1}$$

$$= P_{k+1} \begin{bmatrix} H_k \\ h_{k+1}^T \end{bmatrix}^T \begin{bmatrix} R_k & 0 \\ 0 & \sigma_{k+1}^2 \end{bmatrix}^{-1} \begin{bmatrix} Z_k \\ z_{k+1} \end{bmatrix}$$

$$= P_{k+1} \left(H_k^T R_k^{-1} Z_k + \frac{h_{k+1}}{\sigma_{k+1}^2} z_{k+1} \right).$$

By (1.4-5) this becomes

$$\hat{X}_{k+1} = P_{k+1} \left(P_k^{-1} \hat{X}_k + \frac{h_{k+1}}{\sigma_{k+1}^2} z_{k+1} \right). \tag{1.4-8}$$

This is the *estimate update*. To find a more familiar form for this update, note that

$$P_{k+1} P_k^{-1} = P_{k+1} \left(P_{k+1}^{-1} - \frac{h_{k+1} h_{k+1}^T}{\sigma_{k+1}^2} \right)$$

$$= I - P_{k+1} \frac{h_{k+1} h_{k+1}^T}{\sigma_{k+1}^2},$$

hence (1.4-8) is equivalent to

$$\hat{X}_{k+1} = \hat{X}_k + P_{k+1} \frac{h_{k+1}}{\sigma_{k+1}^2} (z_{k+1} - h_{k+1}^T \hat{X}_k). \tag{1.4-9}$$

Note that $\tilde{z}_{k+1} = z_{k+1} - h_{k+1}^T \hat{X}_k$ is the residual.

The recursions (1.4-6), (1.4-9) should be started using $P_0^{-1} = 0$, $\hat{X}_0 = 0$, which models complete ignorance of the a priori statistics of X.

To summarize our results then, if the measurements are independent, the maximum-likelihood (batch) estimator (1.2-10), (1.2-11) is equivalent to the sequential maximum-likelihood estimator given by (1.4-7), (1.4-9), which requires no matrix inversions but only scalar divisions. The sequential

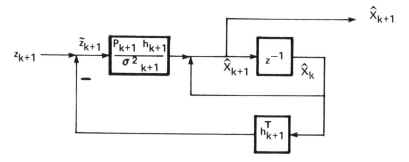

FIGURE 1.4-2   Linear system for sequential maximum-likelihood estimation.

estimator can be implemented as in Fig. 1.4-2. Note that due to the unity feedback it has the structure of a summer or integrator. It is a time-varying system where the σ^2_{k+1}, h_{k+1} are given directly by the structure of the measurement process and the P_{k+1} are computed off-line (i.e., before we begin to take measurements) from knowledge of the measurement process. Compare with Fig. 1.4-1.

If a priori statistics $\bar{X}$ and P_X on X are available, then we should use $\hat{X}_0 = \bar{X}$, $P_0 = P_X$ to initialize (1.4-7) and (1.4-9), which then amount to a recursive mean-square estimator.

Prewhitening of Data

Now suppose that the assumptions required to derive (1.4-6) and (1.4-9) all hold except that R is nonsingular but arbitrary. For sequential processing of Z one component at a time we require a diagonal R, that is, the measurement components must be independent. The data can be *prefiltered* to make R diagonal as follows. Let S be any square root of R^{-1} so that $R^{-1} = S^T S$. Now premultiply (1.4-3) by S to obtain

$$SZ = SHX + SV,$$

and define new quantities by

$$Z^\omega = SZ$$
$$H^\omega = SH$$
$$V^\omega = SV \qquad\qquad (1.4\text{-}10)$$

so that

$$Z^\omega = H^\omega X + V^\omega. \qquad\qquad (1.4\text{-}11)$$

The covariance of the prefiltered noise V^ω is

$$R^\omega = \overline{SV(SV)^T} = S\overline{VV^T}S^T$$
$$= SRS^T$$
$$= S(S^TS)^{-1}S^T = I,$$

so that the components of V^ω are independent and V^ω is a white noise vector. Hence the superscripts. To implement (1.4-10) we define a new measurement matrix H^ω and process all our measurements Z through a linear *prewhitening filter S*. We can now proceed as above to derive (1.4-6), (1.4-9) using Z^ω, H^ω, and $R^\omega = I$ instead of Z, H, and R.

1.5 WIENER FILTERING

Wiener developed the ideas we are about to discuss in the early 1940s for application to antiaircraft fire control systems. His monograph in 1949 brought stochastic notions into control and communication theory, and was the basis for a great deal of subsequent work. Wiener dealt with the continuous-time case; Kolmogorov (1939) applied mean-square theory to discrete stochastic processes. See Van Trees (1968) and the excellent survey by Kailath (1974).

Recall (Papoulis 1965) that if $x(t)$ is a stationary stochastic process with autocorrelation function

$$R_X(\tau) = \overline{x(t+\tau)x^T(t)}, \tag{1.5-1}$$

then its spectral density $\Phi_X(\omega)$ is

$$\Phi_X(s) = \int_{-\infty}^{\infty} R_X(t)e^{-st}\,dt \tag{1.5-2}$$

evaluated at $s = j\omega$. In the discrete case we have

$$R_X(k) = \overline{x_{i+k}x_i^T} \tag{1.5-3}$$

and the spectral density is

$$\Phi_X(z) = \sum_{k=-\infty}^{\infty} R_X(k)z^{-k} \tag{1.5-4}$$

evaluated at $z = e^{j\omega}$.

If $u(t)$ and $y(t)$ are the input and output of a linear system with impulse response $h(t)$ and transfer function $H(s)$, then the autocorrelation functions and spectral densities are related by

$$R_{YU}(t) = h(t) * R_U(t) \tag{1.5-5}$$

$$R_Y(t) = h(t) * R_U(t) * h^T(-t) \tag{1.5-6}$$

$$\Phi_{YU}(s) = H(s)\Phi_U(s) \tag{1.5-7}$$

$$\Phi_Y(s) = H(s)\Phi_U(s)H^T(-s), \tag{1.5-8}$$

where $*$ represents convolution. In the discrete counterparts, $-s$ is replaced by z^{-1}.

We shall have occasion to use the *spectral factorization* theorem, which can be stated as follows (Åstrom 1970, Anderson and Moore 1979).

Let $|\Phi_Y(s)| \neq 0$ for almost every s (such a process is said to be of *full rank*). Then $\Phi_Y(s)$ can be factored uniquely as

$$\Phi_Y(s) = H(s)\Omega H^T(-s) \tag{1.5-9}$$

where $H(s)$ is a square, real, rational transfer function with all poles in the open left-half plane; $\lim_{s\to\infty} H(s) = I$; $H^{-1}(s)$ has all poles in the left-half plane or on the $j\omega$ axis; and $\Omega = \Omega^T > 0$ is a real matrix.

If $|\Phi_Y(s)| \neq 0$ for all s satisfying $\mathrm{Re}(s) = 0$, then $H^{-1}(s)$ has all poles in the open left-half plane.

For a square multivariable transfer function $H(s)$ as described above, the *zeros* of $H(s)$ are the poles of $H^{-1}(s)$. A transfer function with all poles *and* zeros stable is said to be of *minimum phase*.

A similar result holds for discrete systems, with $-s$ replaced by z^{-1} and using the discrete version of stability (i.e., with respect to the unit circle).

According to (1.5-8) and the spectral factorization theorem, the process $y(t)$ can be regarded as the output of the stable system $H(s)$ when the input is white noise with spectral density Ω. This is the *representation theorem*.

Example 1.5-1: ***Discrete Spectral Factorization***

The rational (in cos ω) spectral density

$$\Phi(\omega) = \frac{1.25 + \cos\omega}{1.0625 + 0.5\cos\omega} = \frac{(e^{j\omega} + 0.5)(e^{-j\omega} + 0.5)}{(e^{j\omega} + 0.25)(e^{-j\omega} + 0.25)}$$

can be factored in four ways:

$$H_1(z) = \frac{z + 0.5}{z + 0.25}$$

$$H_2(z) = \frac{z + 0.5}{1 + 0.25z}$$

$$H_3(z) = \frac{1 + 0.5z}{z + 0.25}$$

$$H_4(z) = \frac{1 + 0.5z}{1 + 0.25z}.$$

Only $H_1(z)$ has no poles and zeros outside the unit circle. Thus if the stable system $H_1(z)$ has a white noise input with unit spectral density, the output will have spectral density given by $\Phi(\omega)$. ∎

Example 1.5-2: ***Continuous Spectral Factorization***

The rational function of ω

$$\Phi(\omega) = \frac{\omega^2 + 4}{\omega^2 + 1} = \frac{(j\omega + 2)(-j\omega + 2)}{(j\omega + 1)(-j\omega + 1)}$$

can be factored in four ways, one of which has

$$H(s) = \frac{s + 2}{s + 1}$$

which is minimum phase. The system with this transfer function will have output with spectral density $\Phi(\omega)$ when driven by continuous-time white noise with unit spectral density. ∎

The Linear Estimation Problem

Suppose that $x(t) \in R^n$ is an unknown stochastic process. A related process $z(t) \in R^p$ is measured, and based on this data it is desired to reconstruct

$x(t)$. We want to use a linear filter, in general time varying, for this purpose; so suppose that the estimate for $x(t)$ is

$$\hat{x}(t) = \int_{t_0}^{T} h(t, \tau) z(\tau) \, d\tau. \qquad (1.5\text{-}10)$$

The interval $[t_0, T]$ is the time interval over which the data are available. It is desired to find the optimal filter impulse response $h(t, \tau)$, which is called the time-varying *Wiener filter*.

If $T < t$, the determination of $h(t, \tau)$ is called the linear *prediction* problem; if $T = t$ it is the linear *filtering* problem; and if $T > t$ it is called linear *smoothing* or interpolation. The general problem is known as *linear estimation*.

To find the optimal linear filter $h(t, \tau)$ which reconstructs unknown $x(t)$ given the data $z(\tau)$ for $t_0 \leq \tau \leq T$, minimize the mean-square error at time t

$$J(t) = E[\tilde{x}(t)^T \tilde{x}(t)] = \text{trace } P_{\tilde{x}}(t), \qquad (1.5\text{-}11)$$

where the estimation error is $\tilde{x}(t) = x(t) - \hat{x}(t)$. The error covariance is

$$P_{\tilde{x}}(t) = \overline{[x(t) - \hat{x}(t)][x(t) - \hat{x}(t)]^T}$$

$$E\left\{ \left[x(t) - \int_{t_0}^{T} h(t, \tau) z(\tau) \, d\tau \right] \left[x(t) - \int_{t_0}^{T} h(t, \tau) z(\tau) \, d\tau \right]^T \right\}. \qquad (1.5\text{-}12)$$

According to the orthogonality principle, $J(t)$ is minimized when $\tilde{x}(t)$ is orthogonal to the data over the interval (t_0, T), that is,

$$\overline{\tilde{x}(t) z(u)^T} = E\left\{ \left[x(t) - \int_{t_0}^{T} h(t, \tau) z(\tau) \, d\tau \right] z(u)^T \right\}$$

$$= 0; \qquad t_0 < u < T. \qquad (1.5\text{-}13)$$

This results in the nonstationary *Wiener–Hopf* equation

$$\overline{x(t) z(u)^T} = \int_{t_0}^{T} h(t, \tau) \overline{z(\tau) z(u)^T} \, d\tau; \qquad t_0 < u < T, \quad (1.5\text{-}14)$$

In terms of correlation functions, this can be written as

$$R_{XZ}(t, u) = \int_{t_0}^{T} h(t, \tau) R_Z(\tau, u) \, d\tau; \qquad t_0 < u < T. \qquad (1.5\text{-}15)$$

It is important to note that the optimal *linear* filter is found by knowing only the first- and second-order joint statistics of $x(t)$ and $z(t)$.

If the statistics are Gaussian, then (1.5-15) provides the *optimal* filter for reconstructing $x(t)$ from $z(t)$. For arbitrary statistics, it provides the optimal *linear* filter. (In the Gaussian case, the optimal filter is linear.)

With impulse response given by the Wiener–Hopf equation, the error covariance is equal to [use (1.5-12) and the orthogonality of $\tilde{x}(t)$ and $z(\tau)$]

$$P_{\tilde{X}}(t) = E\left\{\left[x(t) - \int_{t_0}^{T} h(t,\tau)z(\tau)\,d\tau\right]x(t)^T\right\}$$

or

$$P_{\tilde{X}}(t) = R_X(t,t) - \int_{t_0}^{T} h(t,\tau)R_{ZX}(\tau,t)\,d\tau. \qquad (1.5\text{-}16)$$

If $x(t)$ and $z(t)$ are stationary and $t_0 = -\infty$ so that there is an infinite amount of data, then the optimal filter turns out to be time invariant, $h(t,\tau) = h(t-\tau)$. In this *steady-state* case (1.5-15) becomes

$$R_{XZ}(t-u) = \int_{-\infty}^{T} h(t-\tau)R_Z(\tau-u)\,d\tau; \qquad -\infty < u < T,$$

or by changing variables

$$R_{XZ}(\tau) = \int_{t-T}^{\infty} h(u)R_Z(\tau-u)\,du; \qquad t-T < \tau < \infty. \qquad (1.5\text{-}17)$$

This stationary version is the equation Wiener actually worked with, and it is the one he and Hopf provided an elegant frequency domain solution for in 1931 in another context. Wiener was quite delighted in 1942 to discover that the linear estimation problem depended on the solution of this equation he had studied 11 years earlier!

In the steady-state case the error covariance is a constant given by

$$P_{\tilde{X}} = R_X(0) - \int_{t-T}^{\infty} h(\tau)R_{ZX}(-\tau)\,d\tau$$

$$= R_X(0) - \int_{t-T}^{\infty} h(\tau)R_{XZ}^{T}(\tau)\,d\tau. \qquad (1.5\text{-}18)$$

Solution of the Wiener–Hopf Equation

Infinite-Delay Steady-State Smoothing

Let us begin by solving (1.5-17) with $T = \infty$. This means that we must wait until all the data $z(t)$, $-\infty \le t \le \infty$, are received before we begin processing, and so it amounts to non-real-time data processing. It corresponds to the infinite-delay smoothing problem.

In this case the Wiener–Hopf equation (1.5-17) becomes

$$R_{XZ}(t) = \int_{-\infty}^{\infty} h(\tau)R_Z(t-\tau)\,d\tau, \qquad -\infty < t < \infty. \qquad (1.5\text{-}19)$$

The solution is trivial, for according to Fourier transform theory we can write this convolution in terms of spectral densities as

$$\Phi_{XZ}(\omega) = H(j\omega)\Phi_Z(\omega). \qquad (1.5\text{-}20)$$

The optimal linear filter is therefore given by

$$H(j\omega) = \Phi_{XZ}(\omega)\Phi_Z^{-1}(\omega). \qquad (1.5\text{-}21)$$

We can quickly find an expression for the error covariance (1.5-18) in terms of spectral densities, for if we define an auxiliary function by

$$f(t) = R_X(t) - \int_{-\infty}^{\infty} h(\tau) R_{ZX}(t - \tau) \, d\tau,$$

then $P_{\tilde{x}} = f(0)$. But, using the convolution property of Fourier transforms

$$F(\omega) = \Phi_X(\omega) - H(j\omega)\Phi_{ZX}(\omega). \qquad (1.5\text{-}22)$$

Substituting for the optimal $H(j\omega)$ and taking the inverse transform there results

$$P_{\tilde{x}} = \frac{1}{2\pi} \int_{-\infty}^{\infty} [\Phi_X(\omega) - \Phi_{XZ}(\omega)\Phi_Z^{-1}(\omega)\Phi_{ZX}(\omega)] \, d\omega. \qquad (1.5\text{-}23)$$

Compare this to the time-domain expression (1.1-41).

Example 1.5-3: *Infinite-Delay Smoothing for Unknown Markov Process*

Suppose the unknown process $x(t)$ is first-order Markov. Then it can be modeled as being generated by the stable shaping filter

$$\dot{x}(t) = -ax(t) + w(t) \qquad (1)$$

where $a > 0$ and the *process noise* $w(t)$ is white with spectral density $2a\sigma_w^2$. For simplicity we consider the scalar case. In the steady-state case $x(t)$ has spectral density (1.5-8) and autocorrelation (1.5-6) given by (prove this)

$$\Phi_X(\omega) = \frac{2a\sigma_w^2}{\omega^2 + a^2} \qquad (2)$$

$$R_X(t) = \sigma_w^2 e^{-a|t|}. \qquad (3)$$

To find Φ_Z and Φ_{XZ}, let us assume the special case of linear measurements with additive noise. Hence, let

$$z(t) = cx(t) + v(t) \qquad (4)$$

where $c \in R$ and the *measurement noise* $v(t)$ is white with spectral density r. Suppose here that $x(t)$ and $v(t)$ are orthogonal. Then

$$\begin{aligned} R_{XZ}(t) &= \overline{x(\tau + t)z(\tau)} \\ &= cR_X(t) \end{aligned} \qquad (5)$$

and

$$\begin{aligned} R_Z(t) &= \overline{z(\tau + t)z(\tau)} \\ &= \overline{[cx(\tau + t) + v(\tau + t)][cx(\tau) + v(\tau)]} \\ &= c^2 R_X(t) + r\delta(t), \end{aligned} \qquad (6)$$

with $\delta(t)$ the Kronecker delta. Hence

$$\Phi_{XZ}(\omega) = c\Phi_X(\omega) \tag{7}$$

and

$$\Phi_Z(\omega) = c^2\Phi_X(\omega) + r. \tag{8}$$

In this linear measurement case then, the optimal linear filter (1.5-21) is

$$H(j\omega) = \frac{\Phi_{XZ}(\omega)}{\Phi_Z(\omega)} = \frac{c\Phi_X(\omega)}{c^2\Phi_X(\omega) + r}$$

$$= \frac{2ac\gamma}{\omega^2 + (a^2 + 2ac^2\gamma)} \tag{9}$$

where signal-to-noise ratio $\gamma = \sigma_w^2/r$. In this problem it is convenient to define another parameter (Van Trees 1968)

$$\lambda = \frac{2c^2\gamma}{a}, \tag{10}$$

which can also be thought of as a signal-to-noise ratio. Note that $\lambda > 0$. Then

$$H(j\omega) = \frac{2ac\gamma}{\omega^2 + a^2(1 + \lambda)}, \tag{11}$$

so the impulse response of the optimal linear filter is

$$h(t) = \frac{c\gamma}{\sqrt{1 + \lambda}} e^{-a\sqrt{1+\lambda}|t|}. \tag{12}$$

This impulse response is sketched in Fig. 1.5-1.

The error variance (1.5-23) becomes

$$P_{\tilde{x}} = \frac{1}{2\pi} \int_{-\infty}^{\infty} \frac{r\Phi_X(\omega)}{c^2\Phi_X(\omega) + r} \, d\omega$$

$$= \frac{1}{2\pi} \int_{-\infty}^{\infty} \frac{2a\sigma_w^2}{\omega^2 + (a^2 + 2ac^2\gamma)} \, d\omega.$$

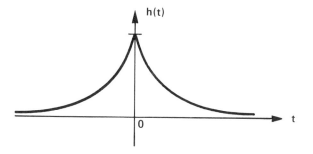

FIGURE 1.5-1 Optimal infinite-delay smoother.

Letting $\alpha = a\sqrt{1 + \lambda}$ we get

$$P_{\tilde{x}} = \frac{2 a \sigma_w^2}{\pi \alpha} \int_0^\infty \frac{\alpha}{\omega^2 + \alpha^2} \, d\omega = a \sigma_w^2 / \alpha,$$

or

$$P_{\tilde{x}} = \frac{\sigma_w^2}{\sqrt{1 + \lambda}}. \tag{13}$$

If $\gamma = 0$, so that the measurements provide no information, then $h(t) = 0$ so that the estimate is $\hat{x}(t) = 0$, the expected value of $x(t)$. The error variance is then

$$P_{\tilde{x}} = \sigma_w^2 = R_x(0), \tag{14}$$

which is independent of the measurement noise.

If $\gamma = \infty$, then the measurements are perfect. Examine (9) to see that in this case

$$h(t) = \frac{1}{c} \delta(t) \tag{15}$$

so that

$$\hat{x}(t) = \frac{1}{c} z(t). \tag{16}$$

Then the error variance is zero.

If

$$\gamma = a / 2c^2 \tag{17}$$

(i.e., $\lambda = 1$) then the optimal infinite-delay smoother is

$$h(t) = \frac{c}{\sqrt{2}\,r} \sigma_w^2 e^{-\sqrt{2}\,a|t|} \tag{18}$$

with error variance

$$P_{\tilde{x}} = \frac{\sigma_w^2}{\sqrt{2}} = \frac{R_x(0)}{\sqrt{2}}. \qquad \blacksquare$$

This example points out a problem with the simple-minded approach of letting $T = \infty$. Although this assumption leads to an easy solution, the resulting impulse response is noncausal and so unrealizable.

Causal Steady-State Filtering

To find a stable causal $h(t)$ which can be used for real-time processing of the data, we present the solution of Wiener and Hopf in a modification owing to Bode and Shannon (1950). Accordingly, consider the steady-state Wiener–Hopf equation (1.5-17) with $T = t$, which corresponds to the filtering problem,

$$R_{XZ}(t) = \int_0^\infty h(\tau) R_Z(t - \tau) \, d\tau; \qquad 0 < t < \infty. \tag{1.5-24}$$

Recall that the steady-state case means that $x(t)$ and $z(t)$ are stationary, and that there is available a data signal $z(t)$ with infinite past history (i.e., $t_0 = -\infty$).

Because of the constraint $t > 0$ the Fourier transform theorems can no longer be directly used to provide a simple solution $h(t)$ to the Wiener–Hopf equation. However, if the data signal has a rational spectral density $\Phi_Z(\omega)$ we can proceed as follows. For simplicity we initially consider the scalar case. Our presentation is modified from that of Van Trees (1968).

Note first that if the data $z(t)$ are white then (1.5-24) has a trivial solution, for then $R_Z(t) = r\,\delta(t)$ and

$$h(t) = \begin{cases} \dfrac{1}{r}\,R_{XZ}(t), & t > 0 \\[2mm] 0, & t < 0. \end{cases} \tag{1.5-25}$$

This will not happen except in pathological cases, but we can always use spectral factorization to *prewhiten* the data!

To wit, we know that the rational spectral density of $z(t)$ can always be factored as

$$\Phi_Z(\omega) = W^{-1}(j\omega)\,W^{-1}(-j\omega) \tag{1.5-26}$$

where $W^{-1}(s)$ is stable with zeros in the left-half plane. The representation theorem says that $z(t)$ can be considered as the output of the linear system $W^{-1}(s)$ driven by a white noise process $v(t)$. It is clear then that the "inverse" of this theorem says that $W(s)$ describes a stable linear system which results in a white noise output $v(t)$ when the input is the data $z(t)$. Thus $W(s)$ is a *whitening filter* for $z(t)$. See Fig. 1.5-2.

The fact that there exists a realizable whitening filter with a realizable inverse for any process with rational spectral density is called the *whitening property*. It can be considered as the "inverse" of the representation theorem.

Since $v(t)$ is white, it is uncorrelated with itself from instant to instant. The information contained in the data $z(t)$ is present also in $v(t)$, but the uncorrelatedness of $v(t)$ means that each value brings us new information, unlike the values of $z(t)$ which are correlated with each other. The information in $z(t)$ is therefore available in the whitened process $v(t)$ in particularly convenient form; this process is called the *innovations*, or new information process. See Kailath (1981).

The prewhitening of the data $z(t)$ by spectral factorization of $\Phi_Z^{-1}(\omega)$

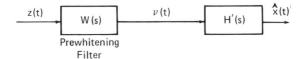

FIGURE 1.5-2 Solution of the Wiener–Hopf equation by prewhitening of the data.

should be compared to the prewhitening we did in the random variable case in Section 1.4. There we factored the inverse of the noise covariance as $R^{-1} = S^T S$.

There is now only one step left in our solution of (1.5-24). From Fig. 1.5-2 we see that it is necessary to design an optimal filter $h'(t)$ which reconstructs $x(t)$ from the innovations $\nu(t)$. Thus, $h'(t)$ satisfies

$$R_{X\nu}(t) = \int_0^\infty h'(\tau) R_\nu(t - \tau) \, d\tau; \qquad 0 < t < \infty. \qquad (1.5\text{-}27)$$

Since $R_\nu(t) = \delta(t)$ [note that $\nu(t)$ has spectral density equal to 1], the optimal solution is given by

$$h'(t) = \begin{cases} R_{X\nu}(t), & t > 0 \\ 0, & t < 0. \end{cases} \qquad (1.5\text{-}28)$$

It is not hard to find $\Phi_{X\nu}(\omega)$ in terms of $\Phi_{XZ}(\omega)$ and $W(j\omega)$ and so provide a frequency-domain solution. In fact, using a modification of (1.5-7)

$$\Phi_{X\nu}(\omega) = \Phi_{XZ}(\omega) W(-j\omega). \qquad (1.5\text{-}29)$$

Since $h'(t)$ is causal, we retain only the portion of $R_{X\nu}(t)$ for $t \geq 0$. This means that we must do a partial fraction expansion of $\Phi_{X\nu}(s)$ and retain the *realizable portion* $[\Phi_{X\nu}(s)]_+$ which has no poles in the open right-half plane,

$$\Phi_{X\nu}(s) = [\Phi_{X\nu}(s)]_+ + [\Phi_{X\nu}(s)]_-. \qquad (1.5\text{-}30)$$

The second term has poles in the right-half plane and corresponds to the anticausal (i.e., $t < 0$) portion of $R_{X\nu}(t)$. Any constant terms in $\Phi_{X\nu}(s)$ are included in the realizable portion. In this notation

$$H'(j\omega) = [\Phi_{XZ}(\omega) W(-j\omega)]_+. \qquad (1.5\text{-}31)$$

The overall Wiener filter $h(t)$ which solves (1.5-24) is given by the cascade of the whitening filter and $H'(s)$, or

$$H(j\omega) = [\Phi_{XZ}(\omega) W(-j\omega)]_+ W(j\omega), \qquad (1.5\text{-}32)$$

where

$$\Phi_Z^{-1}(\omega) = W(-j\omega) W(j\omega). \qquad (1.5\text{-}33)$$

This should be compared with our solution (1.5-21) to the infinite-delay smoothing problem. The transfer function $H(s)$ manufactured by (1.5-32) represents a realizable system and allows real-time processing of the data $z(t)$ to find the optimal estimate $\hat{x}(t)$ of the unknown process $x(t)$.

To find an expression for the error covariance [i.e., equation (1.5-18) with $T = t$], substitute (1.5-32) into (1.5-22) to get (scalars!)

$$F(\omega) = \Phi_X(\omega) - [\Phi_{XZ}(\omega) W(-j\omega)]_+ W(j\omega) \Phi_{XZ}(-\omega)$$

$$= \Phi_X(\omega) - [\Phi_{X\nu}(\omega)]_+ \Phi_{X\nu}(-\omega), \qquad (1.5\text{-}34)$$

where (1.5-29) was used. If the cross-correlation of unknown $x(t)$ and innovations $\nu(t)$ is $R_{X\nu}(t)$, we can retain only the causal portion by noting that

$$\mathcal{F}[R_{X\nu}(t)u_{-1}(t)] = [\Phi_{X\nu}(\omega)]_+. \tag{1.5-35}$$

where $\mathcal{F}[\cdot]$ denotes Fourier transform and $u_{-1}(t)$ is the unit step. Hence (1.5-34) can be transformed to obtain [recall the transform of $\Phi_{X\nu}(-\omega)$ is $R_{X\nu}(-t)$]

$$f(t) = R_X(t) - \int_{-\infty}^{\infty} R_{X\nu}(\tau)u_{-1}(\tau)R_{X\nu}(\tau - t)\, d\tau, \tag{1.5-36}$$

so that $P_{\bar{X}} = f(0)$ or

$$P_{\bar{X}} = R_X(0) - \int_0^{\infty} R_{X\nu}^2(\tau)\, d\tau. \tag{1.5-37}$$

We can also use these constructions to obtain a time-domain expression for the optimal error covariance obtained by using the unrealizable filter (1.5-21). Considering the scalar case and using superscripts u for clarity there results, from (1.5-22), (1.5-21), and (1.5-33),

$$F^u(\omega) = \Phi_X(\omega) - \Phi_{XZ}(\omega)W(-j\omega)W(j\omega)\phi_{XZ}(-\omega). \tag{1.5-38}$$

Now use (1.5-29) to write

$$F^u(\omega) = \Phi_X(\omega) - \Phi_{X\nu}(\omega)\Phi_{X\nu}(-\omega). \tag{1.5-39}$$

Transforming this,

$$f^u(t) = R_X(t) - \int_{-\infty}^{\infty} R_{X\nu}(\tau)R_{X\nu}(\tau - t)\, d\tau \tag{1.5-40}$$

so that

$$P_{\bar{X}}^u = R_X(0) - \int_{-\infty}^{\infty} R_{X\nu}^2(\tau)\, d\tau. \tag{1.5-41}$$

Therefore the (unattainable!) error using the unrealizable filter is in general less than the error (1.5-37) using the realizable Wiener filter.

Example 1.5-4: Causal Steady-State Filter for Unknown Markov Process

Let us find the realizable Wiener filter (1.5-32) for Example 1.5-3. From that example

$$\Phi_X(\omega) = \frac{2a\sigma_w^2}{\omega^2 + a^2} \tag{1}$$

$$\Phi_{XZ}(\omega) = c\phi_X(\omega) = \frac{2ac\sigma_w^2}{\omega^2 + a^2} \tag{2}$$

$$\Phi_Z(\omega) = c^2\Phi_X(\omega) + r = \frac{r[\omega^2 + a^2(1 + \lambda)]}{\omega^2 + a^2}. \tag{3}$$

To find the whitening filter, factor the inverse of the spectral density of $z(t)$ to get

$$\Phi_z^{-1}(\omega) = W(-j\omega)\,W(j\omega)$$

$$= \frac{(-j\omega + a)}{\sqrt{r}(-j\omega + a\sqrt{1+\lambda})} \cdot \frac{(j\omega + a)}{\sqrt{r}(j\omega + a\sqrt{1+\lambda})}. \tag{4}$$

Therefore,

$$\Phi_{X\nu}(\omega) = \Phi_{XZ}(\omega)\,W(-j\omega)$$

$$= \frac{2ac\sigma_w^2}{\sqrt{r}}\frac{(-j\omega + a)}{(-j\omega + a\sqrt{1+\lambda})(\omega^2 + a^2)}. \tag{5}$$

To facilitate things, convert this to a function of $s = \sigma + j\omega$. Then a partial fraction expansion yields

$$\Phi_{X\nu}(s) = \frac{-2ac\sigma_w^2}{\sqrt{r}}\frac{s-a}{(s - a\sqrt{1+\lambda})(s-a)(s+a)} \tag{6}$$

$$= \frac{2c\sigma_w^2}{\sqrt{r}(1+\sqrt{1+\lambda})}\left(\frac{1}{s+a} + \frac{-1}{s - a\sqrt{1+\lambda}}\right). \tag{7}$$

The causal portion of $R_{X\nu}(t)$ is given by the first term, which contains the left-half plane pole. [Note that the region of convergence of (7) contains the imaginary axis.] Therefore, by (1.5-32) (recall $\gamma = \sigma_w^2/r$)

$$H(s) = [\Phi_{X\nu}(s)]_+\,W(s)$$

$$= \frac{2c\gamma}{1+\sqrt{1+\lambda}} \cdot \frac{1}{s + a\sqrt{1+\lambda}}, \tag{8}$$

so that

$$h(t) = \frac{2c\gamma}{1+\sqrt{1+\lambda}}\,e^{-a\sqrt{1+\lambda}\,t}u_{-1}(t). \tag{9}$$

This is the impulse response of the Wiener filter for reconstructing $x(t)$ from the data $z(t)$. It is causal and stable, and should be compared to the unrealizable filter (12) found in Example 1.5-3. Note that we can *not* obtain (9) by taking the $t \geq 0$ portion of the unrealizable filter!

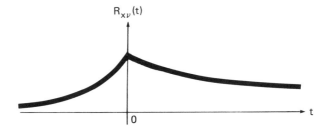

FIGURE 1.5-3 Cross-correlation of unknown with innovations.

From (8) it is clear that the Wiener filter is a low-pass filter whose bandwidth depends on a and λ.

The optimal error covariance is found by using $R_{X_\nu}(t)$, which by (7) is

$$R_{X_\nu}(t) = \begin{cases} \dfrac{2c\sigma_w^2}{\sqrt{r}(1+\sqrt{1+\lambda})} \, e^{-at}, & t \geq 0 \\[3mm] \dfrac{2c\sigma_w^2}{\sqrt{r}(1+\sqrt{1+\lambda})} \, e^{a\sqrt{1+\lambda}\,t}, & t < 0. \end{cases} \tag{10}$$

This cross-correlation is sketched in Fig. 1.5-3. Now, by (1.5-37)

$$P_{\tilde{X}} = \sigma_w^2 - \int_0^\infty \frac{4c^2\sigma_w^2\gamma}{(1+\sqrt{1+\lambda})^2} \, e^{-2at} \, dt \tag{11}$$

$$= \sigma_w^2 \frac{[(1+\sqrt{1+\lambda})^2 - \lambda]}{(1+\sqrt{1+\lambda})^2} \tag{12}$$

$$= \frac{2\sigma_w^2}{1+\sqrt{1+\lambda}}. \tag{13}$$

This error covariance and $P_{\tilde{X}}^u$ given by (13) in Example 1.5-3 are both sketched in Fig. 1.5-4 as a function of λ. ∎

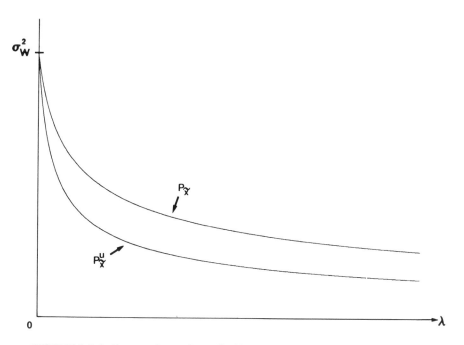

FIGURE 1.5-4 Error variances for realizable and unrealizable optimal linear filters.

Example 1.5-5:   The Wiener Filter for Linear Measurements

Here we derive expressions for the Wiener filter and error covariance for the special case of scalar linear measurements

$$z(t) = cx(t) + v(t), \tag{1}$$

where $c \in R$ and measurement noise $v(t)$ is white with spectral height r. We assume $v(t)$ is uncorrelated with the unknown $x(t)$.

From Example 1.5-3

$$\Phi_{XZ}(\omega) = c\Phi_X(\omega) \tag{2}$$

and

$$\Phi_Z(\omega) = c^2\Phi_X(\omega) + r. \tag{3}$$

To find the Wiener filter we factor $\Phi_Z(\omega)$ into unrealizable and realizable spectral factors as

$$\Phi_Z(\omega) = W^{-1}(j\omega) W^{-1}(-j\omega), \tag{4}$$

where $W^{-1}(s)$ has poles in the open left-half plane (i.e., not on the $j\omega$ axis). From the spectral factorization theorem we know that $W^{-1}(s)$ can only be guaranteed to have zeros in the left-half plane, including possibly on the $j\omega$ axis. Thus if we write

$$\Phi_Z^{-1}(\omega) = W(-j\omega) W(j\omega), \tag{5}$$

we can not in general guarantee that $W(-j\omega)$ has all poles in the s plane with real parts strictly greater than zero (i.e., unstable). However, it is easy to show that if $\Phi_Z(\omega)$ has the form of (3) with $r \neq 0$, then $W(-j\omega)$ does have all poles in the open right-half plane. [We shall soon see why we want $W(-j\omega)$ to be completely unstable!] For

$$\Phi_Z^{-1}(s) = \frac{1/r}{1 + (c^2/r)\Phi_X(s)} \tag{6}$$

is in exactly the canonical form for the root-locus theory. Hence as the "feedback gain" c^2/r increases, the poles of $\Phi_Z^{-1}(s)$ migrate from the poles of $\Phi_X(s)$ to the zeros of $\Phi_X(s)$. Since $\Phi_X(s)$ corresponds to a spectral density, it may have zeros but not poles on the imaginary axis. It does not have poles or zeros in the right-half plane. Hence if $r \neq 0$ the poles of $\Phi_Z^{-1}(s)$ do not occur on the $j\omega$ axis and the spectral factors given by

$$\Phi_Z^{-1}(\omega) = W(-j\omega) W(j\omega) \tag{7}$$

are such that $W(s)$ has all poles in the open left-half plane and $W(s^*)$ has all poles in the open right-half plane (where * denotes complete conjugate).

We can use this fact to find an easy expression for the realizable part of $W(-j\omega)$ which we need shortly. Let

$$\Phi_X(s) = \frac{n(s)}{d(s)} \tag{8}$$

for polynomials $n(s)$ and $d(s)$ in s^2. If $x(t)$ is band limited, the degree of $d(s)$ in s^2 is at least one greater than the degree of $n(s)$ (why?). Therefore $\Phi_Z^{-1}(s)$ has relative degree of zero since

$$\Phi_Z^{-1}(s) = \frac{(1/r) d(s)}{d(s) + (c^2/r)n(s)}. \tag{9}$$

In (7), a multiplier of $1/\sqrt{r}$ is assigned to each of the spectral factors, which also have relative degrees of zero. $W(-j\omega)$ thus has the form

$$W(-j\omega) = \frac{1}{\sqrt{r}} + \text{unrealizable part} \tag{10}$$

and so

$$[W(-j\omega)]_+ = \frac{1}{\sqrt{r}}. \tag{11}$$

This can be used to find a very simple expression for the Wiener filter in the case of linear measurements. From (1.5-32) the Wiener filter is

$$H(j\omega) = \frac{1}{c}[c^2\Phi_X(\omega) W(-j\omega)]_+ W(j\omega), \tag{12}$$

(note that the "realizable part" operator is linear), or

$$H(j\omega) = \frac{1}{c}[(c^2\Phi_X(\omega) + r) W(-j\omega) - rW(-j\omega)]_+ W(j\omega)$$

$$= \frac{1}{c}[W^{-1}(j\omega) - rW(-j\omega)]_+ W(j\omega)$$

$$= \frac{1}{c}[W^{-1}(j\omega)]_+ W(j\omega) - \frac{r}{c}[W(-j\omega)]_+ W(j\omega).$$

Now, since $W^{-1}(s)$ is completely realizable (all poles in the open left-half plane) the first term is $1/c$. Therefore [see (11)]

$$H(j\omega) = \frac{1}{c}[1 - \sqrt{r}\, W(j\omega)], \tag{1.5-42}$$

which is the desired expression for the Wiener filter in the linear measurement case.

It is not too difficult to show (Yovits and Jackson 1955) that the error covariance is given by

$$P_{\tilde{x}} = \frac{r}{2\pi c^2} \int_{-\infty}^{\infty} \ln\left[1 + \frac{c^2\phi_X(\omega)}{r}\right] d\omega \tag{1.5-43}$$

Using (1.5-42) and (1.5-43) the results of Example 1.5-4 can be obtained at somewhat less expense!

A useful identity when using (1.5-43) is

$$\frac{1}{2\pi} \int_{-\infty}^{\infty} \ln\left(\frac{\omega^2 + \alpha^2}{\omega^2 + \beta^2}\right) d\omega = \alpha - \beta. \tag{13}$$

■

It should be noted that although the whitening filter $W(s)$ is the central concept in our derivation of the Wiener filter, only the latter is actually implemented in an application.

The Wiener filter can be generalized to multivariable processes and a counterpart also exists in the discrete case. In fact, if the unknown x_k and the data z_k are discrete vector random processes, then (1.5-32), (1.5-33) become

$$H(z) = [\Phi_{XZ}(z) W^T(z^{-1})]_+ W(z) \qquad (1.5\text{-}44)$$

where

$$\Phi_Z^{-1}(z) = W^T(z^{-1}) W(z). \qquad (1.5\text{-}45)$$

In this case the realizable part operator means to retain the portion with poles inside or on the unit circle. If $|\Phi_Z(z)| \neq 0$ on $|z| = 1$, then $W(z)$ is minimum phase and the realizable poles are strictly within the unit circle.

In the multivariable case with linear measurements

$$z(t) = x(t) + v(t) \qquad (1.5\text{-}46)$$

where $v(t) \sim (0, R)$, the results of Example 1.5-5 can be generalized. The Wiener filter is given in this case by the simple formula

$$H(s) = I - R^{1/2} W(s), \qquad (1.5\text{-}47)$$

where $W(s)$ is the stable spectral factor of $\phi_Z^{-1}(s)$, and $R = R^{1/2} R^{T/2}$ with $R^{1/2}$ the square root of R (Kailath 1981). We shall derive the discrete counterpart to this result in Section 2.6.

PROBLEMS

Section 1.1

1.1-1: Least-Squares Solution of Linear Equations. The connotation of *least-squares* estimation is a setting in which both the unknown X and the data Z are deterministic, so that no a priori information is available on either one.

Let

$$Z = HX \qquad (1)$$

with H of full row rank. Minimize the performance index

$$J = \tfrac{1}{2}(Z - HX)^T R^{-1}(Z - HX), \qquad (2)$$

where R^{-1} is a weighting matrix, to show that the (weighted) least-squares estimate is

$$\hat{X} = (H^T R^{-1} H)^{-1} H^T R^{-1} Z. \qquad (3)$$

Note that the least-squares performance index is expressed in terms of the *residual*.

1.1-2: Nonlinear LS Estimation. Let the data Z and the unknown X be related by

$$Z = h(X) + e, \qquad (1)$$

where e is a residual and $h(\cdot)$ is a (possibly nonlinear) function.

a. Show that the estimate minimizing the least-squares error

$$J = \tfrac{1}{2}\langle e, e\rangle = \tfrac{1}{2}\langle Z - h(X), Z - h(X)\rangle, \tag{2}$$

where $\langle \cdot, \cdot \rangle$ represents inner product, is the value of X satisfying the *least-squares normal equations*

$$\left\langle \frac{\partial h}{\partial X}, h(X)\right\rangle = \left\langle \frac{\partial h}{\partial X}, Z\right\rangle. \tag{3}$$

b. Expand $h(X)$ in a Taylor series about a given estimate $\hat{X}$ to show that

$$\langle H, HX\rangle = \langle H, H\hat{X}\rangle + \langle H, Z - h(\hat{X})\rangle, \tag{4}$$

where $H = \partial h / \partial X$ is the Jacobian. This provides an iterative method for solving (3), for it shows how to determine a more accurate updated estimate for X from a previous estimate $\hat{X}$.

1.1-3: Continuous Estimate Updating. Let the inner product in Problem 1.1-2 be defined as

$$J = \frac{1}{2}\int_0^T e^T R^{-1} e\, dt \tag{1}$$

where $R > 0$.

a. Write the linearized normal equations (4).

b. Note that $H = -\partial e / \partial X$ and define

$$M = \int_0^T H^T R^{-1} H\, dt. \tag{2}$$

Using your result in part a, let T become small to show that the estimate satisfies

$$\dot{X} = -M^{-1}\left(\frac{\partial e}{\partial X}\right)^T R^{-1} e, \tag{3}$$

where $e = z - h(x)$.

Note that

$$\dot{M} = H^T R^{-1} H. \tag{4}$$

Equations (3) and (4) provide a method for continuously updating the least-squares estimate X if data are taken continuously over a time interval.

1.1-4

a. Suppose we have three measurements Z_i of a constant nonrandom voltage X. The Z_i are corrupted by additive noises $V_1 \sim (0, 1)$, $V_2 \sim (0, 2)$, $V_3 \sim (0, 4)$, respectively. If we observe $Z_1 = 4$ V, $Z_2 = 3$ V, $Z_3 = 5$ V, find the mean-square estimate for X.

b. Find the minimum error variance $\sigma_{\hat{X}}^2$.

1.1-5: Estimate for Linear Function of Unknown. If $\hat{X}$ is the linear mean-square estimate for unknown X, show that the linear mean-square estimate for the linear function $f(X) = CX + d$ is given by $f(\hat{X}) = C\hat{X} + d$.

1.1-6: Fill in the steps in showing (1.1-18).

1.1-7: A stochastic unknown X and the data Z are jointly distributed according to

$$f_{XZ}(x, z) = \tfrac{6}{7}(x + z)^2; \qquad 0 \le x \le 1 \quad \text{and} \quad 0 \le z \le 1$$

a. Show that

$$f_X(x) = \tfrac{6}{7}(x^2 + x + \tfrac{1}{3}); \qquad 0 \le x \le 1$$

$$f_{X/Z}(x/z) = \frac{x^2 + 2xz + z^2}{\tfrac{1}{3} + z + z^2}; \qquad 0 \le x \le 1$$

b. Find the best estimate for X and its error covariance in the absence of data.

c. Find the best mean-square estimate for X given the data Z.

d. Find the best linear mean-square estimate for X and its error covariance given the data Z.

1.1-8: Multiplicative Noise. An unknown X is measured in the presence of multiplication noise so that

$$Z = (1 + W)X$$

where $X \sim (0, \sigma_X^2)$ and the noise is $W \sim (0, \sigma_W^2)$. The noise and unknown are uncorrelated.

a. Find the optimal linear estimate $\hat{X} = aZ$.

b. Find the error variance.

1.1-9: Linear Measurements with Correlated Noise and Unknown. Repeat Example 1.1-3 if $\overline{XV^T} = P_{XV} \ne 0$.

1.1-10: Experiment Design. An unknown voltage S is distributed according to

$$f_S(s) = \frac{1}{2\sqrt{2\pi}} e^{-s^2/8}.$$

Consider two ways to measure S.

a. Use two expensive meters that add to S noise which is $(0, 2)$ so that

$$Z_1 = S + V_1, \qquad V_1 \sim (0, 2)$$

$$Z_2 = S + V_2, \qquad V_2 \sim (0, 2),$$

with V_1 and V_2 uncorrelated. Find linear mean-square estimate and error covariance.

b. Use four inexpensive meters that add to S noise which is $(0, 3)$. Find linear mean-square estimate and error covariance.

c. For the best estimate should we use scheme a or scheme b?

d. Find numerical values for $\hat{S}_{LMS}$: in a. if $Z_1 = 1$, $Z_2 = 2$; in b. if $Z_1 = 1$, $Z_2 = 2$, $Z_3 = 0.5$, $Z_4 = 1.2$.

Section 1.2

1.2-1: Inaccurately Known Noise Covariance. To examine the effects of inaccurately known noise parameters, suppose we have

$$Z = \begin{bmatrix} 1 \\ 1 \end{bmatrix} X + v, \qquad v \sim N\left(0, \begin{bmatrix} \sigma^2 & 0 \\ 0 & 1 \end{bmatrix}\right).$$

a. Compute $\hat{X}_{ML}$, $P_{\tilde{x}}$ in terms of σ^2 if the measurement is

$$Z = \begin{bmatrix} 5 \\ 8 \end{bmatrix}.$$

Sketch $P_{\tilde{x}}$ and $\hat{X}_{ML}$ versus σ^2.

b. Evaluate $\hat{X}_{ML}$, $P_{\tilde{X}}$ for $\sigma^2 = \frac{1}{2}$, 1, 2.

1.2-2: Let

$$Z = \begin{bmatrix} 1 & 2 \\ 0 & -1 \end{bmatrix} X + V, \qquad V \sim \left(\begin{bmatrix} 0 \\ 0 \end{bmatrix}, \begin{bmatrix} 1 & 1 \\ 1 & 3 \end{bmatrix}\right)$$

$$X \sim \left(\begin{bmatrix} 1 \\ -1 \end{bmatrix}, \begin{bmatrix} 1 & 2 \\ 2 & 5 \end{bmatrix}\right),$$

and let the measured data be

$$z = \begin{bmatrix} 2 \\ 0.5 \end{bmatrix}.$$

a. Ignore statistics and use matrix inverse to find $\hat{X}$ (i.e. $\hat{X} = H^{-1}Z$).

b. Ignore statistics and use weighted least-squares estimation (Problem 1.1-1) to find $\hat{X}$. Use weighting matrix R equal to the covariance of V. Can you find an error covariance?

c. Include statistics on V but not on X. Find maximum-likelihood estimate. Find $P_{\tilde{x}}$.

d. Include statistics on X and V. Find linear mean-square estimate. Find $P_{\tilde{x}}$.

e. What is your estimate for X before Z is measured? Find error covariance $P_{\tilde{x}}$ in this case.

1.2-3: Consider two measurements of a scalar unknown random variable X

$$Z_1 = X + V_1, \qquad V_1 \sim (0, \sigma_1^2)$$

$$Z_2 = X + V_2, \qquad V_2 \sim (0, \sigma_2^2)$$

with noises V_1 and V_2 uncorrelated.

a. X is known a priori to be $(\bar{X}, \sigma_X^2)$. Find the linear mean-square estimate $\hat{X}$ and error covariance $P_{\tilde{X}}$. Compare with Example 1.2-3.

b. Find linear mean-square estimate if no statistics are available on X.

1.2-4: Let the data Z be related to an unknown X by

$$Z = \ln X + V, \tag{1}$$

with scalar random variables X and V independent. Suppose X is uniformly distributed between 0 and b, and noise V is Laplacian with $a = 1$ [i.e., $f_V(v) = (a/2)e^{-a|v|}$].

a. Find the optimal a priori estimate and error covariance.

b. Find maximum-likelihood estimate.

c. Find conditional mean estimate.

1.2-5: Arbitrary Linear Combination of Data. Let a constant scalar unknown X be measured with two meters corrupted by additive noise so that

$$Z_1 = X + V_1$$

$$Z_2 = X + V_2 \tag{1}$$

where $V_1 \sim N(0, \sigma_1^2)$, $V_2 \sim N(0, \sigma_2^2)$, and V_1 and V_2 are independent. We can estimate X by an arbitrary linear combination of the data,

$$\hat{X} = aZ_1 + bZ_2. \tag{2}$$

a. Show that for this estimate to be unbiased we require

$$b = 1 - a. \tag{3}$$

b. In the unbiased case, find the error covariance $P_{\tilde{X}}$. For what value of a is this estimate equal to $\hat{X}_{\mathrm{ML}}$?

c. Is it possible to find a biased estimate (2) with smaller error covariance than that in b?

1.2-6: Optimal Combination of Two Independent Estimates. Suppose two estimates $\hat{X}_1$ and $\hat{X}_2$ of an unknown random variable X are obtained by independent means. The associated error covariances are P_1 and P_2. It is desired to find the overall optimal linear estimate of X, that is, the optimal estimate of the form

$$\hat{X} = A_1 \hat{X}_1 + A_2 \hat{X}_2 \tag{1}$$

where A_1 and A_2 are constant matrices.

a. Show that if $\hat{X}_1$ and $\hat{X}_2$ are unbiased [i.e., $E(\hat{X}_i) = \bar{X}$], then for $\hat{X}$ to be unbiased also it is required that

$$A_2 = I - A_1. \tag{2}$$

b. Define estimation errors as $\tilde{X}_1 = X - \hat{X}_1$, $\tilde{X}_2 = X - \hat{X}_2$, $\tilde{X} = X - \hat{X}$. Show that, using (2),

$$\tilde{X} = A_1\tilde{X}_1 + (I - A_1)\tilde{X}_2. \tag{3}$$

c. Let the error covariance of $\hat{X}$ be P. Differentiate trace(P) to find that the optimal A_1 is

$$A_1 = P_2(P_1 + P_2)^{-1} \tag{4}$$

and hence

$$A_2 = P_1(P_1 + P_2)^{-1}. \tag{5}$$

d. Show that the overall error covariance P is given by

$$P^{-1} = P_1^{-1} + P_2^{-1}. \tag{6}$$

e. Show that

$$P^{-1}\hat{X} = P_1^{-1}\hat{X}_1 + P_2^{-1}\hat{X}_2. \tag{7}$$

(This is a generalized Millman's theorem. See Examples 1.2-3 and 1.2-4.)

1.2-7: Alternate Form of Estimate Update with Linear Measurements. Let unknown X be distributed as $(\bar{X}, P_X)$, and measurements be taken so that

$$Z = HX + V, \tag{1}$$

with noise $V \sim (0, R)$ uncorrelated with X. Show that the linear mean-square estimate $\hat{X}$ is given by

$$P^{-1}\hat{X} = P_X^{-1}\bar{X} + H^T R^{-1} Z, \tag{2}$$

with error covariance

$$P^{-1} = P_X^{-1} + H^T R^{-1} H. \tag{3}$$

[If X and V are normal, then (2) and (3) yield *the* optimal mean-square estimate.]

Section 1.4

1.4-1: Derive (1.4-2).

1.4-2: Sequential Estimation with Block Diagonal R. Derive (1.4-6), (1.4-7), (1.4-9) for the case $X \in R^n$ and R block diagonal, so that $R = \text{diag}\{S_i\}$, where S_i is $p_i \times p_i$ and nonsingular. Thus the components of Z are sequentially processed in groups of size p_i.

Section 1.5

1.5-1: An Alternative Design Method for Optimal Filters. A measurement $z(t)$ is taken of a signal $x(t)$ corrupted by additive noise $n(t)$. The signal and noise are uncorrelated, and

$$\Phi_X(\omega) = \frac{2\sigma^2}{1 + \omega^2}, \tag{1}$$

$$\Phi_N(\omega) = \frac{r\omega^2}{1 + \omega^2}. \tag{2}$$

An estimate $\hat{x}(t)$ is formed by passing $z(t)$ through the ideal low-pass filter

$$H(j\omega) = \begin{cases} 1, & |\omega| < \omega_c \\ 0, & \text{otherwise.} \end{cases} \tag{3}$$

a. Sketch $\Phi_X(\omega)$ and $\Phi_N(\omega)$. Show that $\Phi_X(\omega) = \Phi_N(\omega)$ at

$$\omega = \omega_0 = \sqrt{\frac{2\sigma^2}{r}}. \tag{4}$$

b. If $\tilde{x}(t) = x(t) - \hat{x}(t)$, show that

$$\tilde{X}(s) = [I - H(s)]X(s) - H(s)N(s), \tag{5}$$

and

$$\Phi_{\tilde{X}}(\omega) = |I - H(j\omega)|^2 \Phi_X(\omega) + |H(j\omega)|^2 \Phi_N(\omega). \tag{6}$$

Sketch $|H(j\omega)|^2$ and $|I - H(j\omega)|^2$.

c. Find the mean-square error $P_{\tilde{X}} = \overline{\tilde{x}^2(t)}$. [Recall that $P_{\tilde{X}} = R_{\tilde{X}}(0)$.]

d. Differentiate $P_{\tilde{X}}$ with respect to ω_c to find the optimal value of ω_c for the filter $H(\omega)$.

e. For $r = 1$, $\sigma^2 = 2$, evaluate the optimal ω_c and error covariance $P_{\tilde{X}}$. (Gail White)

1.5-2: Assume measurement $z(t)$ is derived from unknown $x(t)$ as in the previous problem.

a. Find the Wiener filter $H(j\omega)$ to estimate $x(t)$. Sketch $H(j\omega)$. Note that its cutoff frequency is the same as the cutoff frequency in the previous problem.

b. Find the error covariance $P_{\tilde{X}}$.

c. For $r = 1$, $\sigma^2 = 2$, find $H(j\omega)$ and $P_{\tilde{X}}$. (Gail White)

1.5-3: Wiener Filter with Prediction. If it is desired to estimate an unknown scalar signal $x(t)$ at time $t + d$ using data $z(t)$ up through time t, the Wiener–Hopf equation becomes

$$R_{XZ}(t+d) = \int_0^\infty h(\tau)R_Z(t-\tau)\,d\tau; \qquad 0 < t < \infty. \tag{1}$$

Show that in this case the Wiener filter is given by

$$H(s) = [\phi_{XZ}(s)e^{sd}W(-s)]_+\, W(s), \tag{2}$$

with $W(s)$ the stable factor given by

$$\phi_Z^{-1}(s) = W(-s)\,W(s). \tag{3}$$

(Kailath 1981).

1.5-4: Wiener Filter for Second-Order Markov Process. The unknown $x(t)$ is a second-order Markov process satisfying

$$\dot{x} = x_2 + w \tag{1a}$$

$$\dot{x}_2 = -2x - 2x_2 \tag{1b}$$

with white process noise $w(t) \sim (0, 2)$. The data are generated using measurements given by

$$z(t) = x(t) + v(t) \tag{2}$$

with white measurement noise $v(t) \sim (0, 1)$ uncorrelated with $x(t)$.

a. Find spectral densities $\phi_X(s)$, $\phi_{XZ}(s)$, and $\phi_Z(s)$.

b. Find optimal noncausal filter for reconstructing $x(t)$. Compute and sketch the impulse response.

   ANSWER: $H(s) = \dfrac{2(-s^2 + 4)}{s^4 - 2s^2 + 12}.$

c. Find a whitening filter for the data $z(t)$.

   ANSWER: $W(s) = \dfrac{s^2 + 2s + 2}{s^2 + 2.988s + 3.4641}.$

d. Find the Wiener filter. Compute and sketch impulse response.

2

DISCRETE-TIME
KALMAN FILTER

The object of this chapter is to estimate the state of a discrete linear system developing dynamically through time. We shall see that the optimal estimator, known as the Kalman filter, is itself a linear dynamical system, and that it is a natural extension of the estimators discussed in the previous chapter.

To introduce the basic ideas used in the discrete Kalman filter, we first review the deterministic state observer.

2.1 DETERMINISTIC STATE OBSERVER

Suppose there is prescribed a dynamical plant

$$x_{k+1} = Ax_k + Bu_k \qquad\qquad (2.1\text{-}1a)$$

$$z_k = Hx_k \qquad\qquad (2.1\text{-}1b)$$

where state $x_k \in R^n$, control input $u_k \in R^m$, output $z_k \in R^p$; and A, B, and H are known constant matrices of appropriate dimension. All variables are deterministic, so that if initial state x_0 is known then (2.1-1) can be solved exactly for x_k, z_k for $k \geq 0$.

The *deterministic asymptotic estimation* problem is as follows. Design an estimator whose output $\hat{x}_k$ converges with k to the actual state x_k of (2.1-1) when the initial state x_0 is unknown, but u_k and z_k are given exactly.

An estimator or *observer* which solves this problem has the form

$$\hat{x}_{k+1} = A\hat{x}_k + L(z_k - H\hat{x}_k) + Bu_k, \qquad\qquad (2.1\text{-}2)$$

which is shown in Fig. 2.1-1. A, B, and H are given, so it is only necessary

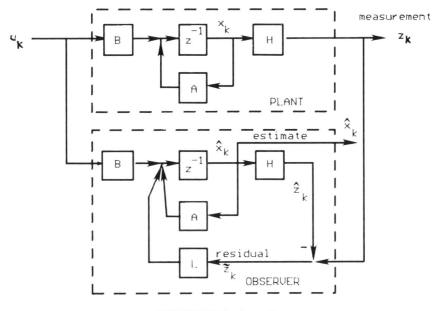

FIGURE 2.1-1   State observer.

to find the constant observer matrix L. Note that the observer consists of two parts, a replica (A, B, H) of the dynamics of the plant whose state is to be estimated, and an error-correcting portion described by $L(z_k - \hat{z}_k) = L\tilde{z}_k$, where the output estimate is $\hat{z}_k = H\hat{x}_k$ and $\tilde{z}_k = z_k - \hat{z}_k$ is the *residual*. The observer should be compared with the recursive estimators shown in Figs. 1.4-1 and 1.4-2. Note, however, that here the gain L that multiplies the residual is a constant. Since we built the observer, its state $\hat{x}_k$ is available to us. (Plant state x_k is not available except through measurement z_k; therein lies the problem.)

To choose L so that the *estimation error* $\tilde{x}_k = x_k - \hat{x}_k$ goes to zero with k for all x_0, it is necessary to examine the dynamics of $\tilde{x}_k$. Write

$$\tilde{x}_{k+1} = x_{k+1} - \hat{x}_{k+1}$$

$$= Ax_k + Bu_k - [A\hat{x}_k + L(z_k - H\hat{x}_k) + Bu_k]$$

$$= A(x_k - \hat{x}_k) - L(Hx_k - H\hat{x}_k);$$

so

$$\tilde{x}_{k+1} = (A - LH)\tilde{x}_k \qquad (2.1\text{-}3)$$

describes the dynamics of the estimation error. System (2.1-3) is called the *error system*. It does not actually exist, but is a mathematical fiction which allows us to study the convergence properties of the observer. The initial condition of (2.1-3) is $\tilde{x}_0 = x_0 - \hat{x}_0$, the unknown initial error.

It is now apparent that in order that $\tilde{x}_k$ go to zero with k for any $\tilde{x}_0$, observer gain L must be selected so that $(A - LH)$ is stable. Thus the output injection problem is the key to deterministic observer design. L can be chosen so that $\tilde{x}_k \to 0$ if and only if (A, H) is detectable (i.e., the unstable poles can be arbitrarily assigned), for then $(A - LH)$ is stable for some L. If we want the error $\tilde{x}_k$ not simply to go to zero with k, but to die out as fast as we desire, then all the poles of $(A - LH)$ must be arbitrarily assignable, so that (A, H) must be observable.

There are many ways to compute L. In the single-output case, *Ackermann's formula*

$$L = \Delta^d(A) V_n^{-1} e_n \qquad (2.1\text{-}4)$$

can be used. Here e_n is the nth column of the $n \times n$ identity matrix, $\Delta^d(s)$ is the desired error system characteristic polynomial, and

$$V_n = \begin{bmatrix} C \\ CA \\ \vdots \\ CA^{n-1} \end{bmatrix} \qquad (2.1\text{-}5)$$

is the observability matrix. Alternatively, we can use the next result, which applies for multioutput systems.

Theorem 2.1-1

Let G and D be any matrices such that (A, G) is reachable and rank $([H \quad D]) = \text{rank } (D)$. Then (A, H) is detectable if and only if there exists a positive definite solution P_+ to the *algebraic Riccati equation*

$$P = APA^T - APH^T(HPH^T + DD^T)^{-1}HPA^T + GG^T. \qquad (2.1\text{-}6)$$

In this event

$$L = AP_+H^T(HPH^T + DD^T)^{-1} \qquad (2.1\text{-}7)$$

stabilizes $(A - LH)$.

The proof of this result is fairly easy; it is part of a theorem we will prove in Section 2.5.

There is as yet no basis for selecting D and G; any matrices that satisfy the hypotheses will do. Different D and G will result in different locations for the error system poles. (One choice that satisfies the hypotheses is $D = I$, $G = I$.)

Example 2.1-1:   Ricatti Equation Design of Scalar Deterministic Observer

We can gain some insight on observer design using (2.1-6) by considering the scalar case. Let the given plant be

$$x_{k+1} = ax_k + bu_k \qquad (1)$$

with measurement

$$z_k = hx_k, \tag{2}$$

where u_k, x_k, and z_k are scalars. If we let $q = g^2$ and $r = d^2$ then (2.1-6) becomes

$$p = a^2 p - \frac{a^2 h^2 p^2}{h^2 p + r} + q, \tag{3}$$

or

$$p = \frac{a^2 r p}{h^2 p + r} + q, \tag{4}$$

which can be written as

$$h^2 p^2 + [(1 - a^2)r - h^2 q]p - qr = 0. \tag{5}$$

This equation has two solutions given by

$$p = \frac{q}{2\lambda} \left[\pm \sqrt{(1 - \lambda)^2 + \frac{4\lambda}{(1 - a^2)}} - (1 - \lambda) \right], \tag{6}$$

where

$$\lambda = \frac{h^2 q}{(1 - a^2) r}. \tag{7}$$

[Compare λ to (10) in Example 1.5-3.]
   We must now consider two cases.

a. Original System Stable

If $|a| < 1$, then $(1 - a^2) > 0$ and $\lambda > 0$. In this case the positive (definite) solution to (5) is

$$p_+ = \frac{q}{2\lambda} \left[\sqrt{(1 - \lambda)^2 + \frac{4\lambda}{(1 - a^2)}} - (1 - \lambda) \right], \tag{8}$$

and the observer gain (2.1-7) is selected as

$$l = \frac{ahp_+}{h^2 p_+ + r}. \tag{9}$$

Then the error system matrix is

$$a^{cl} = a - lh = \frac{a}{1 + (h^2/r)p_+}, \tag{10}$$

or

$$a^{cl} = \frac{a}{1 + \frac{1 - a^2}{2} \left[\sqrt{(1 - \lambda)^2 + \frac{4\lambda}{(1 - a^2)}} - (1 - \lambda) \right]}. \tag{11}$$

(Superscript "cl" means "closed loop.")
   Since the denominator is greater than or equal to 1, we have $|a^{cl}| \le |a| < 1$, so the observer error system is stable.

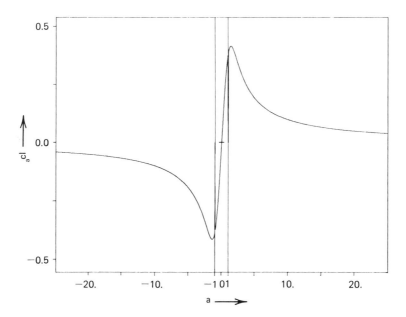

FIGURE 2.1-2 Closed-loop versus open-loop plant matrix.

b. Original System Unstable

If $|a| > 1$, then $(1 - a^2) < 0$ and $\lambda < 0$. In this case the positive solution to (5) is

$$p_+ = \frac{-q}{2\lambda} \left[\sqrt{(1 - \lambda)^2 + \frac{4\lambda}{(1 - a^2)}} + (1 - \lambda) \right]. \tag{12}$$

Then the error system is given by

$$a^{cl} = \frac{a}{1 - \frac{1 - a^2}{2} \left[\sqrt{(1 - \lambda)^2 + \frac{4\lambda}{(1 - a^2)}} + (1 - \lambda) \right]}. \tag{13}$$

It is not quite so easy this time to show that $|a^{cl}| < 1$ (although it is if $\lambda \neq 0$). Note, however, that if $|a| \gg 1$ then $\lambda \approx 0$ and it is not difficult to show that

$$a^{cl} \approx \frac{1}{a}. \tag{14}$$

This is certainly stable if $|a| \gg 1$!

A plot of a^{cl} versus a for the case $h^2 q / r = 1$ is shown in Fig. 2.1-2. Note that a^{cl} is always stable. ∎

Now consider the situation shown in Fig. 2.1-3. To the plant has been added *process noise* w_k, and to the measurement has been added *measurement noise* v_k. Under these more realistic circumstances the observer designed deterministically as above cannot be expected to make $\tilde{x}_k \to 0$. It

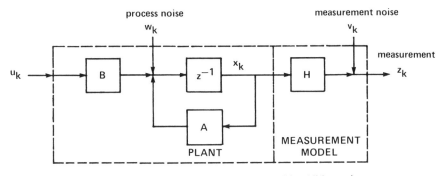

process noise

w_k

measurement noise

v_k

measurement

u_k

B

z^{-1} x_k

H z_k

A

PLANT

MEASUREMENT
MODEL

FIGURE 2.1-3   Plant and measurement with additive noise.

is somewhat surprising, however, that by using a structure virtually identical to Fig. 2.1-1 the error can be forced to zero even in this noisy case. What is required in general is to find a time-varying observer gain L_k based on stochastic considerations such as those of the previous chapter. The resulting state estimator depends on a time-varying version of the Riccati equation (2.1-6) and is known as the Kalman filter.

2.2   LINEAR STOCHASTIC SYSTEMS

To describe a system with noise like the one shown in Fig. 2.1-3, write the *discrete stochastic dynamical equation*

$$x_{k+1} = Ax_k + Bu_k + Gw_k \qquad (2.2\text{-}1a)$$

$$z_k = Hx_k + v_k, \qquad (2.2\text{-}1b)$$

where $x_k \in R^n$, $u_k \in R^m$, $z_k \in R^p$, $w_k \in R^l$, $v_k \in R^p$. Let *process noise* w_k have mean $\bar{w}_k = 0$ and covariance $\overline{w_k w_k^T} = Q$, which is symbolized as $w_k \sim (0, Q)$. The overbar denotes expected value. Let *measurement noise* v_k be $(0, R)$. Both $\{w_k\}$ and $\{v_k\}$ are assumed to be stationary *white noise processes*, that is, their covariance functions satisfy

$$P_{w_{k_1} w_{k_2}} = \overline{w_{k_1} w_{k_2}^T} = Q\,\delta(k_1 - k_2) \qquad (2.2\text{-}2)$$

where $\delta(k)$ is the Kronecker delta, and the corresponding version for v_k.

Suppose that initial state x_0 is unknown, but that there is available a priori knowledge that x_0 is $(\bar{x}_0, P_{x_0})$. We assume that x_0, w_k, and v_k are mutually *uncorrelated*, so that $\overline{w_k v_j^T} = 0$ for all j and k, and so on. Of course, P_{x_0}, Q, and R are symmetric and positive semidefinite. Signal u_k is a *deterministic* control input.

Under these circumstances, state x_k is a random variable with mean $\bar{x}_k$ and covariance P_{x_k}. Similarly, output z_k is random $(\bar{z}_k, P_{z_k})$. Notice that $\{x_k\}$ and $\{z_k\}$ are stochastic processes, in general nonstationary.

Exercise 2.2-1: *Discrete Markov Process*

A discrete first-order Markov process x_k can be represented as the state of the linear system (2.2-1a). Using this system, we can manufacture stationary or nonstationary Markov processes. Let x_k be a scalar for simplicity. Let $u_k = 0$, $g = 1$. (We use lowercase letters for scalar system parameters.)

a. *Stationary Markov Process*

Let the initial condition at $k = -\infty$ equal zero and $w_k \sim (0, (1-a^2)\sigma_w^2)$. Show that x_k is a stationary Markov process with spectral density

$$\Phi_X(\omega) = \frac{(1-a^2)\sigma_w^2}{1+a^2-2a\cos\omega} \tag{1}$$

and autocorrelation function

$$R_X(k) = \sigma_w^2 a^{|k|}. \tag{2}$$

In this case the system must be stable.

b. *Nonstationary Markov Process*

Let $w_k = 0$, and initial condition $x_0 \sim (0, \sigma_X^2)$. Show that x_k is a nonstationary Markov process with autocorrelation function

$$R_X(j, k) = \sigma_X^2 a^{(j+k)}. \tag{3}$$

This result is valid for a stable or unstable system.

Note: If w_k is Gaussian, then so is x_k. In that case the statistics of x_k are completely described by the autocorrelation function.

Propagation of Means and Covariances

It is not difficult to find how the means and covariances of x_k and z_k propagate through time under the influence of the dynamics (2.2-1a).

For the mean of the state, simply write

$$\bar{x}_{k+1} = \overline{Ax_k + Bu_k + Gw_k}$$

$$= A\bar{x}_k + B\bar{u}_k + G\bar{w}_k,$$

or,

$$\bar{x}_{k+1} = A\bar{x}_k + Bu_k. \tag{2.2-3}$$

The initial condition of (2.2-3) is $\bar{x}_0$, which is given. Therefore, the mean propagates exactly according to the deterministic dynamics (2.1-1a)!

To find how the state covariance propagates, write

$$P_{x_{k+1}} = \overline{(x_{k+1} - \bar{x}_{k+1})(x_{k+1} - \bar{x}_{k+1})^T}$$

$$= \overline{[A(x_k - \bar{x}_k) + Gw_k][A(x_k - \bar{x}_k) + Gw_k]^T}$$

$$= A\overline{(x_k - \bar{x}_k)(x_k - \bar{x}_k)^T}A^T$$

$$\quad + G\overline{w_k(x_k - \bar{x}_k)^T}A^T + A\overline{(x_k - \bar{x}_k)w_k^T}G^T$$

$$\quad + G\overline{w_k w_k^T}G^T,$$

or

$$P_{x_{k+1}} = AP_{x_k}A^T + GP_{w_kx_k}A^T + AP_{x_kw_k}G^T + GQG^T.$$

Hence,

$$P_{x_{k+1}} = AP_{x_k}A^T + GQG^T. \tag{2.2-4}$$

The last step follows from the uncorrelatedness of x_0 and w_k and the whiteness of w_k through the following argument. State x_k depends on x_0 and the process noise w_j for $j \le k - 1$. But w_k is uncorrelated with x_0 and all previous w_j, hence $P_{w_kx_k} = 0$.

Equation (2.2-4) is a *Lyapunov equation* for P_{x_k}. The initial condition is P_{x_0}, which is given. (A Lyapunov equation is a linear matrix equation.)

Next, we want to find the mean and covariance of the output. Taking expectations in (2.2-1b) yields

$$\bar{z}_k = H\bar{x}_k. \tag{2.2-5}$$

The cross covariance between state and output is

$$P_{x_kz_k} = \overline{(x_k - \bar{x}_k)(z_k - \bar{z}_k)^T}$$
$$= \overline{(x_k - \bar{x}_k)[H(x_k - \bar{x}_k) + v_k]^T},$$

or

$$P_{x_kz_k} = P_{x_k}H^T \tag{2.2-6}$$

due to the uncorrelatedness of x_k and v_k (why are they uncorrelated?).

For the covariance of the output,

$$P_{z_k} = \overline{(z_k - \bar{z}_k)(z_k - \bar{z}_k)^T}$$
$$= \overline{[H(x_k - \bar{x}_k) + v_k][H(x_k - \bar{x}_k) + v_k]^T}$$

so

$$P_{z_k} = HP_{x_k}H^T + R, \tag{2.2-7}$$

where we again used the uncorrelatedness assumptions.

Now note, and this is the most important point of this section, that x_k and z_k are jointly distributed random variables (RV) with moments given by

$$\begin{bmatrix} x_k \\ z_k \end{bmatrix} \sim \left(\begin{bmatrix} \bar{x}_k \\ \bar{z}_k \end{bmatrix}, \begin{bmatrix} P_{x_k} & P_{x_kz_k} \\ P_{z_kx_k} & P_{z_k} \end{bmatrix} \right)$$

$$= \left(\begin{bmatrix} \bar{x}_k \\ H\bar{x}_k \end{bmatrix}, \begin{bmatrix} P_{x_k} & P_{x_k}H^T \\ HP_{x_k} & HP_{x_k}H^T + R \end{bmatrix} \right), \tag{2.2-8}$$

where $\bar{x}_k$ and P_{x_k} are determined recursively by (2.2-3), (2.2-4). If RV sequence z_k is measured and the related RV x_k is unknown, then we can apply the results of the previous chapter to find an estimate $\hat{x}_k$ for the state given the data z_k. (At this point, x_0, w_k, v_k are assumed to have arbitrary statistics and only the first- and second-joint moments are given above, so the linear mean-square approach would be used.) In the next section we pursue this to find the optimal linear mean-square estimate of x_k given the measurement sequence z_k.

The quantities $\bar{x}_k$ and P_{x_k} found using (2.2-3) and (2.2-4) should actually be considered as the *conditional* mean and covariance given $\bar{x}_0$, P_{x_0}, and inputs $u_0, u_1, \ldots, u_{k-1}$.

All of these derivations have been for the case of general statistics. If it is assumed that x_0, w_k, and v_k are Gaussian, then since system (2.2-1) is linear, x_k and z_k are also Gaussian for all k. Furthermore, x_k and z_k are jointly Gaussian, with moments given by (2.2-8). In this case the first and second moments totally specify the joint PDF, so that complete information on the interdependence of state and output is available in (2.2-8).

If all statistics are Gaussian, then x_k is described by the probability density function (PDF)

$$f_{x_k}(x, k) = \frac{1}{\sqrt{(2\pi)^n |P_{x_k}|}} \exp\left[-\frac{1}{2}(x - \bar{x}_k)^T P_{x_k}^{-1}(x - \bar{x}_k) \right]. \qquad (2.2\text{-}9)$$

Note that this is actually a conditional PDF given $\bar{x}_0$, P_{x_0}, and inputs $u_0, u_1, \ldots, u_{k-1}$. Define the *hyperstate* as the conditional PDF of the state. Then in the Gaussian case, if no measurements are taken the hyperstate is completely described by the pair

$$(\bar{x}_k, P_{x_k}), \qquad (2.2\text{-}10)$$

the conditional mean and covariance, and so (2.2-3), (2.2-4) provide an updating algorithm for the entire hyperstate in the absence of measurements.

Example 2.2-2: *Propagation of Scalar Mean and Covariance*

Let $x_0 \sim N(0, 1)$ (i.e., normal) and x_k, z_k be given by (2.2-1) where all quantities are scalars and $w_k \sim N(0, 1)$, $v_k \sim N(0, \frac{1}{2})$. Suppose that all of our assumptions hold. Let $u_k = u_{-1}(k)$ the unit step. Take $b = 1$, $g = 1$ for simplicity.

Then the mean $\bar{x}_k$ is given by (2.2-3), so

$$\bar{x}_{k+1} = a\bar{x}_k + u_k,$$

which has solution

$$\bar{x}_k = a^k \bar{x}_0 + \sum_{i=0}^{k-1} a^{k-1-i}$$

$$= a^{k-1} \sum_{i=0}^{k-1} \left(\frac{1}{a}\right)^i$$

$$= a^{k-1} \left[\frac{1 - (1/a)^k}{1 - 1/a}\right],$$

$$\bar{x}_k = \frac{a^k - 1}{a - 1}. \qquad (1)$$

This has a limiting value $\bar{x}_\infty$ if and only if $|a| < 1$, that is, if (1) is stable. It is

$$\bar{x}_\infty = \frac{1}{1-a}. \tag{2}$$

To find the covariance of the state use (2.2-4),

$$P_{x_{k+1}} = a^2 P_{x_k} + 1, \qquad P_{x_0} = 1.$$

Thus

$$P_{x_k} = (a^2)^k + \sum_{i=0}^{k-1} (a^2)^{k-1-i}$$

$$= a^{2k} + \frac{a^{2k} - 1}{a^2 - 1}$$

$$= \frac{a^{2(k+1)} - 1}{a^2 - 1}. \tag{3}$$

There is a limiting solution P_{x_∞} if and only if (1) is stable, and then

$$P_{x_\infty} = \frac{1}{1 - a^2}. \tag{4}$$

Since x_k, z_k are normal, the conditional PDF $f_{x_k}(x, k)$ is given by (2.2-9), and we can sketch it as a function of k. This is done in Fig. 2.2-1 for $a = \frac{1}{2}$ and $a = 2$. It is evident that the effect of the process noise is to "flatten" the PDF for x_k as k

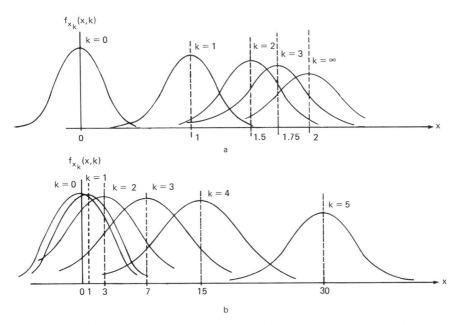

FIGURE 2.2-1 Propagation of the hyperstate. (a) Stable plant. (b) Unstable plant.

increases, thus decreasing our confidence that x_k is near $\bar{x}_k$. There are two cases. If the system is stable a limiting PDF $f_{x_\infty}(x)$ is reached and subsequent propagation does not inject further uncertainty. This limiting case is known as *statistical steady state*. In this case we can say that $\lim_{k\to\infty} x_k$ is near $\bar{x}_\infty$ to some tolerance described by P_{x_∞}. On the other hand, if (1) is unstable there is no limiting PDF and as k increases both the value of $\bar{x}_k$ and the probable spread of x_k about $\bar{x}_k$ go to infinity.

∎

The analysis of the time-varying stochastic system

$$x_{k+1} = A_k x_k + B_k u_k + G_k w_k$$

$$z_k = H_k x_k + v_k,$$

where $x_0 \sim (\bar{x}_0, P_{x_0})$, $w_k \sim (0, Q_k)$, $v_k \sim (0, R_k)$ is identical to the above. The results differ only in that subscripts k must be added to A, B, G, H, Q, and R.

Statistical Steady-State and Spectral Densities

In this subsection we are interested in examining the state and output of a discrete system when they are *stationary* random processes. We shall set the deterministic input u_k to zero.

The solution P_k to (2.2-4) is in general time varying. However, if A is asymptotically stable, then the Lyapunov equation has a steady-state solution $P \triangleq P_{k+1} = P_k$ for large k. This steady-state covariance satisfies the *algebraic Lyapunov equation*

$$P = APA^T + GQG^T. \tag{2.2-11}$$

Under these circumstances, as k increases the state x_k tends to a stationary Markov process with mean $\bar{x}_k = 0$ and covariance P. This limiting case is called *statistical steady state*. If A is unstable, then in general the covariance P_{x_k} increases without bound.

If the system has reached statistical steady state then it is possible to discuss spectral densities. Using (2.2-1a) and the identities at the beginning of Section 1.5, the spectral density of the state x_k is

$$\Phi_X(z) = (zI - A)^{-1} GQG^T (z^{-1}I - A)^{-T}, \tag{2.2-12}$$

where superscript $-T$ denotes the transpose of the inverse. [Note that $(zI - A)^{-1}G$ is the transfer function from w_k to x_k, and $\Phi_W(z) = Q$.]

Likewise, the spectral density of the output z_k in statistical steady state is

$$\Phi_Z(z) = H(zI - A)^{-1} GQG^T (z^{-1}I - A)^{-T} H^T + R. \tag{2.2-13}$$

(If we add two uncorrelated processes their spectral densities add.) This can be written as

$$\Phi_Z(z) = H(z)H^T(z^{-1}) + R \tag{2.2-14}$$

where

$$H(z) = H(zI - A)^{-1}G\sqrt{Q} \qquad (2.2\text{-}15)$$

is the transfer function from w'_k to z_k, where $w_k = \sqrt{Q}\,w'_k$ and w'_k is white noise with unit spectral density. $H(z)$ will be very useful to us later.

2.3 THE DISCRETE-TIME KALMAN FILTER

Consider the discrete-time stochastic dynamical system (2.2-1) with attendant assumptions. All matrices describing the system, and $\bar{x}_0$ and all covariances are given, with $Q \geq 0$, $P_{x_0} \geq 0$. We shall assume that $R > 0$. The statistics are not necessarily Gaussian, but can be arbitrary.

According to our results from Section 1.1, in the absence of measurements the best mean-square estimate for the unknown state x_k is just the conditional mean $\bar{x}_k$ of the state given $\bar{x}_0$ and $u_0, u_1, \ldots, u_{k-1}$, which propagates according to the deterministic version of the system (2.2-3). Then the error covariance is just the (conditional) covariance of the state, and it propagates according to the Lyapunov equation (2.2-4). To derive these results we used only the state equation (2.2-1a).

On the other hand, according to (1.1-41), (1.1-39), and Example 1.1-3, given the linear measurement model (2.2-1b), the best *linear* estimator for the RV x_k given measurement z_k and a priori statistics $\bar{x}_k$, P_{x_k} on x_k is given by

$$P_{\tilde{x}_k} = (P_{x_k}^{-1} + H^T R^{-1} H)^{-1} \qquad (2.3\text{-}1)$$

$$\hat{x}_k = \bar{x}_k + P_{\tilde{x}_k} H^T R^{-1}(z_k - H\bar{x}_k), \qquad (2.3\text{-}2)$$

where the estimation error is $\tilde{x}_k = (x_k - \hat{x}_k)$. In the case of Gaussian statistics this is the conditional mean estimator, which is the *optimal* estimator.

In this section we must reconcile the *time update* (2.2-3), (2.2-4), which describes the effects of the system dynamics (2.2-1a) with the *measurement update* (2.3-1), (2.3-2) which describes the effects of measuring the output z_k according to (2.2-1b). Since measurements z_k occur *at* each time k and the system dynamics describe the transition *between* times, we can do this as follows.

Kalman Filter Formulations

Define at each time k the *a priori* (before including measurement z_k) estimate $\hat{x}_k^-$ and error covariance P_k^-, and the *a posteriori* (after including measurement z_k) estimate $\hat{x}_k$ and error covariance P_k. The relation between P_k^- and P_k is illustrated in Fig. 2.3-1. Then the optimal estimator for the state of (2.2-1) with stated assumptions and given measurements z_k for all $k \geq 1$ is realized as follows, assuming that there is no measurement z_0 at time zero.

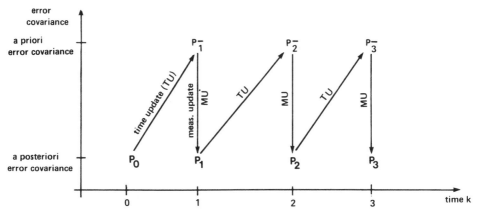

FIGURE 2.3-1   Error covariance update timing diagram.

Discrete Kalman Filter Algorithm:

Set $\hat{x}_0 = \bar{x}_0$, $P_0 = P_{x_0}$. Given $\hat{x}_{k-1}$ and P_{k-1}, for each value of $k \geq 1$ apply time update (2.2-3), (2.2-4) to obtain $\hat{x}_k^-$, P_k^-. Then apply measurement update (2.3-1), (2.3-2) to obtain $\hat{x}_k$, P_k.

The resulting estimator is called the *discrete-time Kalman filter* and it is the best *linear* estimator if x_0, w_k, v_k have arbitrary statistics. If x_0, w_k, v_k are normal then it is the optimal estimator. For convenient reference, the discrete-time Kalman filter is summarized in Table 2.3-1. The time-varying case, where all system matrices and noise covariances are functions of k, is presented for the sake of generality. Note that even for time-invariant system and statistics, the Kalman filter itself is still time varying since P_{k+1} depends on k!

The time update portion of the algorithm gives a *prediction* $\hat{x}_{k+1}^-$ of the state at time $k+1$, along with the associated error covariance P_{k+1}^-. The measurement update provides a *correction* based on the measurement z_{k+1} at time $k+1$ to yield the net a posteriori estimate $\hat{x}_{k+1}$ and its error covariance P_{k+1}. For this reason the equations in Table 2.3-1 are called the *predictor–corrector* formulation of the discrete Kalman filter.

The correction term in (2.3-7) depends on the residual

$$\tilde{z}_k = z_k - H_k \hat{x}_k^-. \tag{2.3-8}$$

The residual weighting coefficient is called the *Kalman gain*

$$K_k = P_k H_k^T R_k^{-1}. \tag{2.3-9}$$

It is often useful to have an expression for K_k in terms of the a priori

TABLE 2.3-1 Discrete-Time Kalman Filter

System model and measurement model:

$$x_{k+1} = A_k x_k + B_k u_k + G_k w_k \tag{2.3-3a}$$

$$z_k = H_k x_k + v_k \tag{2.3-3b}$$

$$x_0 \sim (\bar{x}_0, P_{x_0}), \qquad w_k \sim (0, Q_k), \qquad v_k \sim (0, R_k)$$

Assumptions:

$\{w_k\}$ and $\{v_k\}$ are white noise processes uncorrelated with x_0 and with each other. $R_k > 0$.

Initialization:

$$P_0 = P_{x_0}, \qquad \hat{x}_0 = \bar{x}_0.$$

Time update: (effect of system dynamics)

error covariance: $\qquad P_{k+1}^- = A_k P_k A_k^T + G_k Q_k G_k^T \tag{2.3-4}$

estimate: $\qquad \hat{x}_{k+1}^- = A_k \hat{x}_k + B_k u_k \tag{2.3-5}$

Measurement update: (effect of measurement z_k)

error covariance: $\qquad P_{k+1} = [(P_{k+1}^-)^{-1} + H_{k+1}^T R_{k+1}^{-1} H_{k+1}]^{-1} \tag{2.3-6}$

estimate: $\qquad \hat{x}_{k+1} = \hat{x}_{k+1}^- + P_{k+1} H_{k+1}^T R_{k+1}^{-1} (z_{k+1} - H_{k+1} \hat{x}_{k+1}^-) \tag{2.3-7}$

---

error covariance P_k^-. To derive this expression, write

$$\begin{aligned}
K_k &= P_k H_k^T R_k^{-1} \\
&= P_k H_k^T R_k^{-1} (R_k + H_k P_k^- H_k^T)(R_k + H_k P_k^- H_k^T)^{-1} \\
&= P_k (H_k^T + H_k^T R_k^{-1} H_k P_k^- H_k^T)(R_k + H_k P_k^- H_k^T)^{-1} \\
&= P_k [(P_k^-)^{-1} + H_k^T R_k^{-1} H_k] P_k^- H_k^T (R_k + H_k P_k^- H_k^T)^{-1} \\
&= P_k P_k^{-1} P_k^- H_k^T (R_k + H_k P_k^- H_k^T)^{-1},
\end{aligned}$$

so that

$$K_k = P_k^- H_k^T (H_k P_k^- H_k^T + R_k)^{-1}. \tag{2.3-10}$$

There are many equivalent formulations for equations (2.3-4) and (2.3-6). Using the matrix inversion lemma, the latter can be written as

$$P_{k+1} = P_{k+1}^- - P_{k+1}^- H_{k+1}^T (H_{k+1} P_{k+1}^- H_{k+1}^T + R_{k+1})^{-1} H_{k+1} P_{k+1}^-. \tag{2.3-11}$$

The advantage of (2.3-11) over (2.3-6) is that inversion of only one $p \times p$ matrix is required. (2.3-6) requires the inversion of two $n \times n$ matrices, and the number of measurements p is usually less than n. If $|P_{k+1}^-| = 0$ then (2.3-11) must be used in any event.

TABLE 2.3-2   Alternative Measurement Update Equations

$$K_{k+1} = P_{k+1}^- H_{k+1}^T (H_{k+1} P_{k+1}^- H_{k+1}^T + R_{k+1})^{-1} \quad (2.3\text{-}12)$$

$$P_{k+1} = (I - K_{k+1} H_{k+1}) P_{k+1}^- \quad (2.3\text{-}13)$$

$$\hat{x}_{k+1} = \hat{x}_{k+1}^- + K_{k+1}(z_{k+1} - H_{k+1}\hat{x}_{k+1}^-) \quad (2.3\text{-}14)$$

It is often convenient to replace the measurement update equations (2.3-6), (2.3-7) by the alternative equations in Table 2.3-2.

Equation (2.3-13) is equivalent (see the problems) to the *Joseph stabilized version*

$$P_{k+1} = (I - K_{k+1} H_{k+1}) P_{k+1}^- (I - K_{k+1} H_{k+1})^T + K_{k+1} R_{k+1} K_{k+1}^T. \quad (2.3\text{-}15)$$

This symmetric version guarantees the positive semidefiniteness of P_{k+1} in the presence of roundoff error and is often used for actual software implementation of the filter.

Note that the Kalman gain is the weighting which determines the influence of the residual in updating the estimate. Thus it plays the same role as the $1/(k+1)$ in the recursive sample mean equation in Example 1.4-1.

It is worth noting that although we demanded $|R_k| \neq 0$, this is not required since we only need $|H_k P_k^- H_k^T + R_k| \neq 0$.

An important point is that error covariance update equations (2.3-4), (2.3-6) do not depend on the measurements. Thus, if A_k, G_k, H_k, Q_k, R_k are known beforehand for all k, these equations can be used to find P_k off-line *before* the experiment is performed.

The time and measurement updates can be combined. The recursion for the a posteriori error covariance P_k involves a somewhat lengthy equation, so it is customary to use the recursion for the a priori error covariance P_k^-. To derive this, substitute (2.3-11) into (2.3-4). If we also use (2.3-7) in (2.3-5), there results the alternative filter formulation in Table 2.3-3, which manufactures the a priori estimate and error covariance. We have suppressed the time index on the plant matrices, as we shall usually do henceforth for convenience.

TABLE 2.3-3   Kalman Filter A Priori Recursive Formulation

Kalman filter:

$$\hat{x}_{k+1}^- = A(I - K_k H)\hat{x}_k^- + Bu_k + AK_k z_k \quad (2.3\text{-}16)$$

$$K_k = P_k^- H^T (H P_k^- H^T + R)^{-1} \quad (2.3\text{-}17)$$

Error covariance:

$$P_{k+1}^- = A[P_k^- - P_k^- H^T (H P_k^- H^T + R)^{-1} H P_k^-] A^T + G Q G^T \quad (2.3\text{-}18)$$

The a priori filter formulation is identical in structure to the deterministic observer in Section 2.1, and Fig. 2.1-1 provides a block diagram if L is replaced by the time-varying gain AK_k. Even if plant (2.3-3) is time invariant, the Kalman filter is time varying since the gain K_k depends on time. Some authors define AK_k as the Kalman gain. Equation (2.3-18) is a *Riccati equation*, which is a matrix quadratic equation. As was promised at the end of Section 2.1, it is just a time-varying version of the equation (2.1-6) which can be used to design the deterministic state observer!

In terms of the a posteriori estimate, we can write the recursion

$$\hat{x}_{k+1} = (I - K_{k+1}H)A\hat{x}_k + (I - K_{k+1}H)Bu_k + K_{k+1}z_{k+1}. \quad (2.3\text{-}19)$$

It is worth remarking that if the unknown is constant, then (2.3-3a) becomes

$$x_{k+1} = x_k \quad (2.3\text{-}20)$$

so the time update is simply $\hat{x}^-_{k+1} = \hat{x}_k$, $P_{k+1} = P_k$. Then the error covariance measurement update in Table 2.3-1 is

$$P^{-1}_{k+1} = P^{-1}_k + H^T_{k+1}R^{-1}_{k+1}H_{k+1}, \quad (2.3\text{-}21)$$

which has the simple solution for the information matrix

$$P^{-1}_k = P^{-1}_0 + \sum_{i=1}^{k} H^T_i R^{-1}_i H_i. \quad (2.3\text{-}22)$$

The Kalman filter is

$$\hat{x}_{k+1} = \hat{x}_k + K_{k+1}(z_{k+1} - H_{k+1}\hat{x}_k). \quad (2.3\text{-}23)$$

To get a maximum-likelihood estimate we would set $P^{-1}_0 = 0$, $\hat{x}_0 = 0$ to model a complete lack of a priori information. Use the matrix inversion lemma to compare (2.3-21), (2.3-23) to (1.4-7), (1.4-9).

Software for implementing the Kalman filter efficiently may be found in Bierman (1977). The number of computations required per iteration by the filter is on the order of n^3. As we shall see in subsequent examples, it is often possible in a particular application to do some *preliminary analysis* to simplify the equations that must actually be solved. Some short computer programs showing implementations of presimplified filters are included in the examples.

Examples and Exercises

To gain some insight into the Kalman filter let us consider a few examples.

Example 2.3-1: Ship Navigational Fixes

Suppose a ship is moving east at 10 mph. This velocity is assumed to be constant except for the effects of wind gusts and wave action. An estimate of the easterly position d and velocity $s = \dot{d}$ is required every hour. The navigator guesses at time zero that $d_0 = 0$, $s_0 = 10$, and his "guesses" can be modeled as having independent Gaussian distributions with variances $\sigma_{d_0}^2 = 2$, $\sigma_{s_0}^2 = 3$. Hence the initial estimates are $\hat{d}_0 = 0$, $\hat{s}_0 = 10$. The north–south position is of no interest. If d_k, s_k indicate position and velocity at hour k, we are required to find estimates $\hat{d}_k$, $\hat{s}_k$.

The first task is to model the system dynamics. During hour k the ship moves with velocity s_k mph, so its position will change according to

$$d_{k+1} = d_k + s_k. \tag{1}$$

The quantity s_{k+1} should ideally be equal to s_k since velocity is constant at 10 mph; however to model the unknown effects of wind and waves add a noise term w_k so that

$$s_{k+1} = s_k + w_k. \tag{2}$$

To apply the Kalman filter, suppose w_k is white, and independent of x_0 for all k. If, for example, there is a steady easterly current then w_k is not white. (We shall see later how to handle nonwhite noise. Basically, in this case a model of the noise process itself must be included in the system dynamics).

Define the system state at time k as

$$x_k \triangleq \begin{bmatrix} d_k \\ s_k \end{bmatrix},$$

and write (1) and (2) as one state equation

$$x_{k+1} = \begin{bmatrix} 1 & 1 \\ 0 & 1 \end{bmatrix} x_k + \begin{bmatrix} 0 \\ 1 \end{bmatrix} w_k. \tag{3}$$

There is no deterministic input u_k. Suppose that previous observations of ships in the region have shown that w_k has on the average no effect, but that its effect on velocity can be statistically described by a Gaussian PDF with covariance 1. Then $w_k \sim N(0, Q) = N(0, 1)$. From above, the initial condition is

$$x_0 \sim N(\bar{x}_0, P_0) = N\left(\begin{bmatrix} 0 \\ 10 \end{bmatrix}, \begin{bmatrix} 2 & 0 \\ 0 & 3 \end{bmatrix} \right).$$

This completes our modeling of the dynamics (2.3-3a).

Now, suppose that navigational fixes are taken at times $k = 1, 2, 3$. These fixes determine position to within an error covariance of 2, but yield directly no information about velocity. The fixes yield easterly positions of 9, 19.5, and 29 miles, respectively.

Before we can incorporate this new information into the estimates for d_k and s_k we need to model the measurement process. The situation described above corresponds to the output equation.

$$z_k = \begin{bmatrix} 1 & 0 \end{bmatrix} x_k + v_k, \qquad v_k \sim N(0, 2) \tag{4}$$

where only the position is observed and it has been assumed that the fixes are corrupted by additive zero mean Gaussian noise with covariance of 2. Equation (4) corresponds to (2.3-3b). We are now ready to estimate the state using the Kalman filter.

To incorporate the measurements $z_1 = 9$, $z_2 = 19.5$, $z_3 = 29$ into the estimate we use Table 2.3-1 to proceed as follows.

$k = 0$

   initial estimate:

$$\hat{x}_0 = \bar{x}_0 = \begin{bmatrix} 0 \\ 10 \end{bmatrix}$$

$$P_0 = \begin{bmatrix} 2 & 0 \\ 0 & 3 \end{bmatrix}$$

$k = 1$

   time update: propagate estimate to $k = 1$ using system dynamics.

$$\hat{x}_1^- = A\hat{x}_0 + Bu_0 = \begin{bmatrix} 10 \\ 10 \end{bmatrix}, \text{a priori estimate at } k = 1$$

$$P_1^- = AP_0A^T + GQG^T = \begin{bmatrix} 5 & 3 \\ 3 & 4 \end{bmatrix}$$

   measurement update: include the effects of z_1.

$$P_1 = [(P_1^-)^{-1} + H^TR^{-1}H]^{-1} = \begin{bmatrix} 1.429 & 0.857 \\ 0.857 & 2.714 \end{bmatrix}$$

$$\hat{x}_1 = \hat{x}_1^- + P_1H^TR^{-1}(z_1 - H\hat{x}_1^-)$$

$$= \begin{bmatrix} 9.286 \\ 9.571 \end{bmatrix}, \text{a posteriori estimate at } k = 1$$

$k = 2$

   time update: effects of system dynamics.

$$\hat{x}_2^- = A\hat{x}_1 + Bu_1 = \begin{bmatrix} 18.857 \\ 9.571 \end{bmatrix}, \text{a priori estimate at time } k = 2$$

$$P_2^- = AP_1A^T + GQG^T = \begin{bmatrix} 5.857 & 3.571 \\ 3.571 & 3.714 \end{bmatrix}$$

   measurement update: effects of z_2.

$$P_2 = [(P_2^-)^{-1} + H^TR^{-1}H]^{-1} = \begin{bmatrix} 1.491 & 0.909 \\ 0.909 & 2.091 \end{bmatrix}$$

$$\hat{x}_2 = \hat{x}_2^- + P_2H^TR^{-1}(z_2 - H\hat{x}_2^-)$$

$$= \begin{bmatrix} 19.336 \\ 9.864 \end{bmatrix}, \text{a posteriori estimate at time } k = 2$$

$k = 3$

   time update: effects of system dynamics.

$$\hat{x}_3^- = A\hat{x}_2 + Bu_2 = \begin{bmatrix} 29.2 \\ 9.864 \end{bmatrix}, \text{a priori estimate}$$

$$P_3^- = AP_2A^T + GQG^T = \begin{bmatrix} 5.4 & 3 \\ 3 & 3.091 \end{bmatrix}$$

measurement update: effects of z_3.

$$P_3 = [(P_3^-)^{-1} + H^T R^{-1} H]^{-1} = \begin{bmatrix} 1.46 & 0.811 \\ 0.811 & 1.875 \end{bmatrix}$$

$$\hat{x}_3 = \hat{x}_3^- + P_3 H^T R^{-1}(z_3 - H\hat{x}_3^-)$$

$$= \begin{bmatrix} 29.054 \\ 9.783 \end{bmatrix}, \text{ a posteriori estimate}$$

Note that there are in fact two covariance sequences, the a priori sequence P_1^-, P_2^-, P_3^- and the a posteriori sequence P_0, P_1, P_2, P_3 as shown in Fig. 2.3-1. As indicated in the figure, a time update generally increases the error covariance due to the injection of uncertainty by the process noise w_k. A measurement update generally decreases the error covariance due to the injection of information by the measurement.

In this example, the error covariance sequences seem to be approaching a limit even after three iterations. In fact, limiting values to three decimal places for P_k^- and P_k are reached after 10 iterations. They are

$$P_{10}^- = \begin{bmatrix} 4.783 & 2.604 \\ 2.604 & 2.836 \end{bmatrix}$$

$$P_{10} = \begin{bmatrix} 1.410 & 0.768 \\ 0.768 & 1.836 \end{bmatrix}.$$

To give an indication of how the covariances develop, the $(2, 2)$ components of P_k^- and P_k, that is, the a priori and a posteriori velocity error variances $\overline{(s_k - \hat{s}_k)(s_k - \hat{s}_k)^T}$, are plotted in Fig. 2.3-2. It is thus apparent that even for an unstable system, taking measurements can make the error covariance converge to a finite value.

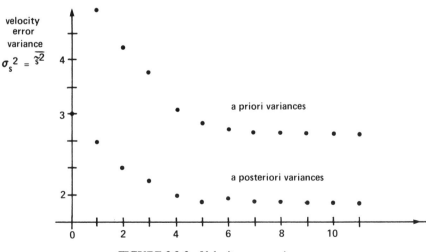

FIGURE 2.3-2 Velocity error variances.

Example 2.3-2: **Recursive Estimation of a Constant Scalar Unknown**

To estimate the value of a constant scalar unknown x from measurements corrupted by additive white noise, the system and measurement models are

$$x_{k+1} = x_k$$

$$z_k = x_k + v_k, \tag{1}$$

where $x_k, z_k \in R$ and $v_k \sim (0, r)$ is white. Let $x_0 \sim (\bar{x}_0, p_0)$.

Before any measurements are taken, determine the error covariance off-line using (2.3-4), (2.3-6) (i.e. (2.3-21)). Thus (Gelb, 1974)

$k = 0$

$$p_0 \text{ is given}$$

$k = 1$

   time update:

$$p_1^- = p_0$$

   measurement update:

$$p_1 = \left(\frac{1}{p_0} + \frac{1}{r} \right)^{-1} = \frac{p_0}{1 + p_0/r}$$

$k = 2$
   time update:

$$p_2^- = p_1 = \frac{p_0}{1 + p_0/r}$$

   measurement update:

$$p_2 = \left(\frac{1}{p_1} + \frac{1}{r} \right)^{-1}$$

$$= \frac{p_1}{1 + p_1/r} = \frac{p_0}{1 + 2p_0/r}$$

$\vdots$

k
   time update:

$$p_k^- = \frac{p_0}{1 + (k-1)p_0/r}$$

   measurement update:

$$p_k = \frac{p_0}{1 + k(p_0/r)} \tag{2}$$

The error covariance is shown in Fig. 2.3-3.

Now, to find the estimate $\hat{x}_k$ incorporating the a priori information on x_0 and all measurements through time k, use (2.3-5), (2.3-7) (or (2.3-23)), so that $\hat{x}_0 = \bar{x}_0$ and

$$\hat{x}_{k+1}^- = \hat{x}_k,$$

$$\hat{x}_{k+1} = \hat{x}_k + \frac{p_{k+1}}{r} (z_{k+1} - \hat{x}_k). \tag{3}$$

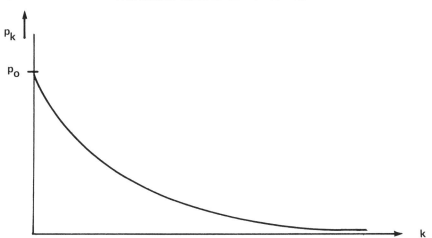

FIGURE 2.3-3   Error covariance for estimation of a constant.

Equations (2) and (3) constitute the Kalman filter for this example. Note that even in this case of time-invariant system and statistics, the filter is still time varying since the Kalman gain (2.3-9) given by

$$K_k = \frac{p_k}{r} = \frac{p_0/r}{1 + k(p_0/r)} \qquad (4)$$

is dependent on k. In fact, the weighting $K_{k+1} = p_{k+1}/r$ of the residual in (3) goes to zero with k. This means that as more measurements are taken and our confidence in the estimate grows, subsequent measurements are increasingly discounted.

It is not difficult to find an explicit expression for $\hat{x}_k$. Putting (4) into (3) yields

$$\hat{x}_{k+1} = \hat{x}_k + \frac{p_0/r}{1 + (k+1)p_0/r}(z_{k+1} - \hat{x}_k). \qquad (5)$$

Therefore

$$\hat{x}_1 = \hat{x}_0 + \frac{p_0/r}{1 + p_0/r}(z_1 - \hat{x}_0) = \frac{\hat{x}_0 + (p_0/r)z_1}{1 + p_0/r}$$

$$\hat{x}_2 = \hat{x}_1 + \frac{p_0/r}{1 + 2p_0/r}(z_2 - \hat{x}_1) = \frac{(1 + p_0/r)\hat{x}_1 + (p_0/r)z_2}{1 + 2p_0/r}$$

$$= \frac{\hat{x}_0 + (p_0/r)(z_1 + z_2)}{1 + 2p_0/r},$$

and in general

$$\hat{x}_k = \frac{1}{1 + kp_0/r}\left[\hat{x}_0 + (p_0/r)\sum_{i=1}^{k} z_i\right]. \qquad (6)$$

As $k \to \infty$, $\hat{x}_k$ becomes

$$\hat{x}_k \to \frac{1}{k}\sum_{i=1}^{k} z_i, \qquad (7)$$

the average of the data points. This is the *law of large numbers* (cf. Bryson and Ho 1975).

Now, let us consider two special cases. First, let $r \to \infty$ so that the measurements are completely unreliable. Then $K_k \to 0$ so that the estimate (3) becomes $\hat{x}_{k+1} = \hat{x}_k = \bar{x}_0$. The filter is thus telling us to disregard all measurements and retain our initial estimate $\hat{x}_0 = \bar{x}_0$. The error covariance p_k is then constant at p_0 for all k.

Suppose on the other hand that $\bar{x}_0 = 0$ and $p_0 \to \infty$, corresponding to the lack of any a priori information on the unknown; the case of deterministic x. The filter (2), (3) now provides the maximum-likelihood estimate (if v_k is normal). Then $K_k \to 1/k$, so that the estimate (3) becomes

$$\hat{x}_0 = 0$$

$$\hat{x}_{k+1} = \hat{x}_k + \frac{1}{k+1}(z_{k+1} - \hat{x}_k). \tag{8}$$

This is exactly the recursive sample mean equation from Example 1.4-1! Note that the same result is obtained if $r \to 0$ (the case of perfect measurements).

If the statistics are time varying so that $v_k \sim (0, r_k)$ then (2), (3) are replaced by

$$p_{k+1} = \frac{p_k}{1 + p_k / r_k}, \qquad \text{given } p_0 \tag{9}$$

$$\hat{x}_{k+1} = \hat{x}_k + \frac{p_{k+1}}{r_k}(z_{k+1} - \hat{x}_k), \qquad \hat{x}_0 = \bar{x}_0. \tag{10}$$

∎

To summarize the insight gained from this example, the Kalman filter embodies many different estimators. If x_0, w_k, v_k have arbitrary statistics it is the best *linear* estimator. If all statistics are normal, it yields the conditional mean estimate. If all statistics are normal and we set $\bar{x}_0 = 0$, $P_0 \to \infty$ then the filter yields the maximum-likelihood estimate.

If $P_0 \to \infty$ then (2.3-6) should be used instead of (2.3-11) for the measurement update, at least initially, since this uses the information matrix $(P_k^-)^{-1}$ which is bounded (i.e., $P_0^{-1} = 0$).

Example 2.3-3: Filtering for a Scalar System

Let there be prescribed the scalar time-invariant system and measurement described by

$$x_{k+1} = ax_k + bu_k + w_k \tag{1}$$

$$z_k = hx_k + v_k, \tag{2}$$

with $x_0 \sim (\bar{x}_0, p_0)$, $w_k \sim (0, q)$, $v_k \sim (0, r)$. Let $\{w_k\}$ and $\{v_k\}$ be white noise processes uncorrelated with each other and with x_0.

Using the formulation in Table 2.3-3, the Riccati equation becomes

$$P_{k+1}^- = a^2 p_k^- - \frac{a^2 h^2 (p_k^-)^2}{h^2 p_k^- + r} + q, \tag{3}$$

or

$$p_{k+1}^- = \frac{a^2 r p_k^-}{h^2 p_k^- + r} + q. \tag{4}$$

The Kalman gain becomes

$$K_{k+1} = \frac{h p_{k+1}^-}{h^2 p_{k+1}^- + r}, \tag{5}$$

so that the (a priori) closed-loop formulation is

$$\hat{x}_{k+1}^- = \frac{ar}{h^2 p_k^- + r} \hat{x}_k^- + b u_k + \frac{a h p_k^-}{h^2 p_k^- + r} z_k. \tag{6}$$

Even in this simple case the closed-form solution to (4) is difficult to deal with. Let us consider three special cases for which it can be solved without too much trouble.

a. Perfect Measurements

Let $r \to 0$ so that our measurements become exact. Then (4) becomes

$$p_{k+1}^- = q \tag{7}$$

and (6) yields for the a priori estimate,

$$\hat{x}_{k+1}^- = \frac{a}{h} z_k + b u_k. \tag{8}$$

We can understand (8) by noting that in this case of no measurement noise, the best (a posteriori) estimate at time k of the state given z_k is just

$$\hat{x}_k = \frac{1}{h} z_k. \tag{9}$$

The predicted value at time $k + 1$ is therefore

$$\hat{x}_{k+1}^- = a \hat{x}_k + b u_k,$$

or (8). Thus, in the case of perfect measurements the estimate depends only on the latest measurement, and the a priori error covariance is just the effect of the process noise over one time step. The initial information $(\bar{x}_0, p_0)$ is not used at all after $k = 0$.

b. Completely Unreliable Measurements

Let $r \to \infty$ so that the measurements are totally unreliable. Then (4) becomes

$$p_{k+1}^- = a^2 p_k^- + q. \tag{10}$$

The solution to this equation is

$$p_k^- = a^{2k} p_0 + q \sum_{i=0}^{k-1} (a^2)^{k-i-1}$$

$$= a^{2k} p_0 + q \left(\frac{1 - a^{2k}}{1 - a^2} \right), \tag{11}$$

or

$$p_k^- = a^{2k} p_0 \left(1 + \frac{q}{p_0} \sum_{i=1}^{k} a^{-2i} \right); \qquad k \ge 1, \tag{12}$$

which converges to a limiting value if and only if a is stable. Equation (6) becomes

$$\hat{x}_{k+1}^- = a\hat{x}_k + bu_k, \qquad \text{with } \hat{x}_0^- = \bar{x}_0. \tag{13}$$

In this case, the estimate depends completely on the system dynamics. In fact, reexamine Example 2.2-2 to see that these two situations are identical!

c.   No Process Noise

Let $q \to 0$. In this situation it is easier to propagate the information matrix than the error covariance. Combine (2.3-4) and (2.3-6) to obtain

$$(p_{k+1}^-)^{-1} = \frac{p_k^{-1}}{a^2} = \frac{1}{a^2} (p_k^-)^{-1} + \frac{h^2}{ra^2}, \tag{14}$$

which has solution

$$(p_k^-)^{-1} = \frac{1}{p_0 a^{2k}} + \sum_{i=1}^{k-1} \frac{h^2}{ra^2} \left(\frac{1}{a^2} \right)^{k-i-1}$$

for $k \ge 2$, with $(p_1^-)^{-1} = 1/a^2 p_0$. This is equivalent to

$$(p_k^-)^{-1} = \frac{1}{p_0 a^{2k}} + \frac{h^2}{r} \frac{(a^2 - a^{2k})}{(1 - a^2)a^{2k}}, \tag{15}$$

which can in turn be manipulated to yield

$$p_k^- = \frac{a^{2k} p_0}{1 + \frac{h^2}{r} p_0 \left(\frac{a^2 - a^{2k}}{1 - a^2} \right)}, \tag{16}$$

or

$$p_k^- = \frac{a^{2k} p_0}{1 + \frac{h^2}{r} p_0 \sum_{i=1}^{k-1} a^{2i}}, \tag{17}$$

for $k \ge 2$. If $a = h = 1$ this is equivalent to Example 2.3-2 equation (2).

Now, compare (17) to (12). Equation (12) converges to a steady-state solution as k increases if and only if $|a| < 1$. However, (17) can converge for unstable a. In fact, it converges for any value of a as long as h^2/r is nonzero!   ∎

Exercise 2.3-4:   Kalman Filter Derivation Using Orthogonality Principle

Assume that the system with measurement (2.3-3) is prescribed, and let all statistics be Gaussian.

Suppose that $z_1, \ldots, z_{k-1}$ have been measured and that the optimal estimate $\hat{x}_k^-$ and error covariance P_k^- have been computed. Now another measurement z_k is made.

   a.   Use the orthogonality principle from Chapter 1 to derive the measurement update equations for the estimate and error covariance.

b. Prove that the residual $\tilde{z}_k$ is orthogonal to all the previous data $z_0, \ldots, z_{k-1}$. It therefore contains the new information in measurement z_k which could not have been anticipated from the previous data. For this reason $\{\tilde{z}_k\}$ is also called the *innovations* process.

c. Now derive the time update equations by finding the best estimate $\hat{x}_{k+1}^-$ for the linear function (2.3-3a) of x_k.

d. Prove that the innovations sequence is a white noise process.

Notes: See Luenberger (1969) for an excellent treatment using these vector space methods.

If the statistics are not Gaussian, then the recursive filter derived in this exercise is the best *linear* estimator.

It is also possible to derive the Kalman filter from the Wiener–Hopf equation. We shall do this in the continuous case in Chapter 3. ■

2.4 DISCRETE MEASUREMENTS OF CONTINUOUS-TIME SYSTEMS

With the increasing sophistication of microprocessors, more and more control schemes are being implemented digitally. In a digital implementation the state of the plant is required at discrete time instants. It therefore behooves us to discuss state estimation for continuous-time systems using discrete, or sampled, data.

Discretization of Continuous Stochastic Systems

Suppose the plant to be controlled is described by

$$\dot{x}(t) = Ax(t) + Bu(t) + Gw(t) \tag{2.4-1a}$$

with measurements

$$z(t) = Hx(t) + v(t). \tag{2.4-1b}$$

Let $x(0) \sim (\bar{x}_0, P_0)$, $w(t) \sim (0, Q)$, $v(t) \sim (0, R)$, where $\{w(t)\}$ and $\{v(t)\}$ are white noise processes uncorrelated with each other and with $x(0)$.

To control the plant using state feedback, we would first need to estimate its state $x(t)$. Suppose the microprocessor samples input $u(t)$ and measurement $z(t)$ every T sec. We can apply the discrete Kalman filter to the continuous plant by first *discretizing* it. To do this, we begin with the solution to (2.4-1a)

$$x(t) = e^{A(t-t_0)}x(t_0) + \int_{t_0}^{t} e^{A(t-\tau)}Bu(\tau)\,d\tau + \int_{t_0}^{t} e^{A(t-\tau)}Gw(\tau)\,d\tau. \tag{2.4-2}$$

To describe the state propagation between samples, let $t_0 = kT$, $t =$

$(k + 1)T$ for an integer k. Defining the sampled state function as $x_k \triangleq x(kT)$, we can write

$$x_{k+1} = e^{AT} x_k + \int_{kT}^{(k+1)T} e^{A[(k+1)T-\tau]} Bu(\tau)\, d\tau$$

$$+ \int_{kT}^{(k+1)T} e^{A[(k+1)T-\tau]} Gw(\tau)\, d\tau. \qquad (2.4\text{-}3)$$

Assuming that control input $u(t)$ is reconstructed from the discrete control sequence u_k by using a zero-order hold, $u(\tau)$ has a constant value of $u(kT) = u_k$ over the integration interval.

The third term is a smoothed (i.e., low-pass filtered) version of the continuous white process noise $w(t)$ weighted by the state transition matrix and the noise input matrix G. It is not difficult to show that this term describes a discrete white noise sequence. Hence define

$$w_k = \int_{kT}^{(k+1)T} e^{A[(k+1)T-\tau]} Gw(\tau)\, d\tau. \qquad (2.4\text{-}4)$$

Now (2.4-3) becomes

$$x_{k+1} = e^{AT} x_k + \int_{kT}^{(k+1)T} e^{A[(k+1)T-\tau]} B\, d\tau \cdot u_k + w_k.$$

On changing variables twice ($\lambda = \tau - kT$ and then $\tau = T - \lambda$) we obtain

$$x_{k+1} = e^{AT} x_k + \int_0^T e^{A\tau} B\, d\tau \cdot u_k + w_k. \qquad (2.4\text{-}5)$$

This is the sampled version of (2.4-1a), which we can write as

$$x_{k+1} = A^s x_k + B^s u_k + w_k \qquad (2.4\text{-}6)$$

with

$$A^s = e^{AT} \qquad (2.4\text{-}7)$$

$$B^s = \int_0^T e^{A\tau} B\, d\tau. \qquad (2.4\text{-}8)$$

To find the covariance Q^s of the new noise sequence w_k in terms of Q, write

$$Q^s = \overline{w_k w_k^T}$$

$$\int\int_{kT}^{(k+1)T} e^{A[(k+1)T-\tau]} G\overline{w(\tau)w(\sigma)^T} G^T e^{A^T[(k+1)T-\sigma]}\, d\tau\, d\sigma.$$

But

$$\overline{w(\tau)w(\sigma)^T} = Q\delta(\tau - \sigma),$$

so

$$Q^s = \int_{kT}^{(k+1)T} e^{A[(k+1)T-\tau]} GQG^T e^{A^T[(k+1)T-\tau]} \, d\tau.$$

By changing variables twice (as above)

$$Q^s = \int_0^T e^{A\tau} GQG^T e^{A^T\tau} \, d\tau. \tag{2.4-9}$$

It is worth noting that even if Q is diagonal, Q^s need not be. Sampling can destroy independence among the components of the process noise!

Discretizing the measurement equation is easy since it has no dynamics:

$$z_k = Hx_k + v_k. \tag{2.4-10}$$

To find the covariance R^s of v_k in terms of the given R we need to think a little. Define the *unit rectangle* as

$$\prod(t) = \begin{cases} 1, & -1/2 \le t \le 1/2 \\ 0, & \text{otherwise} \end{cases} \tag{2.4-11}$$

and note that $\lim_{T\to 0} (1/T)\Pi(t/T) = \delta(t)$, the Kronecker delta. We can write the covariance of v_k in terms of its spectral density R^s as

$$R_v(k) = R^s \, \delta(k). \tag{2.4-12}$$

Since $\delta(k)$ has a value of one at $k = 0$, in this discrete case the covariance is equal to R^s, a finite matrix. The covariance of $v(t)$ on the other hand is given by

$$R \, \delta(t), \tag{2.4-13}$$

where R is its spectral density. For continuous white noise the covariance is unbounded since $\delta(t)$ is unbounded at $t = 0$.

To make (2.4-12) approach (2.4-13) in the limit we must write

$$R \, \delta(t) = \lim_{T\to 0} (R^s T) \frac{1}{T} \prod \left(\frac{t}{T}\right), \tag{2.4-14}$$

so that

$$R^s = \frac{R}{T}. \tag{2.4-15}$$

By examining (2.4-14) it is evident that equation (2.4-15) is the relation required between R and R^s so that noise processes $v(t)$ and v_k have the same spectral densities. Note that if R is diagonal then so is R^s.

Sampling does not affect the a priori information $\bar{x}_0$, P_0.

It is worthwhile to write down the first few terms of the infinite series expansions for the sampled matrices. They are

$$A^s = I + AT + \frac{A^2 T^2}{2!} + \cdots \tag{2.4-16a}$$

$$B^s = BT + \frac{ABT^2}{2!} + \cdots \tag{2.4-16b}$$

$$Q^s = GQG^T T + \frac{(AGQG^T + GQG^T A^T) T^2}{2!} + \cdots . \tag{2.4-16c}$$

If terms of order T^2 are disregarded, then the result is *Euler's approximation* to the sampled system. In Euler's approximation note that process noise covariance Q^s results from multiplication by T, and measurement noise covariance R^s results from division by T.

Examples

Several examples are given to illustrate the application of the discrete Kalman filter to continuous systems. First we discuss the discretization of a continuous system which occurs quite often.

Example 2.4-1:   Sampling Motion Governed by Newton's Laws

Consider the continuous system described by

$$\dot{x} = \begin{bmatrix} 0 & 1 \\ 0 & 0 \end{bmatrix} x + \begin{bmatrix} 0 \\ 1/m \end{bmatrix} u + w. \tag{1}$$

If $u(t) = mg$ is a constant, then by an easy integration the components of $x = [d \quad s]^T$ satisfy

$$s(t) = s(0) + gt \tag{2}$$

$$d(t) = d(0) + s(0)t + \tfrac{1}{2} gt^2. \tag{3}$$

For an input force $u(t) = F = mg$, therefore, (1) is just a formulation of Newton's law $F = ma$, with d the position and s the velocity.

Let $x(0) \sim (\bar{x}_0, p_0)$, $w \sim (0, Q)$, and measurements of position $d(t)$ be made so that

$$z = [1 \quad 0]x + v. \tag{4}$$

Measurement noise v is $(0, r)$ and all uncorrelatedness assumptions hold.

If measurements are made at intervals of T units, the discretized system can quickly be written down by noting that, since A is nilpotent (i.e., $A^2 = 0$), the sequences for the sampled plant matrices are all finite, so

$$A^s = e^{AT} = 1 + AT = \begin{bmatrix} 1 & T \\ 0 & 1 \end{bmatrix}, \tag{5}$$

and

$$B^s = BT + \frac{ABT^2}{2} = \frac{1}{m} \begin{bmatrix} T^2/2 \\ T \end{bmatrix}. \tag{6}$$

Letting

$$Q = \begin{bmatrix} q_1 & q_2 \\ q_2 & q_4 \end{bmatrix},$$

we get

$$Q^s = QT + \begin{bmatrix} q_2 & q_4/2 \\ q_4/2 & 0 \end{bmatrix} T^2 + \begin{bmatrix} q_4/3 & 0 \\ 0 & 0 \end{bmatrix} T^3. \tag{7}$$

Finally,

$$R^s = \frac{R}{T}. \tag{8}$$

The sampled version of (1), (4) is thus

$$x_{k+1} = \begin{bmatrix} 1 & T \\ 0 & 1 \end{bmatrix} x_k + \frac{1}{m} \begin{bmatrix} T^2/2 \\ T \end{bmatrix} u_k + w_k \tag{9}$$

$$z_k = [1 \quad 0] x_k + v_k, \tag{10}$$

with noise covariances Q^s and R^s given above. Compare (9) to equation (3) of Example 2.3-1, where we used Euler's approximation.

Note that even if Q is diagonal, Q^s may not be.   ∎

Example 2.4-2:   Estimation of a Constant

If x is a time-invariant scalar random variable measured in additive white noise then the system and measurement models are

$$\dot{x} = 0 \tag{1}$$

$$z = x + v. \tag{2}$$

Let $x \sim (\bar{x}_0, p_0)$ and $v(t) \sim (0, r^c)$ be uncorrelated. Compare this continuous formulation to the discrete formulation in Example 2.3-2.

To estimate x using measurements provided at sampling interval T, discretize (1) and (2) to get

$$x_{k+1} = x_k \tag{3}$$

$$z_k = x_k + v_k, \tag{4}$$

with $v_k \sim (0, r^c/T)$. These are identical to equations (1) in Example 2.3-2, so from the results of that example the error covariance is given by

$$p_k = \frac{p_0}{1 + kT(p_0/r^c)}. \tag{5}$$

The Kalman gain is

$$K_k = \frac{p_0 T/r^c}{1 + kT(p_0/r^c)}. \tag{6}$$

∎

Example 2.4-3:   *Radar Tracking with Discrete Measurements*

Consider the two dimensional radar tracking problem illustrated in Fig. 2.4-1. For simplicity let velocity V be in the x direction.

a.   State and Measurement Models

To derive the state model write

$$r^2 = x^2 + y^2$$

$$2r\dot{r} = 2x\dot{x} + 2y\dot{y}$$

$$\dot{r} = \frac{x}{r}\dot{x} + \frac{y}{r}\dot{y},$$

but $x = r\cos\theta$, $\dot{x} = V$, and $\dot{y} = 0$; so

$$\dot{r} = V\cos\theta. \tag{1}$$

Differentiating,

$$\ddot{r} = \dot{V}\cos\theta - V\dot{\theta}\sin\theta. \tag{2}$$

For θ, write

$$\tan\theta = \frac{y}{x};$$

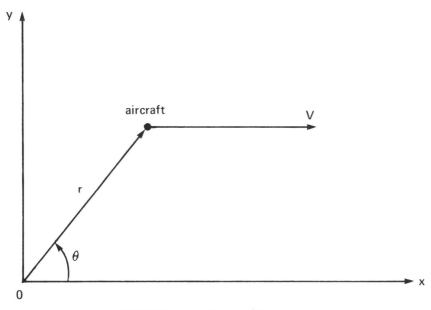

FIGURE 2.4-1 Radar tracking geometry.

so

$$\dot\theta \sec^2 \theta = \frac{x\dot y - \dot x y}{x^2}$$

$$\dot\theta = \frac{x\dot y - \dot x y}{x^2 \sec^2 \theta} = \frac{x\dot y - \dot x y}{r^2},$$

but $y = r \sin \theta$, $\dot x = V$, and $\dot y = 0$; so

$$\dot\theta = -\frac{V}{r} \sin \theta. \tag{3}$$

Differentiating

$$\ddot\theta = -\left(\frac{\dot V r - V\dot r}{r^2}\right) \sin \theta - \frac{V}{r} \dot\theta \cos \theta. \tag{4}$$

State equations (2) and (4) are nonlinear. We shall see in Chapter 5 how to apply the Kalman filter to nonlinear systems. In this example we will apply the Kalman filter in Table 2.3-1 by taking advantage of its robustness and converting to a linear system as follows.

Assume that the aircraft is not accelerating so that $\dot V = 0$. Then, if r and θ do not change much during each scan of the radar (about 1–10 sec), we have $\ddot r \approx 0$ and $\ddot\theta \approx 0$. To account for the small changes in $\dot r$ and $\dot\theta$ during the scan time T, introduce random disturbances $w_1(t)$ and $w_2(t)$ so that

$$\frac{d}{dt} \dot r = w_1(t) \tag{5}$$

$$\frac{d}{dt} \dot\theta = w_2(t). \tag{6}$$

The robustness properties of the Kalman filter in the presence of process noise will now allow tracking of $\dot r$ and $\dot\theta$.

Notice that process noise has been added to compensate for the *unmodeled dynamics* that we introduced by ignoring the nonlinear terms. We shall have more to say about this in Chapter 4.

The state equations are now linear:

$$\frac{d}{dt}\begin{bmatrix} r \\ \dot r \\ \theta \\ \dot\theta \end{bmatrix} = \begin{bmatrix} 0 & 1 & 0 & 0 \\ 0 & 0 & 0 & 0 \\ 0 & 0 & 0 & 1 \\ 0 & 0 & 0 & 0 \end{bmatrix}\begin{bmatrix} r \\ \dot r \\ \theta \\ \dot\theta \end{bmatrix} + \begin{bmatrix} 0 \\ w_1(t) \\ 0 \\ w_2(t) \end{bmatrix}. \tag{7}$$

Symbolize (7) as

$$\dot X(t) = AX(t) + w(t). \tag{8}$$

[Do not confuse the state $X(t)$ with the x coordinate in Fig. 2.4-1!] The process $w(t)$ is known as *maneuver noise*.

The radar measures range r and azimuth θ, so

$$z(t) = \begin{bmatrix} 1 & 0 & 0 & 0 \\ 0 & 0 & 1 & 0 \end{bmatrix} X(t) + v(t) \tag{9}$$

is the measurement model.

To find the maneuver noise covariance, suppose the disturbance accelerations are independent and uniformly distributed between $\pm a$. Then (show this) their variances are $a^2/3$. Note that $w_2(t)$ is an *angular* acceleration, so $w_2(t)r$ is distributed according to the maneuver noise PDF and $w_2(t)$ has variance $a^2/3r^2$. On the other hand, $w_1(t)$ is a radial acceleration, and its variance is $a^2/3$. Therefore $w(t)$ has covariance

$$Q = \begin{bmatrix} 0 & 0 & 0 & 0 \\ 0 & a^2/3 & 0 & 0 \\ 0 & 0 & 0 & 0 \\ 0 & 0 & 0 & a^2/3r^2 \end{bmatrix}. \tag{10}$$

Let the variances of the range and azimuth measurements be σ_r^2 and σ_θ^2 (for radar $\sigma_r^2 \ll \sigma_\theta^2$). Then the measurement noise has covariance

$$R = \begin{bmatrix} \sigma_r^2 & 0 \\ 0 & \sigma_\theta^2 \end{bmatrix}. \tag{11}$$

b. Discretization

The measurements $z(t)$ are available at each scan. Let T be the scan time. Then, since $A^2 = 0$ it is easy to use the series representations for the sampled matrices to get

$$x_{k+1} = \begin{bmatrix} 1 & T & 0 & 0 \\ 0 & 1 & 0 & 0 \\ 0 & 0 & 1 & T \\ 0 & 0 & 0 & 1 \end{bmatrix} x_k + w_k, \tag{12}$$

$$z_k = \begin{bmatrix} 1 & 0 & 0 & 0 \\ 0 & 0 & 1 & 0 \end{bmatrix} x_k + v_k, \tag{13}$$

where

$$w_k \sim \left(0, \frac{a^2}{3} \begin{bmatrix} T^3/3 & T^2/2 & 0 & 0 \\ T^2/2 & T & 0 & 0 \\ 0 & 0 & T^2/3r^2 & T^2/2r^2 \\ 0 & 0 & T^2/2r^2 & T/r^2 \end{bmatrix} \right) \tag{14}$$

$$v_k \sim \left(0, \begin{bmatrix} \sigma_r^2/T & 0 \\ 0 & \sigma_\theta^2/T \end{bmatrix} \right). \tag{15}$$

Suppose the a priori information is $x_0 \sim (\bar{x}_0, p_0)$. To find $\bar{x}_0$ and P_0 we could take one initializing measurement $z_0 = (r_0 \quad \theta_0)^T$. Then

$$\bar{x}_0 = \begin{bmatrix} r_0 \\ 0 \\ \theta_0 \\ 0 \end{bmatrix}, \quad P_0 = \frac{1}{T} \begin{bmatrix} \sigma_r^2 & 0 & 0 & 0 \\ 0 & \sigma_r^2 & 0 & 0 \\ 0 & 0 & \sigma_\theta^2 & 0 \\ 0 & 0 & 0 & \sigma_\theta^2 \end{bmatrix}. \tag{16}$$

In fact it is often more convenient to select $\bar{x}_0 = 0$, $P_0 = I$, since as we shall discover the filter converges to the same stochastic steady state for any $\bar{x}_0$ and P_0, if it is properly designed.

c. Applying the Discrete Kalman Filter

Now all we need to do is enter these system and covariance matrices into our computer program which implements the discrete Kalman filter. At each time kT we give the filter the radar readings $r(kT)$ and $\theta(kT)$ and it will give the best estimates of r, $\dot{r}$, θ, and $\dot{\theta}$.   ∎

The next examples show how to do some preliminary analysis to simplify the software implementation of the filter.

Example 2.4-4: *α–β Tracker for Radar*

a. Parallel Filtering

In the previous example we discussed the radar tracking problem with discrete measurements. By neglecting the nonlinear terms we obtained the discrete equations (12)–(16) in that example. Note that these equations describe two completely decoupled systems, one for the range dynamics $\begin{bmatrix} r \\ \dot{r} \end{bmatrix}$ and one for the angle dynamics $\begin{bmatrix} \theta \\ \dot{\theta} \end{bmatrix}$. This means that we can considerably simplify things by running, not one Kalman filter on a fourth-order system, but two parallel Kalman filters on two second-order systems. See Fig. 2.4-2. The number of computations required per iteration by the Kalman filter is of the order of n^3, so one fourth-order filter requires about $4^3 = 64$ operations while two second-order filters require about $2 \times 2^3 = 16$ operations. Aside from this, using parallel processing a separate microprocessor can be assigned to each filter, resulting in a further saving of time.

b. α–β Tracker

An *α–β tracker*, or *second-order polynomial tracking filter*, is a Kalman filter to estimate the state of the system $\ddot{y}(t) = w$ with scalar measurements $z = y + v$ (Gelb 1974, Schooler 1975). This corresponds exactly to the equations for the range subsystem (5) and angle subsystem (6) in the previous example. The Kalman filters in Fig. 2.4-2 are therefore both α–β trackers.

Let us consider the range subsystem in greater detail. We have the discretized system

$$x_{k+1} = \begin{bmatrix} 1 & T \\ 0 & 1 \end{bmatrix} x_k + w_k \tag{1}$$

$$z_k = [1 \quad 0]x_k + v_k, \tag{2}$$

with x_k now representing

$$\begin{bmatrix} r_k \\ \dot{r}_k \end{bmatrix}.$$

The noises are described by

$$w_k \sim \left(0, \begin{bmatrix} \sigma_p^2 & \mu_{pv} \\ \mu_{pv} & \sigma_v^2 \end{bmatrix}\right) \tag{3}$$

$$v_k \sim (0, \sigma_m^2). \tag{4}$$

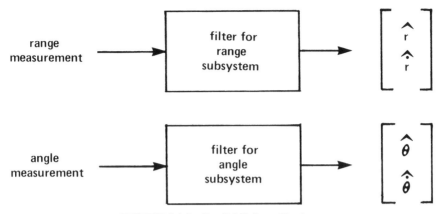

FIGURE 2.4-2   Parallel Kalman filtering.

The range (position) uncertainty is σ_p^2, the range velocity uncertainty is σ_v^2, and the measurement error variance is σ_m^2. We showed a way to find these quantities in the previous example.

With this formulation, a Kalman filter program can now be run to provide filtered estimates of range and range rate.

The Kalman filter for this problem has the form

$$\hat{x}_{k+1} = A\hat{x}_k + K_{k+1}(z_{k+1} - HA\hat{x}_k), \tag{5}$$

where the Kalman gain is traditionally represented in radar problems as

$$K_k = \begin{bmatrix} \alpha_k \\ \dfrac{\beta_k}{T} \end{bmatrix}. \tag{6}$$

Hence the name "α-β tracker." We can write (5) as

$$\hat{r}_{k+1} = \hat{r}_k + T\hat{\dot{r}}_k + \alpha_{k+1}(z_{k+1} - \hat{r}_k - T\hat{\dot{r}}_k) \tag{7a}$$

$$\hat{\dot{r}}_{k+1} = \hat{\dot{r}}_k + \frac{\beta_{k+1}}{T}(z_{k+1} - \hat{r}_k - T\hat{\dot{r}}_k). \tag{7b}$$

c. Zero Process Noise

Setting the process noise to zero in (1) is not a good idea, since as we shall see in Chapter 4 it can be very detrimental to the performance of the filter, especially if the target is maneuvering rapidly. We do it here to get a closed-form solution for K_k in order to obtain further insight.

Thus, suppose that the process noise covariance Q is zero and that there is no a priori state information so that $P_0 = \infty$. ($P_0 = \infty$ corresponds to maximum-likelihood estimation.) Then, since the plant matrix is nonsingular, we have for the information matrix

$$P_{k+1}^{-1} = A^{-T}P_k^{-1}A^{-1} + H^T R^{-1} H. \tag{8}$$

To find a closed-form solution let

$$P_k^{-1} = \begin{bmatrix} p_k^1 & p_k^2 \\ p_k^2 & p_k^4 \end{bmatrix}. \tag{9}$$

Since

$$A^{-1} = \begin{bmatrix} 1 & -T \\ 0 & 1 \end{bmatrix} \tag{10}$$

we have

$$\begin{bmatrix} p_{k+1}^1 & p_{k+1}^2 \\ p_{k+1}^2 & p_{k+1}^4 \end{bmatrix} = \begin{bmatrix} 1 & 0 \\ -T & 1 \end{bmatrix} \begin{bmatrix} p_k^1 & p_k^2 \\ p_k^2 & p_k^4 \end{bmatrix} \begin{bmatrix} 1 & -T \\ 0 & 1 \end{bmatrix} + \begin{bmatrix} 1/\sigma_m^2 & 0 \\ 0 & 0 \end{bmatrix}, \tag{11}$$

or

$$p_{k+1}^1 = p_k^1 + \frac{1}{\sigma_m^2} \tag{12}$$

$$p_{k+1}^2 = p_k^2 - T p_k^1 \tag{13}$$

$$p_{k+1}^4 = p_k^4 - 2 T p_k^2 + T^2 p_k^1. \tag{14}$$

The solution to the first-order difference equation (12) is ($P_0^{-1} = 0$)

$$p_k^1 = \sum_{i=0}^{k-1} \frac{1}{\sigma_m^2} = \frac{k}{\sigma_m^2}. \tag{15}$$

Using this sequence as "input" to (13) we can solve for p_k^2:

$$p_k^2 = -T \sum_{i=0}^{k-1} \left(\frac{i}{\sigma_m^2} \right) = \frac{-T}{2\sigma_m^2} (k-1)k. \tag{16}$$

Finally, using (15) and (16) in (14),

$$p_{k+1}^4 = p_k^4 + \frac{T^2}{\sigma_m^2} (k-1)k + \frac{T^2}{\sigma_m^2} k$$

$$= p_k^4 + \frac{T^2}{\sigma_m^2} k^2, \tag{17}$$

which has solution

$$p_k^4 = \frac{T^2}{\sigma_m^2} \sum_{i=0}^{k-1} i^2 = \frac{T^2}{\sigma_m^2} \frac{(k-1)k(2k-1)}{6}. \tag{18}$$

Therefore

$$P_k^{-1} = \frac{k}{\sigma_m^2} \begin{bmatrix} 1 & (-T/2)(k-1) \\ (-T/2)(k-1) & (T^2/6)(k-1)(2k-1) \end{bmatrix}. \tag{19}$$

Inverting this yields

$$P_k = \frac{2\sigma_m^2}{k(k+1)} \begin{bmatrix} 2k-1 & 3/T \\ 3/T & 6/[T^2(k-1)] \end{bmatrix}. \tag{20}$$

The Kalman gain is given by

$$K_k = P_k H^T R^{-1} = \frac{2}{k(k+1)} \begin{bmatrix} 2k-1 \\ 3/T \end{bmatrix}. \tag{21}$$

In the case of no process noise, the filter gains are thus given by the simple expressions

$$\alpha_k = \frac{2(2k-1)}{k(k+1)}, \tag{22}$$

$$\beta_k = \frac{6}{k(k+1)}. \tag{23}$$

Note that the steady-state gains with no process noise are zero, so that as time passes the measurements are increasingly discounted. This is because $Q = 0$ implies that the model $\ddot{r} = 0$ is exact, so that as time passes we rely to a greater extent on predictions based on this model. In a practical situation this would be undesirable, and Q should not be set to zero.

d. $\alpha-\beta-\gamma$ Tracker

If the aircraft is maneuvering rapidly so that $\ddot{r}$ and $\ddot{\theta}$ are nonzero, then we can obtain more robust performance by using an $\alpha-\beta-\gamma$ *tracker* in each channel in Fig. 2.4-2. This is a Kalman filter to estimate the state of the system $y^{(3)}(t) = w(t)$ with measurements $z = y + v$. (The superscript means "third derivative".) (Marco Mayor, Gail White). ■

Example 2.4-5: *Filter Simplification by Preliminary Analysis*

We can often considerably simplify the implementation of the Kalman filter by doing some *preliminary analysis*. Our object is basically to replace matrix equations by scalar equations. We demonstrate by implementing in FORTRAN the $\alpha-\beta$ tracker in part b of the previous example. We only consider the computation of the error covariances and Kalman gain.

Suppose the a posteriori error covariance P_k is represented as

$$\begin{bmatrix} p_1 & p_2 \\ p_2 & p_4 \end{bmatrix}. \tag{1}$$

For P_k^- we will add superscripts " $-$ " to (1). Let the Kalman gain components be represented as

$$\begin{bmatrix} K_1 \\ K_2 \end{bmatrix}. \tag{2}$$

For the time update, write (2.3-4) as

$$P_{k+1}^- = AP_k A^T + Q$$

$$= \begin{bmatrix} 1 & T \\ 0 & 1 \end{bmatrix} \begin{bmatrix} p_1 & p_2 \\ p_2 & p_4 \end{bmatrix} \begin{bmatrix} 1 & 0 \\ T & 1 \end{bmatrix} + \begin{bmatrix} \sigma_p^2 & 0 \\ 0 & \sigma_v^2 \end{bmatrix}, \tag{3}$$

which results in the scalar replacements

$$p_1^- = p_1 + 2Tp_2 + T^2 p_4 + \sigma_p^2 \tag{4a}$$

$$p_2^- = p_2 + Tp_4 \tag{4b}$$

$$p_4^- = p_4 + \sigma_v^2. \tag{4c}$$

For the measurement update, use Table 2.3-2 to obtain

$$P_{k+1}^- H^T = \begin{bmatrix} p_1^- \\ p_2^- \end{bmatrix}, \tag{5}$$

$$HP_{k+1}^- H^T + R = p_1^- + \sigma_m^2 \tag{6}$$

$$K_{k+1} = \frac{1}{p_1^- + \sigma_m^2} \begin{bmatrix} p_1^- \\ p_2^- \end{bmatrix}, \tag{7}$$

or

$$K_1 = \frac{p_1^-}{p_1^- + \sigma_m^2} \tag{8a}$$

$$K_2 = \frac{p_2^-}{p_1^- + \sigma_m^2}, \tag{8b}$$

and

$$P_{k+1} = (I - K_{k+1}H)P_{k+1}^-$$

$$= \begin{bmatrix} 1 - K_1 & 0 \\ -K_2 & 1 \end{bmatrix} \begin{bmatrix} p_1^- & p_2^- \\ p_2^- & p_4^- \end{bmatrix}, \tag{9}$$

which yields the scalar updates

$$p_1 = (1 - K_1)p_1^-, \tag{10a}$$

$$p_2 = (1 - K_1)p_2^-, \tag{10b}$$

or equivalently,

$$p_2 = p_2^- - K_2 p_1^-, \tag{10b'}$$

and

$$p_4 = p_4^- - K_2 p_2^-. \tag{10c}$$

[Show that (10b) and (10b') are equivalent.]

Equations (4), (8), and (10) are FORTRAN *replacements*, with quantities on the right representing old values, and quantities on the left representing updated values.

A FORTRAN implementation of (4), (8), and (10) appears in Fig. 2.4-3. Note that we have been able to avoid using two sets of covariance elements, one for P_k and one for P_k^-, by careful attention to the order of the operations. For example, in the program (10c) is done *before* (10b) since it uses p_2^-, the a priori cross-covariance, and (10b) updates the (1, 2) element of the error covariance to its a posteriori value, which is *not* what is required in (10c). Using this software and representative values for constants, the filter gains α_k and β_k/T were plotted. See Fig. 2.4-4.

Note that the steady-state values of α_k and β_k/T are reached in five scans. They

C

```
      PROGRAM ALBET

      REAL P(4), K(2)
      DATA P/30.,0.,0.,30./
      REWIND 7
      REWIND 8

      DO 10 I= 1,100
      CALL MEASUP(P,K)
      CALL TIMEUP(P)
      WRITE(7,*) K(1)
      WRITE(8,*) K(2)
10    CONTINUE
      STOP
      END




      SUBROUTINE TIMEUP(P)
      REAL P(*)
      DATA SIGSQP,SIGSQV,T/150.,150.,1./

      P(1)= P(1) + 2.*T*P(2) + T*T*P(4) + SIGSQP
      P(2)= P(2) + T*P(4)
      P(4)= P(4) + SIGSQV
      RETURN
      END




      SUBROUTINE MEASUP(P,K)
      REAL P(*), K(*)
      DATA SIGSQM/30./

      DIV= P(1) + SIGSQM
      K(1)= P(1)/DIV
      K(2)= P(2)/DIV
      OMK1= 1. - K(1)
      P(4)= P(4) - K(2)*P(2)
      P(1)= OMK1*P(1)
      P(2)= OMK1*P(2)
      RETURN
      END
```

FIGURE 2.4-3 Program for alpha-beta tracker error covariances and gains.

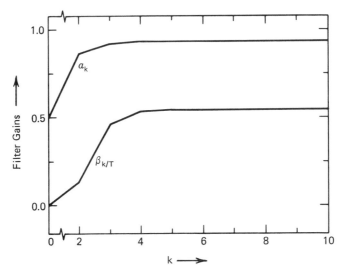

FIGURE 2.4-4   Filter gains for $\alpha-\beta$ tracker.

are nonzero, in contrast with the situation in part c of the previous example, so that measurements are not discounted with increasing k. We shall see in Section 2.5 that this is a consequence of nonzero Q.

In practice it is wise to use a stabilized version of (10) (Bierman 1977), since it can occur that (10b) and (10b') do not give the same values due to roundoff error. For greater numerical stability, the average of these two values can be taken as p_2. ■

Multiple Sampling Rates

We can considerably increase the usefulness of the method presented here by sampling the system at a higher rate than the data sampling rate.

Let the continuous system be discretized using a period of T in (2.4-7)–(2.4-9), where T is much less than the data sampling period T_d. Let T be an integral divisor of T_d. Since the data arrive at the rate $1/T_d$, we should use the data sampling period T_d in (2.4-15).

The Kalman filter would now be applied as follows. No data arrive between the times iT_d, where i is an integer, so during these intervals we should perform *only the time-update portion* of the Kalman filter equations. For this update we would use the system dynamics discretized with period T.

When data are received, at times iT_d, we should perform the measurement update portion of the filter algorithm. For this update we would use the measurement noise covariance $R^s = R/T_d$.

Two advantages of this technique immediately come to mind. First, an optimal estimate is available at every time iT, where i is an integer.

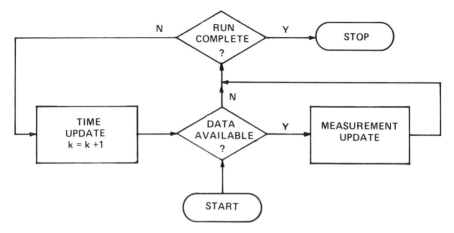

FIGURE 2.4-5   Flowchart for multiple-rate Kalman filter.

Estimates can thus be made available at points on as fine a division of the time line as we desire, independently of the sampling period T_d which may not be up to us to choose. This means, for example, that in a radar tracking problem, range and azimuth estimates can be made available *between* scans. For a slowly scanning radar this can be quite useful!

The second advantage is that the data arrival times need not be uniform. For aperiodically sampled data we merely perform time updates until a data sample is received, then we perform a measurement update at the next integral multiple of T. If the discretization period T is chosen to be small, the data are incorporated virtually as soon as they arrive.

A flowchart/block diagram for the "multiple-rate Kalman filter" is given in Fig. 2.4-5. The next example illustrates the technique.

Example 2.4-6:   Multiple-Rate Kalman Filter for Radar Tracking

Suppose we discretize the range subsystem of the previous examples using a period of $T = 50$ msec. The data sampling period T_d is the radar period of revolution (which we called T in the previous examples). Suppose T_d is 1 sec. Using the multiple-rate Kalman filter will then result in up-to-date optimal estimates every 50 msec although the data arrive only every 1 sec.

If $T = 50$ msec is used in the error covariance time update (4) in Example 2.4-5, and the measurement update (8) and (10) is performed only every $T_d = 1$ sec, then position and velocity error variances like those in Fig. 2.4-6 result. A simple modification of Fig. 2.4-3 was used to obtain these curves.   ■

In Section 3.7 we present another approach to discrete filtering for continuous-time systems which is known as the *continuous-discrete Kalman filter*. No preliminary discretization of the plant is required there. Instead,

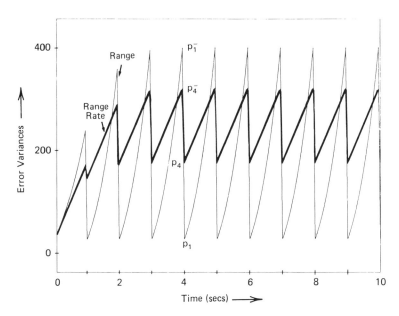

FIGURE 2.4-6 Range and range-rate error variances with time update every 0.05 sec and measurement update every 1 sec.

the estimate and error covariance are propagated between measurements *in continuous time* using equations we shall derive in Chapter 3.

Discretization of Time-Varying Systems

It is straightforward to find equations for discretizing time-varying continuous systems, as the next exercise indicates.

Exercise 2.4-7:   Discretization of Time-Varying Systems

Suppose the prescribed plant is time varying so that

$$\dot{x} = A(t)x + B(t)u + G(t)w \qquad (2.4\text{-}17a)$$

$$z = H(t)x + v, \qquad (2.4\text{-}17b)$$

with $x(0) \sim (\bar{x}(0), P(0))$, $w(t) \sim (0, Q(t))$, $v(t) \sim (0, R(t))$, and the usual whiteness and uncorrelatedness assumptions.

Let $\phi(t, t_0)$ be the state-transition matrix of $A(t)$ defined by

$$\frac{d}{dt}\phi(t, t_0) = A(t)\phi(t, t_0), \qquad \phi(t_0, t_0) = I. \qquad (2.4\text{-}18)$$

Show that the discrete system and noise covariance matrices are time-varying matrices given by

$$A^s(k) = \phi((k+1)T, kT), \tag{2.4-19}$$

$$B^s(k) = \int_{kT}^{(k+1)T} \phi((k+1)T, \tau) B(\tau) \, d\tau \tag{2.4-20}$$

$$G^s(k) = I \tag{2.4-21}$$

$$Q^s(k) = \int_{kT}^{(k+1)T} \Phi((k+1)T, \tau) G(\tau) Q(\tau) G^T(\tau) \phi^T((k+1)T, \tau) \, d\tau \tag{2.4-22}$$

$$H^s(k) = H(kT) \tag{2.4-23}$$

$$R^s(k) = \frac{R(kT)}{T}. \tag{2.4-24}$$

■

2.5 ERROR DYNAMICS AND STATISTICAL STEADY STATE

The Error System

To study the stability and convergence properties of the Kalman filter we need to examine the *error system*. This will give us some of the insight we need to design Kalman filters for practical problems.

Define the *a priori estimation error* as

$$\tilde{x}_k^- = x_k - \hat{x}_k^-, \tag{2.5-1}$$

the error at time k *before* the inclusion of data z_k. This is a one-step ahead prediction error. To find a recursion for $\tilde{x}_k^-$ we use (2.3-3a), (2.3-16), and (2.3-3b) to write

$$\tilde{x}_{k+1}^- = x_{k+1} - \hat{x}_{k+1}^-$$

$$= Ax_k + Bu_k + Gw_k - A(I - K_k H)\hat{x}_k^-$$

$$- Bu_k - AK_k(Hx_k + v_k).$$

This simplifies to

$$\tilde{x}_{k+1}^- = A(I - K_k H)\tilde{x}_k^- + Gw_k - AK_k v_k. \tag{2.5-2}$$

(We assume a time-invariant original plant for ease of notation.) Note that the error system is driven by both the process and measurement noises, but not by the deterministic input u_k.

It is easy to express the residual as a function of $\tilde{x}_k^-$; using (2.3-8) and (2.3-3b) we have

$$\tilde{z}_k = H\tilde{x}_k^- + v_k. \tag{2.5-3}$$

We call (2.5-2), (2.5-3) the *a priori error system*. Note that the residual depends on $\tilde{x}_k^-$ exactly as the data z_k depends on x_k.

To find the covariance of $\tilde{x}_k^-$ note that

$$E(\tilde{x}_{k+1}^-) = A(I - K_k H)E(\tilde{x}_k^-), \qquad (2.5\text{-}4)$$

so that $E(\tilde{x}_k^-) = 0$ if $E(\tilde{x}_0^-) = 0$. This, however, is guaranteed by the way we initialize the filter (Table 2.3-1). Hence the Kalman filter produces an unbiased estimate and so, from (2.5-2),

$$
\begin{aligned}
P_{k+1}^- &= \overline{\tilde{x}_{k+1}^-(\tilde{x}_{k+1}^-)^T} \\
&= A[(I - K_k H)P_k^-(I - K_k H)^T + K_k R K_k^T]A^T + GQG^T. \quad (2.5\text{-}5)
\end{aligned}
$$

By substituting for K_k from (2.3-17) it is easy to see that we have just found another way to derive the Riccati equation (2.3-18). In fact, (2.5-5) is called the Joseph stabilized formulation of the Riccati equation [cf. (2.3-15)].

Note that the Riccati equation is a symmetric equation (transpose both sides), so P_k^- is symmetric for all $k \geq 0$. P_k^- is also positive semidefinite for all k since it is a covariance matrix.

The Innovations Sequence

According to (2.5-3) the residual has zero mean,

$$\bar{\tilde{z}}_k = 0. \qquad (2.5\text{-}6)$$

Therefore, by the uncorrelatedness of $\tilde{x}_k^-$ and v_k the residual has covariance given by

$$P_{\tilde{z}_k} = HP_k^- H^T + R \qquad (2.5\text{-}7)$$

where P_k^- satisfies the Riccati equation. If the measurements are not used to correct $\hat{x}(t)$, then (2.5-7) still holds, but $P_k^- = P_k$ now satisfies the Lyapunov equation (2.2-4). (Why?)

Suppose that all statistics are Gaussian. Let

$$Z_k \triangleq \{z_1, z_2, \ldots, z_k\} \qquad (2.5\text{-}8)$$

represent all the data up through time k. Then the Kalman filter manufactures the conditional mean and covariance

$$\hat{x}_k = \overline{x_k / Z_k}, \qquad (2.5\text{-}9a)$$

$$P_k = P_{x_k/Z_k} \qquad (2.5\text{-}9b)$$

at each time k. (Recall from Section 1.1 that the error covariance P_k is equal to the conditional covariance P_{x_k/Z_k}.) In other words, in the Gaussian case the Kalman filter provides an update for the entire hyperstate. The conditional mean of the data is

$$\overline{z_k / Z_{k-1}} = H\hat{x}_k^-; \qquad (2.5\text{-}10)$$

so that

$$\tilde{z}_k = z_k - \overline{z_k / Z_{k-1}}. \tag{2.5-11}$$

The residual is therefore the *new information* contained in z_k which could not have been anticipated by using the previous data Z_{k-1}. It is called the *innovations* process. See Exercise 2.3-4.

From the orthogonality principle, $\tilde{z}_k$ is orthogonal to (and hence uncorrelated with) Z_{k-1}. Since $\tilde{z}_j$ depends only on x_0, the input, and Z_j, it follows that $\tilde{z}_k$ is uncorrelated with $\tilde{z}_j$ for $j < k$. The innovation is therefore a *zero-mean white noise* process.

One good way to verify that the Kalman filter is performing as designed is to monitor the residual. If it is not white noise, there is something wrong with the design and the filter is not performing optimally.

The Algebraic Riccati Equation

Now consider the limiting behavior of the Kalman filter. As $k \to \infty$, the solution to the Riccati equation tends to a bounded steady-state value P if

$$\lim_{k \to \infty} P_k^- = P \tag{2.5-12}$$

is bounded. In this case, for large k, $P_{k+1}^- = P_k^- \triangleq P$ and (2.3-18) tends to the *algebraic Riccati equation*

$$P = A[P - PH^T(HPH^T + R)^{-1}HP]A^T + GQG^T. \tag{2.5-13}$$

It is clear that any limiting solution to (2.3-18) is a solution to (2.5-13). The algebraic Riccati equation can have more than one solution. Its solutions can be positive or negative semidefinite or definite, indefinite, or even complex. There is quite a body of literature on conditions under which it has a *unique positive definite* solution. In the scalar case of course, a quadratic equation has only two solutions. See Example 2.1-1.

At this point we can ask several questions:

1. When do bounded limiting solutions to (2.3-18) exist for all choices of P_0? (that is, when is the a priori error covariance bounded for all initial conditions?)
2. Clearly the limiting solution P depends on the initial condition P_0. When is P independent of our choice (possibly inaccurate!) of P_0?

And most important of all,

3. When is the error system (2.5-2) asymptotically stable? (that is, when does the estimate $\hat{x}_k$ *converge* to the true state x_k? This is the whole point of the Kalman filter design!)

We can answer these very important questions in terms of the dynamical properties of the original system (2.3-3). Recall that (A, H) is *observable* if the poles of $(A - LH)$ can be arbitrarily assigned by appropriate choice of the output injection matrix L. (A, H) is *detectable* if $(A - LH)$ can be made asymptotically stable by some matrix L. This is equivalent to the observability of the unstable modes of A.

Recall also that (A, B) is *reachable* if the poles of $(A - BK)$ can be arbitrarily assigned by appropriate choice of the feedback matrix K. (A, B) is *stabilizable* if $(A - BK)$ can be made asymptotically stable by some *matrix* K. This is equivalent to the reachability of all the unstable modes of A.

Theorem 2.5-1

Let (A, H) be detectable. Then for every choice of P_0 there is a bounded limiting solution P to (2.3-18). Furthermore, P is a positive semidefinite solution to the algebraic Riccati equation.

Proof:

Since (A, H) is detectable, there exists a constant output injection matrix L so that $(A - LH)$ is asymptotically stable. Define a suboptimal a priori estimate for x_k by

$$\hat{x}_{k+1}^L = (A - LH)\hat{x}_k^L + Bu_k + Lz_k. \tag{1}$$

Then the associated a priori error system

$$\tilde{x}_{k+1}^L = (A - LH)\tilde{x}_k^L + Gw_k - Lv_k \tag{2}$$

is asymptotically stable.

Using a derivation like that of (2.5-5), the error covariance S_k of $\hat{x}_k^L$ satisfies the Lyapunov equation

$$S_{k+1} = (A - LH)S_k(A - LH)^T + LRL^T + GQG^T. \tag{3}$$

Since $(A - LH)$ is stable, the Lyapunov theory shows that S_k is bounded with a finite steady-state value.

However, the *optimal* estimate is given by the Kalman filter with time-varying gain K_k as shown in Table 2.3-3. Optimality implies that $P_k^- < S_k$ [i.e., $(S_k - P_k^-) > 0$] for all k, so that P_k^- is bounded above by a finite sequence.

It can be shown that the Riccati solution P_k^- is smooth, so that if it is bounded above by a finite sequence, then it converges to a finite limit P (see Casti 1977 or Kwakernaak and Sivan 1972).

Since the Riccati equation consists of a sum of positive semidefinite terms [see (2.5-5)], and since $P_0 \geq 0$, we are guaranteed symmetry and positive semidefiniteness of P_k^-, and hence of P. Clearly, P is a solution to (2.5-13). ∎

This result gives a neat test for the boundedness of the error covariance, especially if (A, H) satisfies the stronger condition of observability, which implies detectability. It is easy to test for observability by determining if the rank of the observability matrix (2.1-5) of (A, H) is equal to n.

Since the state estimation problem clearly depends on the observability properties of the original system, the theorem is not surprising. It reassures our intuition. The next result initially seems very strange. It was first derived by Kalman and Bucy (1961). It amounts to a restatement of Theorem 2.5-1 given a reachability hypothesis. This hypothesis makes the results much stronger.

Theorem 2.5-2

Let $\sqrt{Q}$ be a square root of the process noise covariance so that $Q = \sqrt{Q}\sqrt{Q}^T \geq 0$, and let the measurement noise have $R > 0$.

Suppose $(A, G\sqrt{Q})$ is reachable. Then (A, H) is detectable if and only if

a.   There is a unique positive definite limiting solution P to (2.3-18) which is independent of P_0. Furthermore, P is the *unique* positive *definite* solution to the algebraic Riccati equation.

b.   The steady-state error system defined as in (2.5-2) with *steady-state Kalman gain*

$$K = PH^T(HPH^T + R)^{-1} \qquad (2.5\text{-}14)$$

is asymptotically stable.

Proof:

Define D by $R = DD^T$, and note that $R > 0$ implies that $|D| \neq 0$.

We first claim that if $(A, G\sqrt{Q})$ is reachable and $|D| \neq 0$, then for any L

$$((A - LH), [G\sqrt{Q} \quad LD]) \qquad (1)$$

is also reachable. To show this, note that $|D| \neq 0$ implies that $H = DM$ for some M. The PBH test for reachability (Kailath 1980) implies that

$$\text{rank}\,[zI - A \quad G\sqrt{Q}] = n \qquad (2)$$

for all z, and so

$$n = \text{rank}[zI - A \quad G\sqrt{Q} \quad LD]$$

$$= \text{rank}[zI - A \quad G\sqrt{Q} \quad LD]\begin{vmatrix} I & 0 & 0 \\ 0 & I & 0 \\ M & 0 & I \end{vmatrix}$$

$$= \text{rank}[zI - A + LH \quad G\sqrt{Q} \quad LD]. \qquad (3)$$

This proves the claim.

To prove necessity, let gain L define a suboptimal estimate $\hat{x}_k^L$ as in (1) of the proof of Theorem 2.5-1, with $(A - LH)$ asymptotically stable. By

Theorem 2.5-1, (A, H) detectable implies $P_k^- \to P$ with P bounded and at least positive semidefinite. But $E\|\tilde{x}_k^-\|^2 \leq E\|\tilde{x}_k^L\|^2$ for all L because of the optimality of the Kalman filter [note that $E[(\tilde{x}_k^-)^T \tilde{x}_k^-] = \mathrm{trace}(P_k^-)$]. Hence system $A(I - KH)$ is also asymptotically stable with $K = PH^T(HPH^T + R)^{-1}$.

Define $L' = AK$. Then we can write the algebraic Riccati equation as

$$P = (A - L'H)P(A - L'H)^T + L'DD^T(L')^T + G\sqrt{Q}\sqrt{Q}^T G^T$$

$$= (A - L'H)P(A - L'H)^T + [G\sqrt{Q} \quad L'D][G\sqrt{Q} \quad L'D]^T. \qquad (4)$$

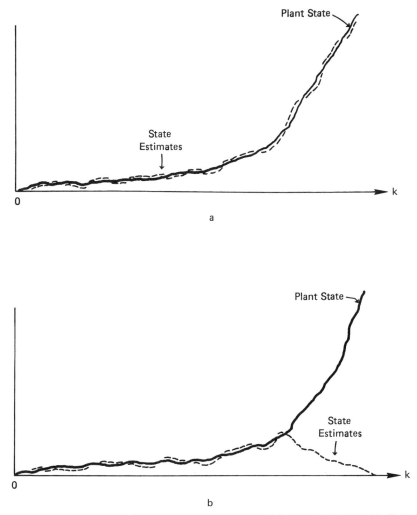

FIGURE 2.5-1 Relation between plant and filter stability. (a) Unstable plant, stable filter. (b) Unstable plant, unstable filter.

We know that P is an at least positive semidefinite solution of (4). But (4) is also a Lyapunov equation, for which we know that $(A - L'H)$ is stable and also that the pair $((A - L'H), [G\sqrt{Q} \quad L'D])$ is reachable. So the solution P is also unique and positive definite, and the gain $K = PH^T(HPH^T + R)^{-1}$ is uniquely defined.

To show sufficiency, note that if $A(I - KH)$ is asymptotically stable, there is an $L = AK$ for which the system $(A - LH)$ is stable and hence the pair (A, H) is detectable. ∎

This theorem is quite surprising. It says that if the state is *reachable by the process noise*, so that every mode is excited by w_k, then the Kalman filter is asymptotically stable if (A, H) is detectable. Thus, we can guarantee a stable filter by selecting the measurement matrix H correctly and ensuring that the process model is sufficiently corrupted by noise! This is the type of thing we need to know to develop sufficient intuition to apply the Kalman filter in practice.

To simplify the proof, we made the hypothesis of the theorem unnecessarily restrictive. In fact, it is only required that $(A, G\sqrt{Q})$ be stabilizable. In this case, though, the unique limiting Riccati solution P can only be guaranteed to be positive semidefinite.

The difference between a stable original plant and a stable Kalman filter should be clearly realized. We are not concerned here with regulating the plant, but only with estimating its state. With a stable filter the estimate closely tracks the state of the plant, whether it is stable or not. Figure 2.5-1 shows the relation between an unstable plant and an unstable filter.

Example 2.5-1: *Steady-State Behavior in the Scalar Case*

Theorem 2.5-2 is illustrated by Example 2.1-1, where the algebraic Riccati equation was solved in the scalar case.

For scalars, detectability is equivalent to the condition $|a| < 1$ or $h \neq 0$. The reachability condition is equivalent to $q \neq 0$. In Example 2.1-1 part a, $|a| < 1$ so detectability is immediate regardless of the value of h. A solution $p > 0$ is given by (8) when $q \neq 0$ and $h \neq 0$. If $q = 0$ and $h \neq 0$, then the solutions to (5) are

$$p = 0, \qquad -(1 - a^2)r/h^2, \tag{15}$$

so there is only a positive semidefinite solution (i.e., $p = 0$). If $h = 0$ but $q \neq 0$ then from (5),

$$p = q/(1 - a^2) \tag{16}$$

is greater than zero.

In Example 2.1-1 part b, $|a| > 1$ so for detectability we require $h \neq 0$ and for stabilizability $q \neq 0$. Indeed, (16) is less than zero if a is unstable. If $h \neq 0$ and $q \neq 0$ then Riccati equation solution (12) is strictly positive. Furthermore, the closed-loop system (13) is stable if and only if $h \neq 0$ and $q \neq 0$. ∎

Example 2.5-2:   Steady-State Filter for Damped Harmonic Oscillator

Many systems can be modeled, at least to a first approximation, by the equation

$$\ddot{y}(t) + 2\alpha\dot{y}(t) + \omega_n^2 = bu(t). \tag{1}$$

For instance, if $y(t)$ is aircraft pitch and $u(t)$ is elevator deflection, then (1) corresponds to the short-period approximation to the longitudinal dynamics (Blakelock 1965).

By selecting the state components as y and $\dot{y}$ the plant (1) can be put into the reachable canonical form

$$\dot{x}(t) = \begin{bmatrix} 0 & 1 \\ -\omega_n^2 & -2\alpha \end{bmatrix} x(t) + \begin{bmatrix} 0 \\ b \end{bmatrix} u(t) + w(t), \tag{2}$$

where we have added $(0, Q)$ process noise $w(t)$ to compensate for unmodeled system dynamics (as we did in Example 2.4-3). Let the measurement be $z = y + v(t)$, with noise $v(t) \sim (0, r)$.

Let sampling period be $T \ll 1$ so that terms of order T^2 can be ignored. Then the sampled plant is approximately

$$x_{k+1} = \begin{bmatrix} 1 & T \\ -\omega_n^2 T & 1 - 2\alpha T \end{bmatrix} x_k + \begin{bmatrix} 0 \\ bT \end{bmatrix} u_k + w_k \tag{3}$$

$$z_k = [1 \quad 0] x_k + v_k \tag{4}$$

with $Q^s = QT$, $r^s = r/T$. (In practice we should use the exact discretized system, not Euler's approximation.)

To study the behavior of the Kalman filter error system, let $\omega_n = 0$ and $\alpha = -0.1$, $r = 0.02$, $T = 20$ msec, and ignore u_k. Then the sampled plant is

$$x_{k+1} = \begin{bmatrix} 1 & 0.02 \\ 0 & 1.004 \end{bmatrix} x_k + w_k \triangleq A^s x_k + w_k, \tag{5}$$

and $r^s = 1$. The poles are at

$$z = 1, 1.004 \tag{6}$$

so the plant is unstable.

Note that the plant is observable through the measurements. Let

$$Q^s = \begin{bmatrix} 0 & 0 \\ 0 & 0.02 \end{bmatrix} \tag{7}$$

so that $(A^s, \sqrt{Q^s})$ is reachable, allowing w_k to excite both modes ($\sqrt{Q^s} = [0 \quad \sqrt{0.02}]^T$).

A computer program was written to solve the Riccati equation (2.3-18). The solution sequence for the case $P_0^- = I$ converges within 4 sec ($k = 200$) to a steady-state value of

$$P = \begin{bmatrix} 0.08247 & 0.16363 \\ 0.16363 & 0.61545 \end{bmatrix}. \tag{8}$$

The same steady-state value is obtained if $P_0^- = 0$.

We can use the Kalman gain (2.3-17) to compute the error system matrix

$$A_k^{cl} \triangleq A^s(I - K_k H) \tag{9}$$

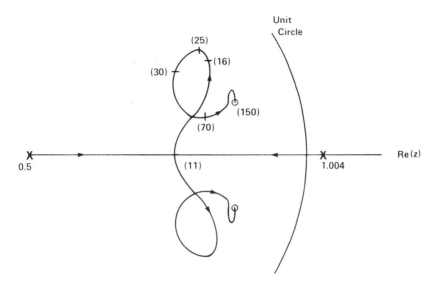

FIGURE 2.5-2 Locus of error system poles for detectable plant reachable by w_k.

for each value of k. Clearly, the poles of the error system are functions of time k, and we can plot a *root locus* for the error system as a function of k. The results are shown in Fig. 2.5-2 for $P_0^- = I$. In this figure, the poles for $k = 0$ are marked by X's and the steady-state poles by 0's. Values of k are shown in parentheses. By $k = 150$ (3 sec) the poles have reached their steady-state values of $0.9624 \pm j0.036117$, corresponding to a steady-state error system of

$$A^{cl} = \begin{bmatrix} 0.92079 & 0.02 \\ -1.15177 & 1.004 \end{bmatrix}. \tag{10}$$

This error system is stable and does not depend on P_0^-, so for any P_0^- the estimation error $\tilde{x}_k^-$ will be bounded with the steady-state variances of its components given by 0.08247 and 0.61545. ∎

Exercise 2.5-3: *Weighted Minimum-Variance Estimate*

Show that under the conditions of Theorem 2.5-2 the steady-state optimal gain minimizes

$$J_\infty = \lim_{k \to \infty} \overline{\tilde{x}_k^T W \tilde{w}_k} \tag{2.5-15}$$

for any $W \geq 0$, and that the minimum value is given by

$$J_\infty = \operatorname{tr}(WP), \tag{2.5-16}$$

where P is the steady-state error covariance. ∎

Exercise 2.5-4:   ***A Posteriori Error System***

Define the *a posteriori estimation error* as

$$\tilde{x}_k = x_k - \hat{x}_k. \tag{2.5-17}$$

a. Show that its dynamics are described by

$$\tilde{x}_{k+1} = (I - K_{k+1}H)A\tilde{x}_k + (I - K_{k+1}H)w_k - K_{k+1}v_{k+1}. \tag{2.5-18}$$

b. Show that the a posteriori error covariance satisfies

$$P_{k+1} = (I - K_{k+1}H)AP_kA^T(I - K_{k+1}H)^T + (I - K_{k+1}H)Q(I - K_{k+1}H)^T$$
$$+ K_{k+1}RK_{k+1}^T \tag{2.5-19}$$

c. When does P_k tend to a steady-state value? When is (2.5-18) asymptotically stable?
d. Show that $A(I - KH)$ and $(I - KH)A$ have the same eigenvalues.   ■

Time-Varying Plant

If the original plant is time varying we need to redefine observability and reachability. Let the plant be given by

$$x_{k+1} = A_kx_k + B_ku_k + G_kw_k \tag{2.5-20}$$

$$z_k = H_kx_k + v_k, \tag{2.5-21}$$

with $w_k \sim (0, Q_k)$ and $v_k \sim (0, R_k)$. Let $\phi(k, i)$ be the state transition matrix.

We say the plant is *uniformly completely observable*, or *stochastically observable*, if for every integer N the *observability gramian* satisfies

$$\alpha_0 I \le \sum_{k=i}^{N-1} \phi^T(k, i)H_k^T R_k^{-1}H_k\phi(k, i) \le \alpha_1 I \tag{2.5-22}$$

for some $i < N$, $\alpha_0 > 0$, and $\alpha_1 > 0$. (This guarantees the positive definiteness of both the gramian and its inverse.) We say the plant is *uniformly completely reachable* or *stochastically reachable*, if for every i the *reachability gramian* satisfies

$$\alpha_0 I \le \sum_{k=i}^{N-1} \phi(N, k + 1)G_kQ_kG_k^T\phi^T(N, k + 1) \le a_1 I, \tag{2.5-23}$$

for some $N > i$, $\alpha_0 > 0$, and $\alpha_1 > 0$.

If the plant is time varying, then there is in general no constant steady-state solution to the Riccati equation. However, stochastic observability and stochastic reachability (and boundedness of A_k, Q_k, R_k) guarantee that for large k the behavior of P_k is unique independent of P_0. They also guarantee uniform asymptotic stability of the Kalman filter error system. See Kalman and Bucy (1961) and Kalman (1963).

It is quite illuminating to compute the Fisher information matrix for the system (2.5-20) with measurements (2.5-21) when $G_k = 0$ (i.e., no process noise) (Sorenson 1970). Let us assume all statistics are Gaussian.

If $u_k = 0$, then the solution to (2.5-20) is $x_k = \phi(k, i)x_i$ for some initial time i, so that the measurements can be written in terms of the initial state as

$$z_k = H_k\phi(k, i)x_i + v_k. \tag{2.5-24}$$

Suppose we take measurements from $k = i$ to $k = N-1$. Let Z represent the entire collection of data, $Z = \{z_i, z_{i+1}, \ldots, z_{N-1}\}$. Then if v_k has the Gaussian PDF

$$f_{v_k}(v_k) = c_v e^{v_k^T R_k^{-1} v_k/2} \tag{2.5-25}$$

with c_v a constant, the independence of the measurements means that

$$f_{Z/x_i}(Z/x_i) = c \exp\left\{ -\frac{1}{2} \sum_{k=i}^{K-1} [z_k - H_k\phi(k, i)x_i]^T R_k^{-1}[z_k - H_k\phi(k, i)x_i] \right\} \tag{2.5-26}$$

with c a constant.

Now we can find the Fisher information matrix J_F (Section 1.3) by writing

$$\frac{\partial}{\partial x_i} \ln f_{Z/x_i}(Z/x_i) = \sum_{k=i}^{N-1} \phi^T(k, i)H_k^T R_k^{-1}[z_k - H_k\phi(k, i)x_i],$$

so

$$J_F = -E\left[\frac{\partial^2 \ln f_{Z/x_i}(Z/x_i)}{\partial x_i^2} \right]$$

$$= \sum_{k=i}^{N-1} \phi^T(k, i)H_k^T R_k^{-1} H_k\phi(k, i). \tag{2.5-27}$$

This is exactly the observability gramian!

2.6 FREQUENCY DOMAIN RESULTS

In the steady-state case we can work in the frequency domain and derive two nice results; one gives an alternative "root locus" approach to the design of the optimal steady-state filter, and the other shows the link between the Kalman filter and the Wiener filter.

A Spectral Factorization Result

The optimal steady-state filter for a time-invariant plant is given by (see Table 2.3-3)

$$\hat{x}_{k+1}^- = A(I - KH)\hat{x}_k^- + Bu_k + AKz_k \tag{2.6-1}$$

where

$$K = PH^T(HPH^T + R)^{-1} \tag{2.6-2}$$

with P the positive semidefinite solution of

$$P = APA^T - APH^T(HPH^T + R)^{-1}HPA^T + GQG^T. \qquad (2.6\text{-}3)$$

It exists and is stable if (A, H) is detectable and $(A, \sqrt{Q})$ is stabilizable.

Note that (Appendix A)

$$\Delta^{cl}(z) \triangleq |zI - A(I - KH)|$$
$$= |I + AKH(zI - A)^{-1}| \cdot |zI - A|$$
$$= |I + H(zI - A)^{-1}AK| \cdot \Delta(z) \qquad (2.6\text{-}4)$$

where $\Delta(z)$, $\Delta^{cl}(z)$ are the open-loop (i.e., plant) and closed-loop (i.e., filter) characteristic polynomials, respectively.

To derive the result on which this section is based, note that

$$P - APA^T = (zI - A)P(z^{-1}I - A)^T + AP(z^{-1}I - A)^T + (zI - A)PA^T.$$

Now use (2.6-3) to write

$$(zI - A)P(z^{-1}I - A)^T + AP(z^{-1}I - A)^T + (zI - A)PA^T$$
$$+ APH^T(HPH^T + R)^{-1}HPA^T = GQG^T.$$

Premultiply this by $H(zI - A)^{-1}$ and postmultiply it by $(z^{-1}I - A)^{-T}H^T$ to get

$$HPH^T + H(zI - A)^{-1}APH^T + HPA^T(z^{-1}I - A)^{-T}H^T$$
$$+ H(zI - A)^{-1}APH^T(HPH^T + R)^{-1}HPA^T(z^{-1}I - A)^{-T}H^T$$
$$= H(zI - A)^{-1}GQG^T(z^{-1}I - A)^{-T}H^T.$$

Substituting from (2.6-2) there results

$$HPH^T + H(zI - A)^{-1}AK(HPH^T + R) + (HPH^T + R)K^TA^T$$
$$\times (z^{-1}I - A)^{-T}H^T + H(zI - A)^{-1}AK(HPH^T + R)K^TA^T(z^{-1}I - A)^{-T}H^T$$
$$= H(zI - A)^{-1}GQG^T(z^{-1}I - A)^{-T}H^T,$$

and by adding R to each side and factoring, we obtain [see (2.2-13)]

$$\Phi_Z(z) = H(zI - A)^{-1}GQG^T(z^{-1}I - A)^{-T}H^T + R$$
$$= [I + H(zI - A)^{-1}AK](HPH^T + R)[I + H(z^{-1}I - A)^{-1}AK]^T. \qquad (2.6\text{-}5)$$

Defining

$$W^{-1}(z) = [I + H(zI - A)^{-1}AK](HPH^T + R)^{1/2} \qquad (2.6\text{-}6)$$

this can be written as

$$\Phi_Z(z) = W^{-1}(z)W^{-T}(z^{-1}). \qquad (2.6\text{-}7)$$

To see why this is an important result, examine (2.6-4). Clearly

$$|W^{-1}(z)| = |HPH^T + R|^{1/2} \cdot \frac{\Delta^{cl}(z)}{\Delta(z)} \qquad (2.6\text{-}8)$$

and

$$|W(z)| = \frac{\Delta(z)}{\Delta^{\text{cl}}(z)|HPH^T + R|^{1/2}}. \tag{2.6-9}$$

The optimal filter (2.6-1) is asymptotically stable, and this means that $|W^{-1}(z)|$ has all zeros inside $|z| = 1$. Thus $W(z)$ has all poles stable. If in addition the original plant is asymptotically stable, then $|W(z)|$ has all zeros inside $|z| = 1$, so that $W^{-1}(z)$ has all poles stable.

Under these circumstances (2.6-7) provides a minimum-phase factorization for the spectral density of z_k. Thus the Kalman filter can be viewed as an algorithm to compute a spectral factorization of the density $\Phi_Z(z)$ of the output process z_k of a stable plant. Solving the algebraic Riccati equation is equivalent to performing a spectral factorization!

The system with transfer function $W^{-1}(z)$ when driven by a white noise input ν_k with unit spectral density yields a process with density $\Phi_Z(z)$, hence system $W(z)$ when driven by z_k results in the white noise process ν_k. This means that $W(z)$ is a minimum-phase *whitening filter* for the data process z_k.

The Kalman filter can be viewed as a system that *whitens* the measurement sequence. The sequence ν_k resulting when z_k is whitened by $W(z)$ is called the *innovations sequence* (cf., Section 1.5).

The Innovations Representation

It is not difficult to relate the innovations ν_k to the residual $\tilde{z}_k$. Using Table 2.3-3 and (2.3-8) we have

$$\hat{x}^-_{k+1} = A(I - K_k H)\hat{x}^-_k + Bu_k + AK_k z_k \tag{2.6-10}$$

$$\tilde{z}_k = -H\hat{x}^-_k + z_k. \tag{2.6-11}$$

This formulation of the filter equations is called the *innovations representation*. It has as input the data z_k and as output the white residual $\tilde{z}_k$. This form makes it especially clear that the Kalman filter is a whitening filter.

Once the filter has reached steady state so that K_k is a constant K, the transfer function from z_k to $\tilde{z}_k$ can be written as

$$W'(z) = \{I - H[zI - A(I - KH)]^{-1}AK\}. \tag{2.6-12}$$

We can use the matrix inversion lemma to see that

$$W'(z) = [I + H(zI - A)^{-1}AK]^{-1}. \tag{2.6-13}$$

It is now clear [see (2.6-6)] that

$$\tilde{z}_k = (HPH^T + R)^{1/2}\nu_k, \tag{2.6-14}$$

so that the residual is simply a scaled version of the innovations. The process $\tilde{z}_k$ has covariance

$$P_{\tilde{z}_k} = HPH^T + R, \qquad (2.6\text{-}15)$$

while ν_k has covariance I.

Although technically the innovations and the residual are not the same, $\tilde{z}_k$ is commonly called the innovations.

Chang–Letov Design Procedure for the Kalman Filter

In this subsection let us not assume that the original plant is stable. Then $W(z)$ is not necessarily minimum phase, but we can still derive a useful result from (2.6-5). In fact, note that this equation along with (2.6-4) yields the *Chang–Letov Equation* (Letov 1960)

$$\Delta^{cl}(z)\Delta^{cl}(z^{-1}) = |H(z)H^T(z^{-1}) + R| \cdot \Delta(z)\Delta(z^{-1}) \cdot |HPH^T + R|^{-1} \quad (2.6\text{-}16)$$

where

$$H(z) = H(zI - A)^{-1}G\sqrt{Q} \qquad (2.6\text{-}17)$$

is the transfer function *in the original plant* from w'_k to z_k, where $w_k = \sqrt{Q}\,w'_k$ and w'_k is white noise with unit covariance. See (2.2-15).

This is an extremely important result, since it allows us an alternative method to steady-state Kalman filter design which does not depend on solving an algebraic Riccati equation. Instead we can compute the gain K by using tools such as Ackermann's formula (2.1-4). See Kailath (1980).

To see this, note that in the single-output case we can represent (2.6-17) as

$$H(z) = \frac{N(z)}{\Delta(z)} \qquad (2.6\text{-}18)$$

with the numerator polynomial $N(z) = H[adj(zI - A)]G\sqrt{Q}$ a row vector. Then (2.6-16) becomes

$$\Delta^{cl}(z)\Delta^{cl}(z^{-1}) = \frac{N(z)N^T(z^{-1}) + r\Delta(z)\Delta(z^{-1})}{HPH^T + r}. \qquad (2.6\text{-}19)$$

Since $\Delta^{cl}(z)$ is asymptotically stable, to find the poles of the optimal steady-state observer we only need to form $N(z)N^T(z^{-1}) + r\Delta(z)\Delta(z^{-1})$ for the given covariances Q and r, and then take the stable roots of this polynomial. (The constant $HPH^T + r$ only plays a normalizing role and can be disregarded.) Then Ackermann's formula or another design technique yields the value for the gain K. By this method, solution of the Riccati equation is completely avoided.

The zeros of (2.6-19) are the same as the zeros of

$$1 + \frac{1}{r}H(z)H^T(z^{-1}). \qquad (2.6\text{-}20)$$

Since this is in exactly the form required for a classical root locus analysis, it is evident that as $1/r$ goes from 0 to ∞, the optimal steady-state filter poles move from the stable poles of

$$G(z) \triangleq H(z) H^T(z^{-1}) \tag{2.6-21}$$

to its stable zeros.

The next example illustrates these notions.

Example 2.6-1: Chang–Letov Design of Steady-State Filter

Consider the discrete system

$$x_{k+1} = \begin{bmatrix} 1 & 0.02 \\ 0.1176 & 1 \end{bmatrix} x_k + \begin{bmatrix} 0 \\ 0.012/m \end{bmatrix} u_k + w_k \tag{1}$$

$$z_k = [5 \quad 0] x_k + v_k, \tag{2}$$

where $x_k = [\theta_k \quad \dot{\theta}_k]^T$ and we are measuring the angle θ_k. Let the process noise have covariance of

$$Q = \begin{bmatrix} 0.04 & 0 \\ 0 & 0.04 \end{bmatrix},$$

and the measurement noise have covariance of r. We want to find the best steady-state filter for estimating θ_k and $\dot{\theta}_k$ from z_k.

The plant is unstable with characteristic equation

$$\Delta(z) = (z - 0.9515)(z - 1.0485). \tag{3}$$

Since (A, H) is observable and $(A, G\sqrt{Q})$ is reachable

$$\left(\text{note } G\sqrt{Q} = \begin{bmatrix} 0.2 & 0 \\ 0 & 0.2 \end{bmatrix} \right),$$

we know that there is an asymptotically stable steady-state Kalman filter. Let us use the Chang–Letov equation to find it.

The transfer function from the process noise to the measurement is

$$H(z) = H(zI - A)^{-1} G\sqrt{Q}$$

$$= \frac{[z-1 \quad 0.02]}{\Delta(z)} \triangleq \frac{N(z)}{\Delta(z)}, \tag{4}$$

so the Chang–Letov equation becomes

$$(HPH^T + r)\Delta^{cl}(z)\Delta^{cl}(z^{-1}) = [z-1 \quad 0.02]\begin{bmatrix} z^{-1}-1 \\ 0.02 \end{bmatrix}$$

$$+ r(z - 0.9515)(z - 1.0485)(z^{-1} - 0.9515)(z^{-1} - 1.0485)$$

or

$$z^2(HPH^T + r)\Delta^{cl}(z)\Delta^{cl}(z^{-1}) = (-z^3 + 2.0004 z^2 - z) + r(0.99765 z^4$$

$$- 3.9953 z^3 + 5.9953 z^2 - 3.9953 z + 0.99765)$$

$$\triangleq n(z) + rd(z). \tag{5}$$

The constant term $(HPH^T + r)$ only plays a normalizing role, and it can be disregarded. The "symmetric" form of the right-hand side of (5) is worth noting. It means that if z is a root, then so is z^{-1}.

The roots of the right-hand side of (5) depend on r. When $r = 0$ they are the roots of $n(z)$; these are (adding an ∞ for reasons we shall soon discover)

$$0, 0.9802, 1.0202, \infty. \tag{6}$$

As $r \to \infty$ the roots of the right-hand side of (5) tend to the roots of $d(z)$; these are

$$0.9515, 0.95375, 1.0485, 1.051. \tag{7}$$

As r increases from 0 to ∞ the roots move from (6) to (7). The root locus is shown in Fig. 2.6-1, where the roots (6) of $n(z) = N(z)N^T(z^{-1})$ are denoted by X's and the roots (7) of $d(z) = \Delta(z)\Delta(z^{-1})$ are denoted by 0's. We added ∞ to the roots of (6) since there are four loci. Note also that both sets of roots (6) and (7) are symmetric about $|z| = 1$, that is, if z is a root then so is z^{-1}. (This is why the factorization of (2.6-19) into $\Delta^{cl}(z)$ and $\Delta^{cl}(z^{-1})$ works.)

According to the Chang–Letov procedure, we should select as the poles of the optimal steady-state filter the stable roots of (5). Suppose $r = 1$. Then the roots are

$$0.38083, 0.98011, 1.0203, 2.6258, \tag{8}$$

so we would select

$$\Delta^{cl}(z) = (z - 0.38083)(z - 0.98011)$$
$$= z^2 - 1.3609z + 0.37325. \tag{9}$$

This is the desired characteristic equation of the closed-loop plant matrix $(A - LH)$. To find the required steady-state gain $L = AK$ for use in the filter equation in Table

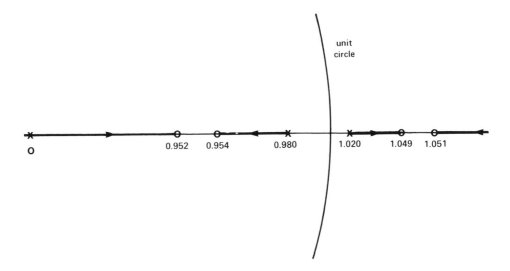

FIGURE 2.6-1   Chang–Letov root locus.

2.3-3, use Ackermann's formula (2.1-4). Thus

$$V_2 = \begin{bmatrix} H \\ HA \end{bmatrix} = \begin{bmatrix} 5 & 0 \\ 5 & 0.1 \end{bmatrix} \tag{10}$$

and

$$L = AK = \Delta^{cl}(A) V_2^{-1} \begin{bmatrix} 0 \\ 1 \end{bmatrix}$$

$$= \begin{bmatrix} 0.12781 \\ 0.1467 \end{bmatrix}. \tag{11}$$

The same gain AK would be found by using the algebraic Riccati equation approach. ∎

In the general multioutput case, it is necessary to use the *multivariable root locus* techniques to study the optimal steady-state filter poles as a function of R. In the case of $R = 0$, meaning perfect measurements, we can see from the Chang–Letov equation that the filter poles are the stable zeros of $|H(z)H^T(z^{-1})|$. In the limit as $R \to \infty$, the case of totally unreliable measurements, the filter poles are the original plant poles simply reflected inside the unit circle (i.e., unstable poles z are replaced by z^{-1}).

Deriving the Discrete Wiener Filter

Consider the Wiener problem of estimating an unknown stationary stochastic process x_k from a stationary data process z_k which is known for all past time. From Section 1.5 we know that the Wiener filter that performs this task is given by

$$F(z) = [\Phi_{XZ}(z) W^T(z^{-1})]_+ W(z) \tag{2.6-22}$$

where

$$\phi_Z^{-1}(z) = W^T(z^{-1}) W(z). \tag{2.6-23}$$

By introducing a *shaping filter* for the unknown process we can easily find $F(z)$ in terms of a solution to the algebraic Riccati equation. To wit, suppose x_k is generated by the shaping filter

$$x_{k+1} = Ax_k + Gw_k, \tag{2.6-24a}$$

with $w_k \sim (0, Q)$ a white process noise. Suppose the data are related to x_k by

$$z_k = Hx_k + v_k \tag{2.6-24b}$$

with $v_k \sim (0, R)$ a white measurement noise. The object is now to estimate x_k in terms of z_k assuming (2.6-24) has reached steady state. This requires of course that A be stable.

The first step in the solution of the problem is to find a whitening filter

$W(z)$ to manufacture the innovations ν_k from the data; but we have already solved this problem! We know that the whitening filter satisfying spectral factorization (2.6-23) is given by (2.6-6), where K is the gain given in (2.6-2).

From Example 1.1-3,

$$\Phi_{XZ}(z) = \Phi_X(z) H^T \qquad (2.6\text{-}25)$$

and

$$\Phi_Z(z) = H\Phi_X(z) H^T + R$$
$$= H\Phi_{XZ}(z) + R; \qquad (2.6\text{-}26)$$

therefore,

$$\Phi_Z(z) = H\Phi_{XZ}(z) + R = W^{-1}(z) W^{-T}(z^{-1})$$

or

$$H\Phi_{XZ}(z) W^T(z^{-1}) = -RW^T(z^{-1}) + W^{-1}(z). \qquad (2.6\text{-}27)$$

To find the Wiener filter we next require the realizable (i.e., stable) part of the left-hand side of this equation. Let us assume that

$$|\Phi_Z(z)| \neq 0 \quad \text{on} \quad |z| = 1 \qquad (2.6\text{-}28)$$

so that $W(z)$ is minimum phase. Then, since the realizable part operator is linear

$$H[\Phi_{XZ}(z) W^T(z^{-1})]_+ = -[RW^T(z^{-1})]_+ + [W^{-1}(z)]_+.$$

$W^{-1}(z)$ is realizable since $W(z)$ is minimum phase, and by using arguments like those in Example 1.5-5 it can be shown that the realizable (i.e., constant) part of $RW^T(z^{-1})$ is $(HPH^T + R)^{1/2}$. Therefore,

$$H[\Phi_{XZ}(z) W^T(z^{-1})]_+ = W^{-1}(z) - (HPH^T + R)^{1/2}. \qquad (2.6\text{-}29)$$

To complete the computation of $F(z)$, note that by the matrix inversion lemma

$$W(z) = (HPH^T + R)^{-1/2}[I - H(zI - A(I - KH))^{-1}AK], \quad (2.6\text{-}30)$$

so that by (2.6-22)

$$HF(z) = H[zI - A(I - KH)]^{-1}AK. \qquad (2.6\text{-}31)$$

This is the Wiener filter required to yield estimates of Hx_k from z_k. By construction it is stable and causal.

Now examine (2.6-1) to see that this is exactly the transfer function from z_k to $H\hat{x}_k^-$ in the steady-state Kalman filter! The Wiener and the Kalman approach therefore give identical solutions for the optimal steady-state filter.

If we examine our formulas we will see that this is not the only interesting conclusion that follows from our line of thought. From (2.6-22) and (2.6-29) we can write

$$HF(z) = I - (HPH^T + R)^{1/2} W(z). \qquad (2.6\text{-}32)$$

This is important because it is possible to find $(HPH^T + R)$ and $W(z)$ without solving the algebraic Riccati equation for P. For, suppose we can factor $\Phi_Z^{-1}(z)$ as in Section 1.5 to obtain

$$\Phi_Z^{-1}(z) = [W'(z^{-1})]^T \Omega W'(z) \qquad (2.6\text{-}33)$$

where $\Omega = \Omega^T > 0$ is a real matrix, $W'(z)$ is minimum phase, and $\lim_{z \to \infty} W'(z) = I$. Then the whitening filter is given by

$$W(z) = \Omega^{1/2} W'(z). \qquad (2.6\text{-}34)$$

Now note that the term in square brackets in the factorization (2.6-5) satisfies the limiting condition which defines (2.6-33) [$\lim_{z \to \infty} W'(z) = I$ is equivalent to $\lim_{z \to \infty} [W'(z)]^{-1} = I$]. Therefore $\Omega = (HPH^T + R)^{-1}$. Hence in the case of linear measurements we have the nice expression for the Wiener filter

$$HF(z) = I - \Omega^{-1/2} W(z), \qquad (2.6\text{-}35)$$

where Ω and $W(z)$ are both found by factoring $\Phi_Z^{-1}(z)$. Alternatively

$$HF(z) = I - W'(z). \qquad (2.6\text{-}36)$$

Compare this to (1.5-47), which was for the continuous case. Note that due to (2.6-12), equation (2.6-36) yields for the Wiener filter

$$HF(z) = H[zI - A(I - KH)]^{-1} AK. \qquad (2.6\text{-}37)$$

This is exactly (2.6-31).

We have been discussing the Wiener filter for estimating $y_k = Hx_k$ from z_k. A natural question arises. If $\hat{y}_k = H\hat{x}_k^-$ is reconstructed by (2.6-31), is $\hat{x}_k^-$ also provided by filtering the data by

$$F(z) = [zI - A(I - KH)]^{-1} AK? \qquad (2.6\text{-}38)$$

We should hope so, since this is the transfer function from z_k to $\hat{x}_k^-$ in the steady-state Kalman filter (2.6-1). Indeed, if (A, H) is observable then this is the case. The demonstration is left to the problems.

The next example uses the simplified equation (2.6-36) to find the Wiener filter, and demonstrates that either the Wiener approach or the Kalman approach can be used to find the optimal steady-state filter.

Example 2.6-2:   Kalman and Wiener Filters for Angle-of-Attack Measurements in the Presence of Wing Flutter

An aircraft wing experiences torsional flutter when it vibrates in a twisting mode. If y_k is the angle of attack, then this vibration might be modeled by the spectral density

$$\Phi_Y(\omega) = \frac{0.36(2 + 2\cos\omega)}{2.04 + 0.8\cos\omega + 2\cos 2\omega}, \tag{1}$$

which corresponds to a resonant frequency ω_n of 1 rad/sec (selected for illustration only!) and a Q factor $Q = \omega_n/2\alpha = 5$. See Fig. 2.7-3.

Let measurements be made with an angle-of-attack indicator with additive white noise

$$z_k = y_k + v_k, \tag{2}$$

$v_k \sim (0, 1)$.

We shall now find the optimal steady-state filter to estimate y_k using both the Wiener and the Kalman approaches.

a. Wiener Filter

We will use (2.6-35) which is based on a spectral factorization of $\Phi_Z^{-1}(z)$. Thus,

$$\Phi_Z(\omega) = \Phi_Y + 1 = \frac{2.76 + 1.52\cos\omega + 2\cos 2\omega}{2.04 + 0.8\cos\omega + 2\cos 2\omega};$$

so

$$\Phi_Z^{-1}(z) = \frac{(e^{-2j\omega} + 0.2e^{-j\omega} + 1)}{(e^{-2j\omega} + 0.239e^{-j\omega} + 0.46)}(0.4597)\frac{(e^{2j\omega} + 0.2e^{j\omega} + 1)}{(e^{2j\omega} + 0.239e^{j\omega} + 0.46)}$$

$$\triangleq [W'(z^{-1})]^T \Omega W'(z) \tag{3}$$

(Note that this factorization is not a trivial step! Much has been written about spectral factorization techniques, and some techniques are based on solving Riccati equations.)

The whitening filter is therefore given by

$$W(z) = 0.678\frac{(z^2 + 0.2z + 1)}{(z^2 + 0.239z + 0.46)}, \tag{4}$$

which is minimum phase, and $\Omega = 0.4597$. According to (2.6-35) [or (2.6-36)], the Wiener filter is given by

$$F(z) = I - \Omega^{-1/2}W(z)$$

$$= \frac{0.039z - 0.54}{z^2 + 0.239z + 0.46}. \tag{5}$$

This filter provides angle-of-attack estimates $\hat{y}_k$ if its input is the data z_k.

b. Steady-State Kalman Filter

The Wiener approach uses a factorization of the spectral density of the data. The Kalman approach uses a state model of the data generation procedure.

From Fig. 2.7-3, one state model corresponding to $\Phi_Y(\omega)$ given by (1) is

$$x_{k+1} = \begin{bmatrix} 0 & 1 \\ -1 & -0.2 \end{bmatrix} x_k + \begin{bmatrix} 1 \\ 0.8 \end{bmatrix} w_k$$

$$\triangleq Ax_k + Gw_k \qquad (6)$$

$$y_k = \begin{bmatrix} 1 & 0 \end{bmatrix} x_k \triangleq Hx_k \qquad (7)$$

with $w_k \sim (0, 0.36)$ a white noise process. This may be verified by finding $H(z) = H(zI - A)^{-1}G$ and using $\Phi_Y(z) = H(z)\Phi_W(z)H^T(z^{-1})$, with $\Phi_W(z) = 0.36$.

State model (6), (7) contains the same stochastic information as the spectral density (1). These are equivalent models for the unknown stochastic process $\{y_k\}$.

To incorporate the measurement noise into our state model, modify (7) to read

$$z_k = Hx_k + v_k, \qquad (8)$$

with $v_k \sim (0, 1)$.

Since (A, H) is observable and $(A, G\sqrt{Q})$ is reachable, there is a unique positive definite solution P to the algebraic Riccati equation, and the resulting error system $A(I - KH)$ is asymptotically stable. To find P, iteratively solve (2.3-18) with any P_0 (we used $P_0 = I$) until a steady-state solution is reached. The result, achieved within 50 iterations, is

$$P = \begin{bmatrix} 1.1758 & 0.0855 \\ 0.0855 & 0.81915 \end{bmatrix}. \qquad (9)$$

Then, by (2.6-2)

$$K = \begin{bmatrix} 0.5404 \\ 0.03931 \end{bmatrix}, \qquad (10)$$

and the error system matrix is

$$A(I - KH) = \begin{bmatrix} -0.03931 & 1 \\ -0.45174 & -0.2 \end{bmatrix}. \qquad (11)$$

The steady-state filter that reconstructs angle-of-attack y_k given z_k is therefore implemented by (2.6-1), or

$$\hat{x}_{k+1}^- = \begin{bmatrix} -0.03931 & 1 \\ -0.45174 & -0.2 \end{bmatrix} \hat{x}_k^- + \begin{bmatrix} 0.03931 \\ -0.5483 \end{bmatrix} z_k$$

$$\hat{y}_k = \begin{bmatrix} 1 & 0 \end{bmatrix} \hat{x}_{k+1}^-. \qquad (12)$$

If we take the transfer function of the Kalman filter (12), we obtain exactly the Wiener filter (5). ∎

2.7 CORRELATED NOISE AND SHAPING FILTERS

We have discussed the optimal filter for the ideal case of white uncorrelated noise sequences. Departure from this case can occur in three ways: the process noise can be nonwhite (i.e., correlated with itself), the measurement and process noises can be correlated with each other, or the measurement noise can be nonwhite.

Colored Process Noise

Suppose we are prescribed the plant

$$x_{k+1} = Ax_k + Bu_k + Gw_k \tag{2.7-1a}$$

$$z_k = Hx_k + v_k \tag{2.7-1b}$$

where $v_k \sim (0, R)$ is white; $x_0 \sim (\bar{x}_0, P_0)$; v_k, w_k, and x_0 are uncorrelated; but process noise w_k is not white.

A noise sequence w_k whose spectral density $\Phi_W(\omega)$ is not a constant is said to be *colored*. From the spectral factorization theorem, if $\phi_W(z)$ is rational and if $|\Phi_W(z)| \neq 0$ for almost every z, then there is a square, rational, asymptotically stable spectral factor $H(z)$ with zeros inside or on the unit circle such that

$$\phi_W(z) = H(z)H^T(z^{-1}). \tag{2.7-2}$$

If $|\Phi_W(z)| \neq 0$ on $|z| = 1$, then $H(z)$ is minimum phase (i.e., all zeros strictly inside $|z| = 1$). If the linear system $H(z)$ is driven by white noise w'_k with unit spectral density, then the output of the system has spectral density $\Phi_W(z)$. The system $H(z)$ that manufactures colored noise w_k with a given spectral density from white noise is called a spectrum *shaping filter*. See Fig. 2.7-1.

In the continuous case (2.7-2) becomes

$$\Phi_W(s) = H(s)H^T(-s). \tag{2.7-3}$$

Given a spectral factorization for the density of w_k, we can find a state realization for $H(z)$,

$$H(z) = H'(zI - A')^{-1}G' + D', \tag{2.7-4}$$

(Kailath 1980). (Note that primes denote additional variables.) Then the noise w_k can be represented by

$$x'_{k+1} = A'x'_k + G'w'_k \tag{2.7-5a}$$

$$w_k = H'x'_k + D'w'_k, \tag{2.7-5b}$$

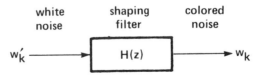

FIGURE 2.7-1 Spectral density shaping filter.

where $w'_k \sim (0, I)$ is white. *Augmenting* the state of (2.7-1) by x'_k we get

$$\begin{bmatrix} x_{k+1} \\ x'_{k+1} \end{bmatrix} = \begin{bmatrix} A & GH' \\ 0 & A' \end{bmatrix} \begin{bmatrix} x_k \\ x'_k \end{bmatrix} + \begin{bmatrix} B \\ 0 \end{bmatrix} u_k + \begin{bmatrix} GD' \\ G' \end{bmatrix} w'_k \qquad (2.7\text{-}6a)$$

$$z_k = [H \quad 0] \begin{bmatrix} x_k \\ x'_k \end{bmatrix} + v_k, \qquad (2.7\text{-}6b)$$

with w'_k and v_k white and uncorrelated. Now the Kalman filter in Table 2.3-1 can be run on this augmented plant. Note that it must estimate the states of the original plant and also of the noise-shaping filter.

In practice, to find $\Phi_W(z)$ in the first place one could assume $\{w_k\}$ is ergodic and use one sample function w_k to determine the autocorrelation function $R_W(k)$ by time averages. A Z transform yields the spectral density, and then a curve fit yields $\Phi_W(z)$.

Example 2.7-1: *Aircraft Longitudinal Dynamics with Gust Noise*

The longitudinal dynamics of an aircraft can be represented in the short period approximation by the harmonic oscillator

$$\dot{x} = \begin{bmatrix} 0 & 1 \\ -\omega_n^2 & -2\delta\omega_n \end{bmatrix} x + \begin{bmatrix} 0 \\ 1 \end{bmatrix} w, \qquad (1)$$

where $x = [\theta \quad \dot{\theta}]^T$ and θ is pitch angle. The process noise $w(t)$ represents wind gusts which change the angle of attack α, and so influence pitch rate. The gust noise might have a spectral density which can be approximated by the low-frequency spectrum

$$\Phi_W(\omega) = \frac{2a\sigma^2}{\omega^2 + a^2}, \qquad a > 0. \qquad (2)$$

This density is shown in Fig. 2.7-2.

Performing a factorization on (2) there results

$$\Phi_W(s) = \frac{\sqrt{2a}\,\sigma}{(s+a)} \cdot \frac{\sqrt{2a}\,\sigma}{(-s+a)}; \qquad (3)$$

so that the minimum-phase shaping filter is

$$H(s) = \frac{1}{s+a}, \qquad (4)$$

which must be driven by input noise $w'(t)$ with covariance $2a\sigma^2$. A state realization of (4) is

$$\dot{x}' = -ax' + w' \qquad (5a)$$

$$w = x'. \qquad (5b)$$

Augmenting the plant (1) by the shaping filter (5) and redefining the state as $x \triangleq [\theta \quad \dot{\theta} \quad x']^T$ there results

$$\dot{x} = \begin{bmatrix} 0 & 1 & 0 \\ -\omega_n^2 & -2\delta\omega_n & 1 \\ 0 & 0 & -a \end{bmatrix} x + \begin{bmatrix} 0 \\ 0 \\ 1 \end{bmatrix} w'. \qquad (6)$$

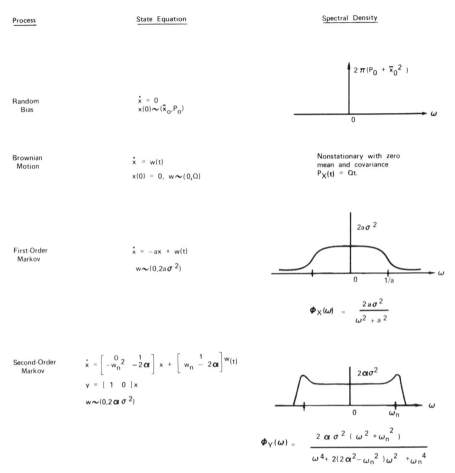

Process	State Equation	Spectral Density

Random
Bias

$\dot{x} = 0$
$x(0) \sim (\bar{x}_0, P_0)$

$2\pi(P_0 + \bar{x}_0^2)$

Brownian
Motion

$\dot{x} = w(t)$

$x(0) = 0, \; w \sim (0, Q)$

Nonstationary with zero
mean and covariance
$P_X(t) = Qt.$

First-Order
Markov

$\dot{x} = -ax + w(t)$

$w \sim (0, 2a\sigma^2)$

$2a\sigma^2$

$\Phi_X(\omega) = \dfrac{2a\sigma^2}{\omega^2 + a^2}$

Second-Order
Markov

$\dot{x} = \begin{bmatrix} 0 & 1 \\ -\omega_n^2 & -2\alpha \end{bmatrix} x + \begin{bmatrix} 1 \\ -2\alpha \end{bmatrix} w(t)$

$y = [\, 1 \quad 0 \,]x$

$w \sim (0, 2\alpha\sigma^2)$

$2\alpha\sigma^2$

$\Phi_Y(\omega) = \dfrac{2\alpha\sigma^2(\omega^2 + \omega_n^2)}{\omega^4 + 2(2\alpha^2 - \omega_n^2)\omega^2 + \omega_n^4}$

FIGURE 2.7-2   Some useful continuous spectrum-shaping filters.

If pitch θ is measured every T sec, then

$$z = [1 \quad 0 \quad 0]x + v; \tag{7}$$

suppose the measurement noise is white with $v \sim (0, r)$. If the sampling period T is small so that T^2 is negligible then the discretized plant is

$$x_{k+1} = \begin{bmatrix} 1 & T & 0 \\ -\omega_n^2 T & 1 - 2\,\delta\omega_n T & T \\ 0 & 0 & 1 - aT \end{bmatrix} x_k + \begin{bmatrix} 0 \\ 0 \\ 1 \end{bmatrix} w_k' \tag{8a}$$

$$z_k = [1 \quad 0 \quad 0]x_k + v_k, \tag{8b}$$

with white noises $w_k' \sim (0, 2a\sigma^2 T)$, $v_k \sim (0, r/T)$.

Now the discrete Kalman filter can be run on (8).

In practice the discretization would be performed using e^{AT}, not by the Euler's

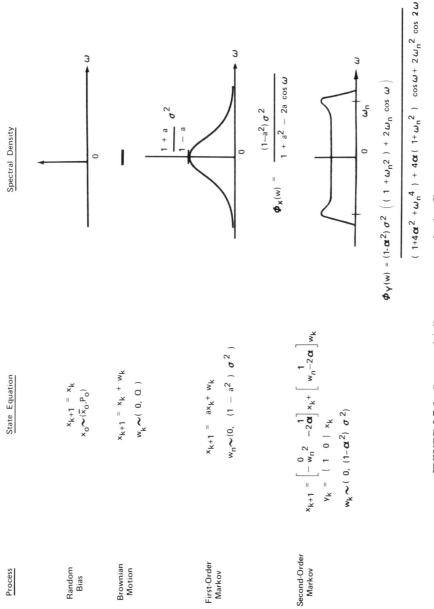

Process	State Equation	Spectral Density
Random Bias	$x_{k+1} = x_k$ $x_o \sim (\bar{x}_o, P_o)$	
Brownian Motion	$x_{k+1} = x_k + w_k$ $w_k \sim (0, Q)$	
First-Order Markov	$x_{k+1} = ax_k^+ + w_k$ $w_n \sim (0, (1 - a^2) \sigma^2)$	$\Phi_x(w) = \dfrac{(1-a^2)\sigma^2}{1 + a^2 - 2a\cos\omega}$
Second-Order Markov	$x_{k+1} = \begin{bmatrix} 0 & 1 \\ -w_n^2 & -2\alpha \end{bmatrix} x_k^+ + \begin{bmatrix} 1 \\ w_n - 2\alpha \end{bmatrix} w_k$ $y_k = \begin{bmatrix} 1 & 0 \end{bmatrix} x_k$ $w_k \sim (0, (1-\alpha^2)\sigma^2)$	$\Phi_Y(w) = (1\text{-}\alpha^2)\sigma^2 \dfrac{\left((1 + \omega_n^2) + 2\omega_n\cos\omega \right)}{\left(1 + 4\alpha^2 + \omega_n^4 \right) + 4\alpha(1 + \omega_n^2)\cos\omega + 2\omega_n^2\cos 2\omega}$

FIGURE 2.7-3 Some useful discrete spectrum-shaping filters.

approximation $I + AT$, BT, QT, R/T. We use the latter method in the examples for simplicity only. The actual sampling process with zero-order-hold corresponds to discretization by e^{AT}, not $I + AT$.   ∎

Some useful continuous and discrete shaping filters are shown in Figs. 2.7-2 and 2.7-3. In these figures, $w(t)$ and w_k are white noise processes.

The random bias could be used in a ship navigation problem where there is a constant unknown current. Brownian motion, or the Wiener process, is useful in describing the motion of particles under the influence of diffusion. It should also be used to model biases that are known to vary with time. First-order Markov processes are useful to model band-limited noises. The second-order Markov process has a periodic autocorrelation, and is useful to describe oscillatory random processes such as fuel slosh or vibration (Maybeck 1979).

Correlated Measurement and Process Noise

Now we consider the plant (2.7-1) with $w_k \sim (0, Q)$ and $v_k \sim (0, R)$ both white, but with

$$\overline{v_j w_k^T} = S\,\delta_{jk}, \tag{2.7-7}$$

where δ_{jk} is the Kronecker delta. In this case we could write

$$\overline{\begin{bmatrix} w_k \\ v_k \end{bmatrix} \begin{bmatrix} w_k \\ v_k \end{bmatrix}^T} = \begin{bmatrix} Q & S^T \\ S & R \end{bmatrix}. \tag{2.7-8}$$

To rederive the Kalman filter equations for this case it is easiest to begin with the measurement update.

Since x_k and v_k in (2.7-1b) are still uncorrelated (x_k depends on w_{k-1} and hence on v_{k-1}), we obtain from Table 2.3-2 at time k

$$\hat{x}_k = \hat{x}_k^- + P_k^- H^T (HP_k^- H^T + R)^{-1}(z_k - H\hat{x}_k^-) \tag{2.7-9}$$

$$P_k = P_k^- - P_k^- H^T (HP_k^- H^T + R)^{-1} HP_k^-. \tag{2.7-10}$$

The correlation (2.7-7) does not affect the measurement update.

For the time update, we recall that the best estimate of a linear function is the same function of the estimate, that is,

$$\hat{f}(x) = f(\hat{x}) \tag{2.7-11}$$

if $f(\cdot)$ is linear. Hence from (2.7-1a)

$$\hat{x}_{k+1}^- = A\hat{x}_k + B\hat{u}_k + G\hat{w}_k. \tag{2.7-12}$$

The effect of the correlatedness of w_k and v_k can be clearly understood at this point, for in contrast to our previous derivations, the estimated value $\hat{w}_k$ of w_k given z_k is no longer zero. To see this, use Example 1.1-2 to write

$$\hat{w}_k = \bar{w}_k + P_{WZ}P_Z^{-1}(z_k - \bar{z}_k), \tag{2.7-13}$$

where

$$P_{WZ} = \overline{w_k(z_k - \bar{z}_k)^T}$$
$$= \overline{w_k(Hx_k + v_k - H\hat{x}_k^-)^T}$$
$$= S^T, \tag{2.7-14}$$

since w_k and x_k are uncorrelated. Therefore, we obtain as the estimate for the process noise given measurement z_k,

$$\hat{w}_k = S^T(HP_k^- H^T + R)^{-1}(z_k - H\hat{x}_k^-). \tag{2.7-15}$$

Now use this and (2.7-9) in (2.7-12) to get

$$\hat{x}_{k+1}^- = A\hat{x}_k^- + Bu_k + (AP_k^- H^T + GS^T)(HP_k^- H^T + R)^{-1}(z_k - H\hat{x}_k^-). \tag{2.7-16}$$

If we define a *modified Kalman gain* by

$$K_k = (AP_k^- H^T + GS^T)(HP_k^- H^T + R)^{-1} \tag{2.7-17}$$

then this becomes

$$\hat{x}_{k+1}^- = A\hat{x}_k^- + Bu_k + K_k(z_k - H\hat{x}_k^-). \tag{2.7-18}$$

Equations (2.7-18) and (2.3-16) are identical except for the gain terms; the only effect of measurement and process noise correlated as in (2.7-7) is to modify the Kalman gain.

To find the effect of the correlation (2.7-7) on the error covariance, write the new a priori error system as

$$\tilde{x}_{k+1}^- = x_{k+1} - \hat{x}_{k+1}^-$$
$$= Ax_k + Bu_k + Gw_k$$
$$\quad - A\hat{x}_k^- - Bu_k - K_k(z_k - H\hat{x}_k^-),$$

or

$$\tilde{x}_{k+1}^- = (A - K_k H)\tilde{x}_k^- + Gw_k - K_k v_k, \tag{2.7-19}$$

with K_k as in (2.7-17). Compare this to (2.5-2). The only difference is the modified gain.

Now, since $\overline{\tilde{x}_{k+1}^-} = 0$, we have

$$P_{k+1}^- = E(\tilde{x}_{k+1}^-(\tilde{x}_{k+1}^-)^T)$$
$$= (A - K_k H)P_k^-(A - K_k H)^T + GQG^T$$
$$\quad + K_k R K_k^T - GS^T K_k^T - K_k SG^T.$$

It takes only a few steps to show that this is equivalent to

$$P_{k+1}^- = AP_k^- A^T - K_k(HP_k^- H^T + R)K_k^T + GQG^T. \tag{2.7-20}$$

Examine (2.3-18) to see that it can be put into this same form, with a difference only in the Kalman gain definition.

All of the results for this correlated noise case are, therefore, identical to our old ones, but the gain term AK_k with K_k as in (2.3-17) must be replaced by the modified gain (2.7-17), and the formulation (2.7-20) of the Riccati equation must be used.

From (2.7-20) it is clear that the error covariance is decreased due to the new information provided by the cross-correlation term S. Even if $S \neq 0$, we could still implement the filter of Table 2.3-1, but it would be suboptimal in this case.

Exercise 2.7-2: *Measurement Noise Correlated with Delayed Process Noise*

Suppose that instead of (2.7-7), we have

$$\overline{v_j w_{k-1}^T} = S \, \delta_{jk}. \tag{2.7-21}$$

This is sometimes useful if (2.7-1) is a sampled continuous system, for it represents correlation between the process noise over a sample period and the measurement noise at the *end* of the period [see (2.4-4)].

Show that in this case:

a. The *time update* is still the same as in the uncorrelated noise case.

b. The measurement update is now given by

$$\hat{x}_k = \hat{x}_k^- + K_k(z_k - H\hat{x}_k^-), \tag{2.7-22}$$

with a modified gain defined by

$$K_k = (P_k^- H^T + GS^T)(HP_k^- H^T + R + HG^T S^T + SG^T H^T)^{-1} \tag{2.7-23}$$

Note that (2.7-22) is identical to (2.3-14).

c. The new error covariance is given by

$$P_{k+1}^- = AP_k^- A^T - AK_k(HP_k^- H^T + R + HG^T S^T + SG^T H^T)K_k^T A^T + GQG^T. \tag{2.7-24}$$

∎

Colored Measurement Noise

Suppose now that we are prescribed the plant (2.7-1) with $w_k \sim (0, Q)$ white and w_k and v_k uncorrelated; but that the measurement noise is

$$v_k = n_k + v_k' \tag{2.7-25}$$

where n_k is colored and v_k' is white.

If $\Phi_N(z)$ is the spectral density of n_k, then a spectral factorization yields

$$\Phi_N(z) = H(z)H^T(z^{-1}), \tag{2.7-26}$$

and we can find a state realization for the shaping filter so that

$$H(z) = H'(zI - A')^{-1}G' + D'. \tag{2.7-27}$$

Then the measurement noise is modeled by

$$x'_{k+1} = A'x'_k + G'w'_k$$
$$v_k = H'x'_k + D'w'_k + v'_k, \tag{2.7-28}$$

where $w'_k \sim (0, I)$ is white. Let $v'_k \sim (0, R')$ with $R' > 0$. The new measurement noise in (2.7-28) is $Dw'_k + v'_k$, and since w'_k and v'_k are uncorrelated,

$$R \triangleq \overline{(D'w'_k + v'_k)(D'w'_k + v'_k)^T} = R' + D'(D')^T \tag{2.7-29}$$

Augmenting the state of (2.7-1) by (2.7-28) there results the new plant

$$\begin{bmatrix} x_{k+1} \\ x'_{k+1} \end{bmatrix} = \begin{bmatrix} A & 0 \\ 0 & A' \end{bmatrix} \begin{bmatrix} x_k \\ x'_k \end{bmatrix} + \begin{bmatrix} B \\ 0 \end{bmatrix} u_k + \begin{bmatrix} G & 0 \\ 0 & G' \end{bmatrix} \begin{bmatrix} w_k \\ w'_k \end{bmatrix}$$

$$z_k = [H \quad H'] \begin{bmatrix} x_k \\ x'_k \end{bmatrix} + (D'w'_k + v'_k), \tag{2.7-30}$$

which describes the original dynamics and the measurement noise process. The new process noise $[w_k^T \quad w_k'^T]^T$ is white with

$$\begin{bmatrix} w_k \\ w'_k \end{bmatrix} \sim \left(0, \begin{bmatrix} Q & 0 \\ 0 & I \end{bmatrix}\right), \tag{2.7-31}$$

and the new measurement noise is white with

$$(D'w'_k + v'_k) \sim (0, R). \tag{2.7-32}$$

Now, however, the measurement and process noises are correlated with

$$E\left[(D'w'_k + v'_k)\begin{bmatrix} w_k \\ w'_k \end{bmatrix}^T\right] = [0 \quad D'], \tag{2.7-33}$$

so the modified gain of the previous subsection should be used for optimality. If the relative degree of $\Phi_N(z)$ is greater than zero, then $D' = 0$ and the new measurement and process noises are uncorrelated.

If $v'_k = 0$ in (2.7-25) then $R' = 0$; however this approach can often still be used, since the discrete Kalman filter requires that $|[H \quad H']P[H \quad H']^T + R| \neq 0$.

In the case of colored measurement noise, there is an alternative to state augmentation. Suppose we are given plant (2.7-1) with $w_k \sim (0, Q)$ white, but with colored v_k generated by the shaping filter

$$x'_{k+1} = A'x'_k + G'w'_k$$
$$v_k = x'_k \tag{2.7-34}$$

with $w'_k \sim (0, I)$ white and uncorrelated with w_k. (Presumably, this shaping filter was found by factoring the spectrum of v_k.)

Define the *derived measurement* as

$$z'_{k+1} = z_{k+1} - A'z_k - HBu_k. \tag{2.7-35}$$

Then

$$z'_{k+1} = Hx_{k+1} + v_{k+1} - A'Hx_k - A'v_k - HBu_k$$
$$= (HA - A'H)x_k + G'w'_k + HGw_k.$$

Defining a new measurement matrix as

$$H' = HA - A'H \qquad (2.7\text{-}36)$$

we can write a new output equation in terms of the derived measurement as

$$z'_{k+1} = H'x_k + (G'w'_k + HGw_k). \qquad (2.7\text{-}37)$$

In this equation the measurement noise $(G'w'_k + HGw_k)$ is *white* with

$$(G'w'_k + HGw_k) \sim [0, (G'(G')^T + HGQG^TH^T)]. \qquad (2.7\text{-}38)$$

Therefore, (2.7-1a) and (2.7-37) define a new plant driven by white noises. In this plant, the measurement and process noises are correlated with

$$E[(G'w'_k + HGw_k)w_k^T] = HGQ, \qquad (2.7\text{-}39)$$

so the modified Kalman gain of the previous subsection should be used for optimality.

The advantage of the derived measurement approach is that the dimension of the state is not increased by adding a shaping filter.

To see how to implement the Kalman filter with derived measurements, by (2.7-18) and (2.7-35) we have

$$\hat{x}^-_{k+1} = A\hat{x}^-_k + Bu_k + K_k(z'_k - H'\hat{x}^-_k)$$
$$= (A - K_kH')\hat{x}^-_k + Bu_k - K_kA'z_{k-1}$$
$$+ K_k(z_k - HBu_{k-1}). \qquad (2.7\text{-}40)$$

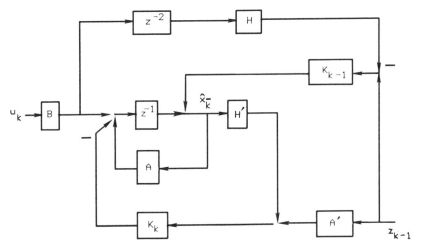

FIGURE 2.7-4 Implementing the Kalman filter with derived measurements.

We claim that an implementation of this recursion is given in Fig. 2.7-4. To see this, note that the signal at the input to the delay z^{-1} is

$$(A - K_k H')\hat{x}_k^- + Bu_k - K_k A' z_{k-1}, \qquad (2.7\text{-}41)$$

so the output of the delay is this signal delayed by 1. Now it is easy to see from the figure that

$$\hat{x}_k^- = (A - K_{k-1}H')\hat{x}_{k-1}^- + Bu_{k-1} - K_{k-1}A' z_{k-2}$$
$$+ K_{k-1}(z_{k-1} - HBu_{k-2}), \qquad (2.7\text{-}42)$$

which is a delayed version of (2.7-40) (see Anderson and Moore 1979). Compare this figure to Fig. 2.1-1.

2.8 OPTIMAL SMOOTHING

This chapter has heretofore been devoted to the filtering problem. Given a linear stochastic system, the solution to the prediction problem is very straightforward, and we deal with it in the problems. In this section we discuss optimal smoothing.

Suppose data are available over a time interval $(0, N]$. The determination of the state estimate $\hat{x}_k$ for $k \in (0, N)$ using *all* the data, past and future, is called the *smoothing problem*. It is a non-real-time operation where the available data are processed to obtain an estimate $\hat{x}_k$ for some past value of k. There are several types of smoothing, and here we shall consider *fixed interval smoothing*, where the final time N is fixed.

An application of smoothing is satellite orbit determination. The satellite's position is measured over a time interval, and smoothed estimates over the interval of its radial and angular positions and velocities are computed. From these estimates the parameters of the orbit can be found. See Example 5.3-2.

Let the unknown x_k be described by the dynamics

$$x_{k+1} = Ax_k + Bu_k + Gw_k \qquad (2.8\text{-}1a)$$

and let the data be

$$z_k = Hx_k + v_k. \qquad (2.8\text{-}1b)$$

White noises $w_k \sim (0, Q)$ and $v_k \sim (0, R)$ are uncorrelated with each other and with $x_0 \sim (\bar{x}_0, P_0)$. Let $Q > 0$, $R > 0$.

To find the optimal state estimate $\hat{x}_k$ at time k in the time interval $(0, N]$, we need to take account of the data z_j for $0 < j \le k$ and also for $k < j \le N$. We know now to incorporate the data for the subinterval prior to and including k. All we must do is use the Kalman filter. If we denote the estimate at time k based only on the data z_j for $0 < j \le k$ by $\hat{x}_k^f$ (for "forward"), then we have $\hat{x}_0^f = \bar{x}_0$, $P_0^f = P_0$ and:

Time Update:

$$\hat{x}_{k+1}^{f-} = A\hat{x}_k^f + Bu_k \tag{2.8-2}$$

$$P_{k+1}^{f-} = AP_k^f A^T + GQG^T \tag{2.8-3}$$

Measurement Update:

$$\hat{x}_k^f = \hat{x}_k^{f-} + K_k^f(z_k - H\hat{x}_k^{f-}) \tag{2.8-4}$$

$$P_k^f = P_k^{f-} - P_k^{f-}H^T(HP_k^{f-}H^T + R)^{-1}HP_k^{f-} \tag{2.8-5a}$$

$$= [(P_k^{f-})^{-1} + H^T R^{-1} H]^{-1} \tag{2.8-5b}$$

$$K_k^f = P_k^{f-}H^T(HP_k^{f-}H^T + R)^{-1} \tag{2.8-6a}$$

$$= P_k^f H^T R^{-1}. \tag{2.8-6b}$$

What we now propose to do is the following. First we will construct a filter that runs *backward* in time beginning at N to make an estimate $\hat{x}_k^b$ incorporating the data z_j for $k < j \le N$. This is the *information formulation* of the Kalman filter. Then we will discuss how to combine $\hat{x}_k^f$ and $\hat{x}_k^b$ into the overall smoothed estimate $\hat{x}_k$ which incorporates all the data over the entire interval.

The smoother we will develop provides the best linear estimate of x_k given the data in $(0, N]$. If all statistics are normal, it provides the optimal estimate. It also extends directly to the time-varying case.

The Information Filter

In the problems for Section 2.3 we discuss a filter based on the information matrix P_k^{-1}. This filter uses A^{-1}. Here we shall see that the natural direction for time flow in the information formulation is backward. If we reverse time, then A appears instead of A^{-1}. The key to solving the smoothing problem by our approach is to derive a convenient backward form of the optimal filter, and this form is provided by the information formulation.

We now derive a filter that provides the best estimate of x_k given the data on the subinterval $k < j \le N$. All $\hat{x}$, K, and P in this subsection of the book should be considered as having a superscript "b" (for "backward"), which we suppress for notational brevity. In the derivation we assume $|A| \ne 0$, though the final form of the information filter does not require this.

Write the system (2.8-1a) as

$$x_k = A^{-1}x_{k+1} - A^{-1}Bu_k - A^{-1}Gw_k. \tag{2.8-7}$$

This is a recursion for decreasing k beginning at $k = N$. At time k, before the data z_k are included, we have the a priori estimate $\hat{x}_k^-$. After z_k is included we have the a posteriori estimate $\hat{x}_k$. Since we are proceeding

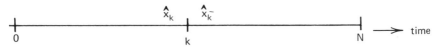

FIGURE 2.8-1 Backward filter update diagram.

backward, the timing diagram is as depicted in Fig. 2.8-1. Note that $\hat{x}_k^-$ occurs immediately to the right of k, while $\hat{x}_k$ occurs to its left.

The error covariance measurement update is

$$P_k^{-1} = (P_k^-)^{-1} + H^T R^{-1} H. \tag{2.8-8}$$

The time update for stochastic system (2.8-7) is easily seen to be

$$P_k^- = A^{-1}(P_{k+1} + GQG^T)A^{-T}, \tag{2.8-9}$$

where $A^{-T} \triangleq (A^{-1})^T$. By using the matrix inversion lemma on (2.8-9) and defining information matrices

$$S_k \triangleq P_k^{-1} \tag{2.8-10a}$$

$$S_k^- \triangleq (P_k^-)^{-1} \tag{2.8-10b}$$

we can write these updates as

$$S_k = S_k^- + H^T R^{-1} H \tag{2.8-11}$$

$$S_k^- = A^T [S_{k+1} - S_{k+1} G(G^T S_{k+1} G + Q^{-1})^{-1} G^T S_{k+1}] A. \tag{2.8-12}$$

If a gain matrix is defined as

$$K_k = S_{k+1} G(G^T S_{k+1} G + Q^{-1})^{-1} \tag{2.8-13}$$

then the time update becomes

$$S_k^- = A^T (I - K_k G^T) S_{k+1} A. \tag{2.8-14}$$

A Joseph stabilized version is

$$S_k^- = A^T (I - K_k G^T) S_{k+1} (I - K_k G^T)^T A + A^T K_k Q^{-1} K_k^T A. \tag{2.8-15}$$

Note that these equations do not involve A^{-1}.

For the estimate updates it is convenient to define intermediate variables as

$$\hat{y}_k = P_k^{-1} \hat{x}_k = S_k \hat{x}_k \tag{2.8-16a}$$

$$\hat{y}_k^- = (P_k^-)^{-1} \hat{x}_k^- = S_k^- \hat{x}_k^-. \tag{2.8-16b}$$

According to the problems for Section 1.2, the measurement update of the Kalman filter can be written as

$$P_k^{-1} \hat{x}_k = (P_k^-)^{-1} \hat{x}_k^- + H^T R^{-1} z_k \tag{2.8-17}$$

or

$$\hat{y}_k = \hat{y}_k^- + H^T R^{-1} z_k. \tag{2.8-18}$$

The backward time update for the estimate for stochastic system (2.8-7) is

$$\hat{x}_k^- = A^{-1}\hat{x}_{k+1} - A^{-1}Bu_k, \tag{2.8-19}$$

$$\hat{y}_k^- = S_k^-\hat{x}_k^- = S_k^-A^{-1}(S_{k+1}^{-1}\hat{y}_{k+1} - Bu_k). \tag{2.8-20}$$

Taking into account (2.8-14) yields

$$\hat{y}_k^- = A^T(I - K_kG^T)(\hat{y}_{k+1} - S_{k+1}Bu_k). \tag{2.8-21}$$

Equations (2.8-11)–(2.8-13), (2.8-18), and (2.8-21) comprise the information filter. Note the duality between this version and the Kalman filter. In the latter, time updates are simple, while in the information formulation measurement updates are simple.

It is sometimes convenient to have another expression for the gain K_k. To find it, write

$$K_k = (AP_k^-A^T)^{-1}(AP_k^-A^T)S_{k+1}G(G^TS_{k+1}G + Q^{-1})^{-1}.$$

But using (2.8-9)

$$(AP_k^-A^T)S_{k+1}G = (P_{k+1} + GQG^T)P_{k+1}^{-1}G$$

$$= GQ(Q^{-1} + G^TP_{k+1}^{-1}G);$$

so that

$$K_k = (AP_k^-A^T)^{-1}GQ. \tag{2.8-22}$$

Compare this to (2.8-6b).

Optimal Smoothed Estimate

Let us now suppose that a Kalman filter has been run on the interval $[0, N]$ to compute for each k an estimate $\hat{x}_k^f$ based on data z_j for $0 < j \le k$. Let us also suppose that a separate information filter has been run backward on the interval to find for each k an estimate $\hat{x}_k^{b-}$ based on data z_j for $k < j \le N$. It is now necessary to combine $\hat{x}_k^f$ and $\hat{x}_k^{b-}$ and their respective error covariances to determine an optimal overall estimate $\hat{x}_k$ and error covariance that incorporate all the available data. See Fig. 2.8-2.

This is very easy to do. Since w_k and v_k are white, estimates $\hat{x}_k^f$ and $\hat{x}_k^{b-}$ are independent and we can use the "Millman's theorem" from the problems in Section 1.2. (For simplicity, we suppress the superscript "b" on $\hat{y}$ and S.)

The a posteriori smoothed error covariance is given by

$$P_k = [(P_k^f)^{-1} + (P_k^{b-})^{-1}]^{-1}, \tag{2.8-23a}$$

$$P_k = [(P_k^f)^{-1} + S_k^-]^{-1}. \tag{2.8-23b}$$

The matrix inversion lemma can be employed to write this in several

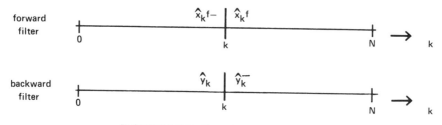

FIGURE 2.8-2 Smoother timing diagram.

alternative ways, two of which are

$$P_k = P_k^f - P_k^f (P_k^f + P_k^{b-})^{-1} P_k^f \qquad (2.8\text{-}24)$$

and

$$P_k = P_k^f - P_k^f S_k^- (I + P_k^f S_k^-)^{-1} P_k^f. \qquad (2.8\text{-}25)$$

Note that the smoothed covariance P_k is smaller than or equal to the Kalman filter covariance P_k^f, so that using future data adds information. See Fig. 2.8-3.

The a posteriori smoothed estimate is (by "Millman's theorem")

$$\hat{x}_k = P_k[(P_k^f)^{-1}\hat{x}_k^f + (P_k^{b-})^{-1}\hat{x}_k^{b-}]. \qquad (2.8\text{-}26)$$

Use (2.8-23b) to write

$$\hat{x}_k = (I + P_k^f S_k^-)^{-1}\hat{x}_k^f + P_k\hat{y}_k^-, \qquad (2.8\text{-}27)$$

so that by the matrix inversion lemma [or we can use (2.8-25) in (2.8-26)]

$$\hat{x}_k = (I - K_k)\hat{x}_k^f + P_k\hat{y}_k^-$$
$$= \hat{x}_k^f + (P_k\hat{y}_k^- - K_k\hat{x}_k^f) \qquad (2.8\text{-}28)$$

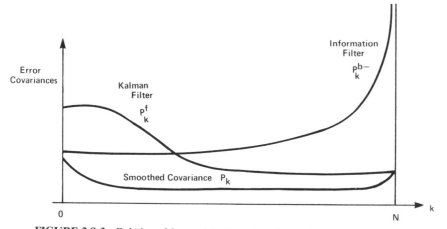

FIGURE 2.8-3 Relation of forward, backward, and smoothed error covariances.

TABLE 2.8-1 Optimal Smoothing Scheme

System and measurement model:

$$x_{k+1} = Ax_k + Bu_k + Gw_k$$

$$z_k = Hx_k + v_k,$$

$$x \sim (\bar{x}_0, P_0),\ w_k \sim (0, Q),\ v_k \sim (0, R).$$

$x_0,\ w_k,\ v_k$ uncorrelated.

Forward filter:
  initialization:

$$\hat{x}_0^f = \bar{x}_0,\ P_0^f = P_o$$

  time update:

$$P_{k+1}^{f-} = AP_k^f A^T + GQG^T$$

$$\hat{x}_{k+1}^{f-} = A\hat{x}_k^f + Bu_k$$

  measurement update:

$$K_k^f = P_k^{f-} H^T (HP_k^{f-} H^T + R)^{-1}$$

$$P_k^f = (I - K_k^f H)P_k^{f-}$$

$$\hat{x}_k^f = \hat{x}_k^{f-} + K_k^f (z_k - H\hat{x}_k^{f-})$$

Backward filter:
  initialization:

$$\hat{y}_N^- = 0,\ S_N^- = 0$$

  measurement update:

$$S_k = S_k^- + H^T R^{-1} H$$

$$\hat{y}_k = \hat{y}_k^- + H^T R^{-1} z_k$$

  time update:

$$K_k^b = S_{k+1} G (G^T S_{k+1} G + Q^{-1})^{-1}$$

$$S_k^- = A^T (I - K_k^b G^T) S_{k+1} A$$

$$\hat{y}_k^- = A^T (I - K_k^b G^T)(\hat{y}_{k+1} - S_{k+1} Bu_k)$$

Smoother:

$$K_k = P_k^f S_k^- (I + P_k^f S_k^-)^{-1}$$

$$P_k = (I - K_k) P_k^f$$

$$\hat{x}_k = (I - K_k)\hat{x}_k^f + P_k \hat{y}_k^-$$

where the "smoother gain" is

$$K_k = P_k^f S_k^-(I + P_k^f S_k^-)^{-1}. \tag{2.8-29}$$

In terms of this gain, smoother update (2.8-25) becomes

$$P_k = (I - K_k)P_k^f. \tag{2.8-30}$$

The entire smoothing scheme, which includes a Kalman filter, an information filter, and a smoother, is shown in Table 2.8-1. A corresponding scheme can be derived for the a priori smoothed estimate $\hat{x}_k^-$.

A word on the boundary conditions is in order. The initial conditions on the forward filter are $\hat{x}_0^f = \bar{x}_0$, $P_0^f = P_0$. At time N we have $P_N = P_N^f$, since the filtered and smoothed estimates for x_N are the same. According to (2.8-23b) then, $S_N^- = 0$. The boundary condition $\hat{x}_N^{b-}$ is not known, but according to (2.8-16b), $\hat{y}_N^- = 0$. These are the boundary conditions needed to start up the backward filter. (Now it is evident why we propagate $\hat{y}_k^-$ instead of $\hat{x}_k^{b-}$ in the information filter; we do not know the boundary condition $\hat{x}_N^{b-}$.)

It is possible in some cases that $P_k = P_k^f$; so that the smoothed estimate is no better than the filtered estimate. It is also possible for only some of the directions of the state $\hat{x}_k$ to be improved by smoothing. These are called *smoothable* states. It can be shown that only those states which are $(A, G\sqrt{Q})$ reachable are smoothable. In particular, when $Q = 0$, smoothing does not improve the estimate. For further discussion, see Gelb (1974) and Fraser (1967). Additional references on smoothing are Anderson and Moore (1979), Bryson and Ho (1975), and Maybeck (1979).

Rauch–Tung–Striebel Smoother

The smoothing scheme in Table 2.8-1 requires three components: a forward filter, a separate backward filter, and a separate smoother. It is possible to devise a scheme with only two components: a forward filter, and a backward smoother algorithm which incorporates both the backward filter and the smoother. This *Rauch–Tung–Striebel formulation* is more convenient than the scheme we have just presented. See Rauch, Tung, and Striebel (1965).

To derive a backward recursion for the smoothed error covariance P_k which depends only on the forward filter covariances, we can eliminate $S_k^- = (P_k^{b-})^{-1}$ and $S_k = (P_k^b)^{-1}$ from the smoother equations as follows.

First note that by using (2.8-8), (2.8-5b), and (2.8-23a) we have

$$\begin{aligned}
P_{k+1}^b &= [(P_{k+1}^{b-})^{-1} + H^T R^{-1} H]^{-1} \\
&= [(P_{k+1}^{b-})^{-1} + (P_{k+1}^f)^{-1} - (P_{k+1}^{f-})^{-1}]^{-1} \\
&= [P_{k+1}^{-1} - (P_{k+1}^{f-})^{-1}]^{-1}. \tag{2.8-31}
\end{aligned}$$

Therefore, by (2.8-9), (2.8-3), and (2.8-31)

$$
\begin{aligned}
\{P_k^f + P_k^{b-}\}^{-1} &= \{P_k^f + A^{-1}(P_{k+1}^b + GQG^T)A^{-T}\}^{-1} \\
&= A^T\{AP_{k+1}^f A^T + GQG^T + P_{k+1}^b\}^{-1}A \\
&= A^T\{P_{k+1}^{f-} + P_{k+1}^b\}^{-1}A \\
&= A^T\{P_{k+1}^{f-} + [P_{k+1}^{-1} - (P_{k+1}^{f-})^{-1}]^{-1}\}^{-1}A \\
&= A^T(P_{k+1}^{f-})^{-1}\{(P_{k+1}^{f-})^{-1} \\
&\quad + (P_{k+1}^{f-})^{-1}[P_{k+1}^{-1} - (P_{k+1}^{f-})^{-1}]^{-1}(P_{k+1}^{f-})^{-1}\}^{-1}(P_{k+1}^{f-})^{-1}A.
\end{aligned}
$$

According to the matrix inversion lemma, there results

$$
\{P_k^f + P_k^{b-}\}^{-1} = A^T(P_{k+1}^{f-})^{-1}\{P_{k+1}^{f-} - P_{k+1}\}(P_{k+1}^{f-})^{-1}A.
$$

Using this in (2.8-24) yields

$$
P_k = P_k^f - P_k^f A^T(P_{k+1}^{f-})^{-1}(P_{k+1}^{f-} - P_{k+1})(P_{k+1}^{f-})^{-1}AP_k^f. \tag{2.8-32}
$$

By defining the gain matrix

$$
F_k = P_k^f A^T(P_{k+1}^{f-})^{-1} \tag{2.8-33}
$$

this becomes

$$
P_k = P_k^f - F_k(P_{k+1}^{f-} - P_{k+1})F_k^T. \tag{2.8-34}
$$

Equations (2.8-33) and (2.8-34) are the desired backward recursion for the smoothed error covariance P_k. They depend only on $\{P_k^f\}$ and $\{P_k^{f-}\}$, so that only the forward filter must be run before using them.

By the same sort of manipulations, the recursive smoother measurement update can be determined to be

$$
\hat{x}_k = \hat{x}_k^f + F_k(\hat{x}_{k+1} - \hat{x}_{k+1}^{f-}). \tag{2.8-35}
$$

The Rauch–Tung–Striebel smoother algorithm is thus as follows. First run the usual Kalman filter on the data. Then run the recursive smoother defined by (2.8-33)–(2.8-35) backward for $k = N-1, N-2, \ldots$. The initial conditions for the smoother are

$$
P_N = P_N^f, \qquad \hat{x}_k = \hat{x}_k^f. \tag{2.8-36}
$$

It is worth remarking that the backward recursive smoother depends neither on the data nor on the deterministic input.

PROBLEMS

Section 2.1

2.1-1: Stabilization via Riccati Equation. The plant is

$$
x_{k+1} = \begin{bmatrix} 1 & 1 \\ 0 & 0 \end{bmatrix} x_k + \begin{bmatrix} 0 \\ 1 \end{bmatrix} u_k, \tag{1}
$$

$$
z_k = [1 \quad 0]x_k
$$

and we want to design a stable observer.

a. Solve the algebraic Riccati equation with $DD^T = 1$, $G^T = [0 \quad \sqrt{q}]$.
    Hint: Substitute

$$P = \begin{bmatrix} p_1 & p_2 \\ p_2 & p_3 \end{bmatrix}$$

and solve for the p_i.

b. Find the observer gain L.

c. Plot the observer poles as q goes from 0 to ∞.

Section 2.2

2.2-1: Writing the Lyapunov Equation as a Vector Equation. Let $\otimes$
denote Kronecker product (Appendix A) and $s(\cdot)$ denote the stacking
operator. Show that the Lyapunov (matrix) equation $P_{k+1} = AP_k A^T + GG^T$
is equivalent to the *vector* equation

$$s(P_{k+1}) = (A \otimes A)s(P_k) + s(GG^T).$$

2.2-2: Solutions to Algebraic Lyapunov Equations. Use the results of
the previous problem to show that $P = APA^T + GG^T$ has a unique solution
if and only if $\lambda_i \lambda_j \neq 1$ for all i and j, where λ_i is an eigenvalue of A. Show
that the stability of A is sufficient to guarantee this.

2.2-3:

a. Find *all* possible solutions to $P = APA^T + GG^T$ if

$$A = \begin{bmatrix} \frac{1}{2} & 1 \\ 0 & -1 \end{bmatrix}, \qquad G = \begin{bmatrix} 2 \\ 0 \end{bmatrix}.$$

Interpret this in the light of the previous problem. Classify the solutions
as positive definite, negative definite, and so on.

b. Repeat if $G = [2 \quad 1]^T$.

c. Repeat if

$$A = \begin{bmatrix} \frac{1}{2} & 1 \\ 0 & -\frac{1}{2} \end{bmatrix}, \qquad G = \begin{bmatrix} 2 \\ 1 \end{bmatrix}.$$

2.2-4: Analytic Solution to the Lyapunov Equation. Show that the
solution to $P_{k+1} = AP_k A^T + GQG^T$ for $k \geq 0$ is

$$P_k = A^k P_0 (A^T)^k + \sum_{i=0}^{k-1} A^i GQG^T (A^T)^i.$$

2.2-5: Stochastic System

$$x_{k+1} = \begin{bmatrix} \frac{1}{2} & 1 \\ 0 & -\frac{1}{2} \end{bmatrix} x_k + \begin{bmatrix} 2 \\ 1 \end{bmatrix} w_k, \tag{1}$$

$$z_k = [1 \quad 0]x_k + v_k,$$

where $w_k \sim (0, 1)$, $v_k \sim (0, 2)$ are white and uncorrelated.

a. Find analytic expressions for $\bar{x}_k$ and P_{x_k} if $\bar{x}_0 = 1$, $P_{x_0} = I$.

b. Find steady-state values $\bar{x}_\infty$, P_{x_∞}.

c. Find spectral densities of x_k and z_k in steady state.

Section 2.3

2.3-1: Recursion for A Posteriori Error Covariance

a. Show that a recursion for P_k is given by

$$P_{k+1} = (AP_kA^T + GQG^T) - (AP_kA^T + GQG^T)H^T$$
$$\times [H(AP_kA^T + GQG^T)H^T + R]^{-1}H(AP_kA^T + GQG^T), \qquad (1)$$

or, if $Q = 0$,

$$P_{k+1} = A(P_k^{-1} + A^TH^TR^{-1}HA)^{-1}A^T. \qquad (2)$$

b. Show that if $|A| \neq 0$ (which is always true if A is obtained by sampling a continuous system) then a recursion for the a posteriori information matrix when $Q = 0$ is given by

$$P_{k+1}^{-1} = A^{-T}P_k^{-1}A^{-1} + H^TR^{-1}H. \qquad (3)$$

(Gail White)

2.3-2: Riccati Equation as Lyapunov Equation for Closed-Loop System. Show that (2.3-15) and (2.3-10) are equivalent to (2.3-11).

2.3-3: Information Matrix Formulation. The Kalman filter measurement update in terms of information matrices is

$$P_k^{-1} = (P_k^-)^{-1} + H^TR^{-1}H. \qquad (1)$$

Show that the information matrix time update is

$$(P_{k+1}^-)^{-1} = S_k - S_kG(G^TS_kG + Q^{-1})^{-1}G^TS_k \qquad (2)$$

where

$$S_k \triangleq A^{-T}P_k^{-1}A^{-1}. \qquad (3)$$

Note that the measurement update is easier to perform in terms of measurement matrices, while the time update is easier in terms of covariance matrices.

2.3-4: Square Root Riccati Formulation. Define square roots S_k, S_k^- by $P_k = S_kS_k^T$, $P_k^- = S_k^-(S_k^-)^T$. The following results show how to propagate S_k, S_k^- instead of P_k, P_k^-, resulting in stable algorithms and a reduced numerical range. See Schmidt (1967, 1970), Businger and Golub (1965), Dyer and McReynolds (1969), Bierman (1977), and Morf and Kailath (1975).

a. Show that time update (2.3-4) is equivalent to

$$[AS_k \quad C]T_1 = [S_{k+1}^- \quad 0], \qquad (1)$$

where $CC^T = GQG^T$ and T_1 is any orthogonal transformation (i.e., $T_1 T_1^T = I$) selected so that S_{k+1}^- has n columns.

b. Show that measurement update (2.3-11) is equivalent to

$$\begin{bmatrix} D & HS_k^- \\ 0 & S_k^- \end{bmatrix} T_2 = \begin{bmatrix} (HP_k^- H^T + R)^{1/2} & 0 \\ \bar{K}_k & S_k \end{bmatrix}, \tag{2}$$

where $R = DD^T$ and T_2 is any orthogonal transformation so that S_k has n columns. (You will find an expression for $\bar{K}_k$ during the demonstration.)

c. Show that both updates can be expressed together [i.e., as in (2.3-18)] in square root form as

$$\begin{bmatrix} D & HS_k^- & 0 \\ 0 & AS_k^- & C \end{bmatrix} T_3 = \begin{bmatrix} (HP_k^- H^T + R)^{1/2} & 0 & 0 \\ A\bar{K}_k & S_{k+1}^- & 0 \end{bmatrix} \tag{3}$$

where T_3 is orthogonal.

2.3-5: Recursive Line of Best Fit. There is prescribed a set of data points (x_k, y_k) as in Fig. P2.3-1 to which it is desired to fit a line $y = mx + c$. We want to use the Kalman filter to determine recursively the best estimates $\hat{m}_k$, $\hat{c}_k$ for the slope and intercept given the data through time k. Define the state as

$$X_k = \begin{bmatrix} c \\ m \end{bmatrix}. \tag{1}$$

Then $X_{k+1} = X_k$.

Suppose the measurements x_k are exact and the y_k are corrupted by $(0, r)$ white noise v_k so that

$$z_k \triangleq y_k = H_k X_k + v_k \tag{2}$$

with $H_k = [1 \quad x_k]$.

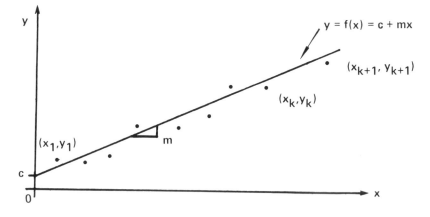

FIGURE P2.3-1 Recursive line of best fit.

a. Write the filter equations.
b. To obtain the maximum-likelihood estimate, set $P_0^{-1} = 0$, $\bar{X}_0 = 0$. Then the information matrix P_k^{-1} is given by (2.3-22). Compute it.
c. For the special case of uniformly spaced abscissas, we may have $x_k = k$. Find the error covariance in this case and show that the Kalman gain is

$$K_k = \frac{2}{k(k+1)} \begin{bmatrix} -(k+1) \\ 3 \end{bmatrix}. \tag{3}$$

Write down the Kalman filter.

2.3-6: Prediction and Intermittent Measurements. Let

$$x_{k+1} = ax_k + w_k \tag{1}$$

$$z_k = x_k + v_k \tag{2}$$

with $w_k \sim (0, q)$, $v_k \sim (0, r_k)$ white and uncorrelated. Let $x_0 \sim (\bar{x}_0, p_0)$.

a. Let $a = 0.9$, $q = 1$, $r_k = \frac{1}{2}$, and $x_0 \sim (1, 2)$. The data are $z_1 = 0.9$, $z_2 = 0.8$, $z_3 = 0.7$, $z_4 = 0.7$, $z_5 = 0.6$. Find filtered estimates $\hat{x}_k$ and error covariances p_k for $k = 0$ through 8. (Note, no data are available beyond $k = 5$.)
b. Find the filtered maximum-likelihood estimate $\hat{x}_k$, $k = 0, 1, 2, 3, 4, 5$, for part a if all statistics are Gaussian.
c. Repeat part a if the available data are $z_1 = 0.9$, $z_3 = 0.7$, $z_4 = 0.7$, $z_6 = 0.53$, $z_8 = 0.43047$.
d. Repeat part a if $r_k = 0.9^k/2$, that is, the data get better with time.
e. Now let $a = 1$, and $q = 0$, so there is no process noise. Show that after two measurements

$$p_2 = \left(\frac{1}{p_0} + \frac{1}{r_1} + \frac{1}{r_2} \right)^{-1} \tag{3}$$

$$\hat{x}_2 = p_2 \left(\frac{\bar{x}_0}{p_0} + \frac{z_1}{r_1} + \frac{z_2}{r_2} \right). \tag{4}$$

Compare to Millman's theorem.

2.3-7: The plant is

$$x_{k+1} = \begin{bmatrix} 0.9 & -1 \\ 0 & 0.8 \end{bmatrix} x_k + \begin{bmatrix} 0 \\ 1 \end{bmatrix} u_k + \begin{bmatrix} 0 \\ 1 \end{bmatrix} w_k$$

$$z_k = [1 \quad 0] x_k + v_k.$$

Let $w_k \sim (0, 2)$ and $v_k \sim (0, 1)$ be white, and $x_0 \sim (0, 1)$, with all three uncorrelated.

a. Find $\bar{x}_k$ if $\bar{x}_0 = 0$, $u_k = u_{-1}(k)$ the unit step for $k = 0$ through 5.
b. Find $\hat{x}_k$ and P_k for $k = 0$ through 5 if $u_k = u_{-1}(k)$ and the data are $z_1 = 0.1$, $z_2 = -1$, $z_3 = -2$, $z_4 = -5$, $z_5 = -7.2$.

c. If all statistics are normal, find the filtered maximum-likelihood estimate $\hat{x}_k$, $k = 0, 1, 2, 3$, for part b.

2.3-8: Time-Varying System. Consider the time-varying system

$$x_{k+1} = 2^k x_k + (-1)^k u_k + w_k,$$

$$z_k = x_k + v_k$$

with $x_0 \sim (1, 2)$, $w_k \sim (0, 1.1^k)$, $v_k \sim (0, 0.9^k)$. (That is, as time passes, the process noise increases and the measurement noise decreases.)

a. Find error covariances for $k = 0, 1, 2, 3$.

b. Suppose $u_k = 1$ for $k = 0, 1, 2, \ldots$ and the measurements are given by $z_1 = -5$, $z_2 = -10$, $z_3 = -30$. Find the state estimates $\hat{x}_k$ for $k = 0, 1, 2, 3$.

2.3-9: Scalar Algebraic Riccati Equation. Consider the scalar system in Examples 2.3-3 and 2.5-1.

a. Let $h = 0$ so there are no measurements. Sketch the steady-state error variance p as a function of the plant matrix a. Interpret this.

b. Let $h = 1$, $r = 1$, $a = \frac{1}{2}$. Write the algebraic Riccati equation. Plot the root locus of this equation as q varies from zero to infinity. For what values of q does there exist a positive root to the algebraic Riccati equation?

Section 2.4

2.4-1: There is prescribed the system

$$\dot{x} = \begin{bmatrix} 0 & 1 \\ -2 & -3 \end{bmatrix} x + \begin{bmatrix} 0 \\ 2 \end{bmatrix} u + w$$

$$z = [1 \quad 1]x + v,$$

with $w(t)$, $v(t)$, and $x(0)$ having means of zero and covariances of

$$Q = \begin{bmatrix} 1 & -2 \\ -2 & 4 \end{bmatrix}, \qquad R = 0.4, \qquad P_0 = \begin{bmatrix} 2 & 1 \\ 1 & 1 \end{bmatrix},$$

respectively. The usual uncorrelatedness assumptions apply.

a. Measurements z_k are taken every $T = 0.1$ sec. Find the discretized system (not Euler's approximation).

b. For the continuous system, find the mean $\bar{z}(t)$ if $u(t) = u_{-1}(t)$ the unit step.

c. For the discretized system, find $\bar{z}_k$ for $k = 0, 1, 2, 3, 4$ if $u_k = u_{-1}(k)$ the unit step. Sketch $\bar{z}_k$ and $\bar{z}(t)$ on the same graph. Do they agree?

d. Find error covariances P_k and Kalman gains K_k for $k = 0, 1, 2$ for the discretized system assuming no measurement at time $k = 0$.

e. If $u_k = u_{-1}(k)$ and the data are $z_1 = 0.181$, $z_2 = 0.330$ find the estimates $\hat{x}_k$ for $k = 0, 1, 2$.

f. Now, after completing part d you discover that there is a change in plan and a measurement z_0 is taken. It is equal to 0.05. The remaining data and the input are given as in part e. Find the new optimal linear estimates $\hat{x}_k$ for $k = 0, 1, 2$.

2.4-2: Vibration Isolation System. A cargo vibration isolation system on an aircraft is shown in Fig. P2.4-1a. The cargo mass is M, k is a spring constant, B is a damper, $y(t)$ is vertical position, and $f(t)$ is an input control force. An electrical analog is shown in Fig. P2.4-1b.

a. Write the state equation if the state is defined as $x = [y \quad \dot{y}]^T$. To compensate for unmodeled dynamics add a white process noise $w(t) \sim (0, \frac{1}{2})$ to $\ddot{y}(t)$.

b. We are interested in the vertical position $y(t)$ of the cargo. Unfortunately, we can only measure the velocity $\dot{y}(t)$, which is done by using a device with a gain of 1 which adds white measurement noise $v(t) \sim (0, \frac{1}{2})$. Write the measurement equation.

c. The onboard flight computer samples the velocity measurement every $T = 50$ msec. Find the discretized system (not Euler's approximation) which we can use to run a Kalman filter on. Let $M = 2$, $k = 4$, $B = 6$. Note that the process noise in the discretized system excites *both* states, although $w(t)$ excites only the velocity in the continuous system.

d. The next step is to find the Kalman gains to store a priori in the computer so that we can save time during the on-line computation of the estimates. Assume no measurement at time $k = 0$. Find P_k and K_k for $k = 0, 1, 2, 3$.

e. Let

$$x(0) \sim \left(\begin{bmatrix} 0 \\ 1 \end{bmatrix}, \begin{bmatrix} 0.5 & 0 \\ 0 & 0.5 \end{bmatrix} \right),$$

and suppose the data samples are $z_1 = 0.429$, $z_2 = 0.366$, $z_3 = 0.311$. Find estimates for the position $y(kT)$ for $k = 0, 1, 2, 3$.

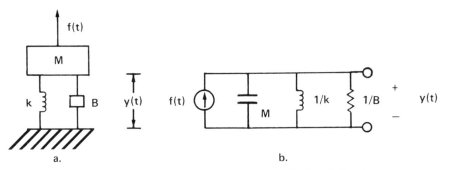

FIGURE P2.4-1 (a) Vibration isolation system. (b) Electrical analog.

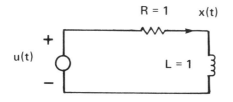

FIGURE P2.4-2   An electric circuit.

2.4-3:

a.   Find the differential equation relating $x(t)$ to $u(t)$ in Fig. P2.4-2.

b.   We take measurements $z(t)$ of $x(t)$ every 50 msec and get the results

Time t (sec)	$u(t)$	$z(t)$
0.05	0.95	0.05
0.10	1.00	0.20
0.15	1.10	0.24
0.20	1.01	0.30

The measurements are corrupted by additive zero-mean white noise with a variance of 0.1. We know that $x(0) \sim (0, 0.2)$. Let $u(0) = 1.05$.

Assume that the plant has process noise which we can model as being zero-mean white noise with variance of 2 added to the derivative of $x(t)$. The usual uncorrelatedness assumptions hold.

Find the best estimates for $x(t)$ at $t = 0, 0.05, 0.10, 0.15$. (Use three decimal places.)

Section 2.5

2.5-1:   Limiting Behavior.   Let

$$x_{k+1} = \begin{bmatrix} 4 & 0 \\ 0 & 1 \end{bmatrix} x_k + w_k, \qquad w_k \sim \left(0, \begin{bmatrix} 0 & 0 \\ 0 & 0.5 \end{bmatrix} \right), \qquad x_0 \sim (0, P_0)$$

$$z_k = H x_k + v_k, \qquad v_k \sim (0, 1).$$

a.   Let $H = [0 \quad 3]$. Find P_k for $k = 0, 1, 2, 3$ if
   i. $P_0 = 0$.
   ii. $P_0 = I$.

b.   Repeat part a if $H = [3 \quad 3]$.

c.   Discuss. What is going on here?

Section 2.6

2.6-1: Wiener Filter for Optimal Estimation of the State. Prove the statements accompanying (2.6-38). [Assume A and H are known. Then suppose that the optimal filter for estimating $y_k = Hx_k$ is given by (2.6-31) using a gain of K_1. Suppose further that $\hat{x}_k$ is provided by (2.6-38) with a gain K_2. Show that observability implies that $K_1 = K_2$. See Anderson and Moore (1979).]

2.6-2: Kalman and Wiener Filters. In this problem we consider a rather artificial discrete system which is simple enough to obtain analytic results. Let

$$x_{k+1} = \begin{bmatrix} 0 & 1 \\ 0 & 0 \end{bmatrix} x_k + \begin{bmatrix} q \\ 1 \end{bmatrix} w_k,$$

$$z_k = [1 \quad 0]x_k + v_k,$$

with $w_k \sim (0, 1)$, $v_k \sim (0, 1)$ white and uncorrelated. Let $q = 1$.

a. Solve the ARE to determine the optimal steady-state Kalman filter. Find the filter poles and the transfer function from data z_k to estimate $\hat{x}_k$.

b. Find spectral density $\Phi_Z(z)$. Factor to obtain the whitening filter $W(z)$. Hence, find the Wiener filter. Compare to part a.

c. Use the Chang–Letov equation to find the steady-state Kalman filter.

d. Find the steady-state Kalman filter if $q = 0$ so that the process noise does not drive state component one.

Section 2.7

2.7-1: Colored Measurement Noise Due to Tachometer Vibration. A DC servo motor is described by the transfer function

$$\frac{\theta}{e} = \frac{1}{s(1 + 0.2s)}, \tag{1}$$

where $e(t)$ is applied voltage and θ is shaft angle. Angular velocity $\dot{\theta}(t)$ is measured using a tachometer which yields a voltage of $y(t) = \dot{\theta}(t)$. Tachometer vibration results in an additive measurement noise $v(t)$ with spectrum

$$\Phi_v(\omega) = \frac{0.2(\omega^2 + 4\pi^2)}{\omega^4 + 4(0.01 - \pi)^2\omega^2 + 16\pi^4}. \tag{2}$$

The sampling period for the measurements is $T = 20$ msec. Assume that terms of the order of T^2 can be ignored.

a. Set up the continuous state and measurement equations.

b. Find a measurement noise shaping filter in state variable form.

c. Use state augmentation to convert to an equivalent problem with white measurement noise. Sample this system and set up the discrete Kalman filter equations. (Use Euler's approximation to simplify this.)

d. Sample the plant and shaping filter in a and b. Now use the derived measurement approach to convert to an equivalent system driven by white noises. Find all covariances and set up the discrete Kalman filtering problem.

2.7-2: Riccati Equation as Lyapunov Equation for Closed-Loop System. Define K_k by (2.7-17). Show that if $Q = I$ and $R = SS^T$ then (2.7-20) can be written as

$$P_{k+1}^{-} = (A - K_k H)P_k^{-}(A - K_k H)^T + [G \quad -K_k S][G \quad -K_k S]^T.$$

This is the Joseph stabilized formulation. Note that this is a *Lyapunov* equation for the closed-loop system formed by applying the time-varying output injection K_k.

2.7-3: Correlated Measurement and Process Noise. In Problem 2.3-7, let $v_k w_k = -0.5$. Repeat part b of the problem.

2.7-4: Colored Process Noise. The plant

$$x_{k+1} = 0.9x_k + w_k$$

$$z_k = x_k + v_k$$

has $v_k \sim (0, 1)$ white, but the process noise has spectral density of

$$\phi_w(\omega) = \frac{2 + 2\cos\omega}{\frac{3}{4} + \cos\omega}.$$

The two noises v_k and w_k are uncorrelated.

a. Find a shaping filter for w_k.

b. Write the augmented state equations and set up the Kalman filter equations.

Section 2.8

2.8-1: A Priori Optimal Smoother. Derive a smoothing scheme like the one in Table 2.8-1 for the a priori smoothed estimate $\hat{x}_k^{-}$.

2.8-2: Find optimal *smoothed* estimates for Problem 2.3-6 part a using:

a. The scheme of Table 2.8-1.

b. The Rauch–Tung–Striebel smoother.

2.8-3: Repeat Problem 2.8-2 for the system of Problem 2.3-7.

3

CONTINUOUS-TIME KALMAN FILTER

If discrete measurements are taken, whether they come from a discrete or a continuous system, the discrete Kalman filter can be used. The continuous Kalman filter is used when the measurements are continuous functions of time. Discrete measurements arise when a system is sampled, perhaps as part of a digital control scheme. Because of today's advanced micro-processor technology and the fact that microprocessors can provide greater accuracy and computing power than analog computers, digital control is being used more and more instead of the classical analog control methods. This means that for modern control applications, the discrete Kalman filter is usually used.

On the other hand, a thorough study of optimal estimation must include the continuous Kalman filter. Its relation to the Wiener filter provides an essential link between classical and modern techniques, and it yields some intuition which is helpful in a discussion of nonlinear estimation.

3.1 DERIVATION FROM DISCRETE KALMAN FILTER

There are several ways to derive the continuous-time Kalman filter. One of the most satisfying is the derivation from the Wiener–Hopf equation presented in Section 3.3. In this section we present a derivation which is based on "unsampling" the discrete-time Kalman filter. This approach provides an understanding of the relation between the discrete and continuous filters. It also provides insight on the behavior of the discrete Kalman gain as the sampling period goes to zero.

Suppose there is prescribed the continuous time-invariant plant

$$\dot{x}(t) = Ax(t) + Bu(t) + Gw(t) \tag{3.1-1a}$$

$$z(t) = Hx(t) + v(t), \tag{3.1-1b}$$

with $w(t) \sim (0, Q)$ and $v(t) \sim (0, R)$ white; $x(0) \sim (\bar{x}_0, P_0)$; and $w(t)$, $v(t)$, and $x(0)$ mutually uncorrelated. Then if sampling period T is small we can use Euler's approximation to write the discretized version of (3.1-1) as

$$x_{k+1} = (I + AT)x_k + BTu_k + Gw_k \tag{3.1-2a}$$

$$z_k = Hx_k + v_k, \tag{3.1-2b}$$

with $w_k \sim (0, QT)$ and $v_k \sim (0, R/T)$ white; $x_0 \sim (\bar{x}_0, P_0)$; and w_k, v_k, and x_0 mutually uncorrelated.

By using Tables 2.3-1 and 2.3-2 we can write the covariance update equations for (3.1-2) as

$$P^-_{k+1} = (I + AT)P_k(I + AT)^T + GQG^TT \tag{3.1-3}$$

$$K_{k+1} = P^-_{k+1}H^T\left(HP^-_{k+1}H^T + \frac{R}{T}\right)^{-1} \tag{3.1-4}$$

$$P_{k+1} = (I - K_{k+1}H)P^-_{k+1}. \tag{3.1-5}$$

We shall manipulate these equations and then allow T to go to zero to find the continuous covariance update and Kalman gain.

Let us first examine the behavior of the discrete Kalman gain K_k as T tends to zero. By (3.1-4) we have

$$\frac{1}{T}K_k = P^-_kH^T(HP^-_kH^TT + R)^{-1}; \tag{3.1-6}$$

so that

$$\lim_{T \to 0} \frac{1}{T}K_k = P^-_kH^TR^{-1}. \tag{3.1-7}$$

This implies that

$$\lim_{T \to 0} K_k = 0; \tag{3.1-8}$$

the discrete Kalman gain tends to zero as the sampling period becomes small. This result is worth remembering when designing discrete Kalman filters for continuous-time systems. For our purposes now it means that the continuous Kalman gain $K(t)$ should not be defined by $K(kT) = K_k$ in the limit as $T \to 0$.

Turning to (3.1-3), we have

$$P^-_{k+1} = P_k + (AP_k + P_kA^T + GQG^T)T + 0(T^2), \tag{3.1-9}$$

where $0(T^2)$ represents terms of order T^2. Substituting (3.1-5) into this equation there results

$$P^-_{k+1} = (I - K_k H)P^-_k$$

$$+ [A(I - K_k H)P^-_k + (I - K_k H)P^-_k A^T + GQG^T]T + 0(T^2),$$

or, on dividing by T,

$$\frac{1}{T}(P^-_{k+1} - P^-_k) = (AP^-_k + P^-_k A^T + GQG^T - AK_k HP^-_k - K_k HP^-_k A^T)$$

$$- \frac{1}{T} K_k HP^-_k + 0(T). \qquad (3.1\text{-}10)$$

In the limit as $T \to 0$ the continuous-error covariance $P(t)$ satisfies

$$P(kT) = P^-_k, \qquad (3.1\text{-}11)$$

so letting T tend to zero in (3.1-10) there results (use (3.1-7) and (3.1-8))

$$\dot{P}(t) = AP(t) + P(t)A^T + GQG^T - P(t)H^T R^{-1} HP(t). \qquad (3.1\text{-}12)$$

This is the *continuous-time Riccati equation* for propagation of the error covariance. It is the continuous-time counterpart of (2.3-18).

The discrete estimate update for (3.1-2) is given by (2.3-19), or

$$\hat{x}_{k+1} = (I + AT)\hat{x}_k + BTu_k$$

$$+ K_{k+1}[z_{k+1} - H(I + AT)\hat{x}_k - HBTu_k], \qquad (3.1\text{-}13)$$

which on dividing by T can be written as

$$\frac{\hat{x}_{k+1} - \hat{x}_k}{T} = A\hat{x}_k + Bu_k + \frac{K_{k+1}}{T}[z_{k+1} - H\hat{x}_k - H(A\hat{x}_k - Bu_k)T]. \qquad (3.1\text{-}14)$$

Since, in the limit as $T \to 0$, $\hat{x}(t)$ satisfies

$$\hat{x}(kT) = \hat{x}_k, \qquad (3.1\text{-}15)$$

we obtain in the limit

$$\dot{\hat{x}}(t) = A\hat{x}(t) + Bu(t) + P(t)H^T R^{-1}[z(t) - H\hat{x}(t)]. \qquad (3.1\text{-}16)$$

This is the estimate update equation. It is a differential equation for the estimate $\hat{x}(t)$ with initial condition $\hat{x}(0) = \bar{x}_0$. If we define the continuous Kalman gain by

$$K(kT) = \frac{1}{T} K_k \qquad (3.1\text{-}17)$$

in the limit as $T \to 0$, then

$$K(t) = P(t)H^T R^{-1}, \qquad (3.1\text{-}18)$$

and

$$\dot{\hat{x}} = A\hat{x} + Bu + K(z - H\hat{x}). \qquad (3.1\text{-}19)$$

TABLE 3.1-1   Continuous-Time Kalman Filter

System model and measure model:

$$\dot{x} = Ax + Bu + Gw \tag{3.1-20a}$$

$$z = Hx + v \tag{3.1-20b}$$

$$x(0) \sim (\bar{x}_0, P_0), \ w \sim (0, Q), \ v \sim (0, R)$$

Assumptions:

$\{w(t)\}$ and $\{v(t)\}$ are white noise processes uncorrelated with $x(0)$ and with each other. $R > 0$.

Initialization:

$$P(0) = P_0, \ \hat{x}(0) = \bar{x}_0$$

Error covariance update:

$$\dot{P} = AP + PA^T + GQG^T - PH^T R^{-1} HP \tag{3.1-21}$$

Kalman gain:

$$K = PH^T R^{-1} \tag{3.1-22}$$

Estimate update:

$$\dot{\hat{x}} = A\hat{x} + Bu + K(z - H\hat{x}) \tag{3.1-23}$$

Equations (3.1-12), (3.1-18), and (3.1-19) are the continuous-time Kalman filter for (3.1-1). They are summarized in Table 3.1-1. If matrices A, B, G, Q, R are time varying, these equations still hold.

It is often convenient to write $P(t)$ in terms of $K(t)$ as

$$\dot{P} = AP + PA^T + GQG^T - KRK^T. \tag{3.1-24}$$

It should be clearly understood that while, in the limit as $T \to 0$, the discrete-error covariance sequence P_k^- is a sampled version of the continuous-error covariance $P(t)$, the discrete Kalman gain is *not* a sampled version of the continuous Kalman gain. Instead, K_k represents the samples of $TK(t)$ in the limit as $T \to 0$.

Figure 3.1-1 shows the relation between $P(t)$ and the discrete covariances P_k and P_k^- for the case when there is a measurement z_0. As $T \to 0$, P_k and P_{k+1}^- tend to the same value since $(I + AT) \to I$ and $QT \to 0$ [see (2.3-4)]. They both approach $P(kT)$.

A diagram of (3.1-23) is shown in Fig. 3.1-2. Note that it is a linear system with two parts: a model of the system dynamics (A, B, H) and an error correcting portion $K(z - H\hat{x})$. The Kalman filter is time varying even when the original system (3.1-1) is time invariant. It has exactly the same form as the deterministic state observer of Section 2.1. It is worth remark-

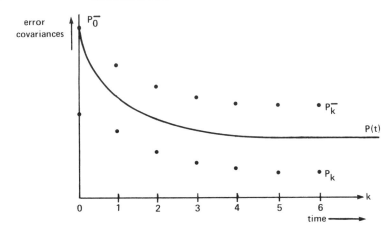

FIGURE 3.1-1 Continuous and discrete error covariances.

ing that according to Fig. 3.1-2 the Kalman filter is a low-pass filter with time-varying feedback.

If all statistics are Gaussian then the continuous Kalman filter provides the optimal estimate $\hat{x}(t)$. In general, for arbitrary statistics it provides the best *linear* estimate.

If all system matrices and noise covariances are known a priori, then $P(t)$ and $K(t)$ can be found and stored *before* any measurements are taken. This allows us to evaluate the filter design before we build it, and also saves computation time during implementation.

The continuous-time *residual* is defined as

$$\tilde{z}(t) = z(t) - H\hat{x}(t), \tag{3.1-25}$$

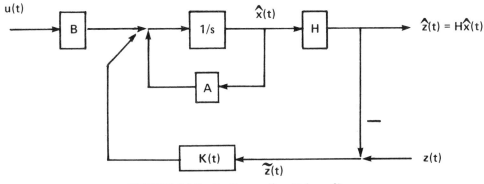

FIGURE 3.1-2 Continuous-time Kalman filter.

since $\hat{z}(t) = H\hat{x}(t)$ is an estimate for the data. We shall subsequently examine the properties of $\tilde{z}(t)$.

It should be noted that Q and R are not covariance matrices in the continuous time case; they are spectral density matrices. The covariance of $v(t)$, for example, is given by $R(t)\delta(t)$. (The continuous Kronecker delta has units of $\sec^{-1}$.)

In the discrete Kalman filter it is not strictly required that R be nonsingular; all we required was $|HP_k^- H^T + R| \neq 0$ for all k. In the continuous case, on the other hand it is necessary that $|R| \neq 0$. If R is singular, we must use the *Deyst filter* (Example 3.6-1).

The continuous Kalman filter can not be split up into separate time and measurement updates; there is no "predictor–corrector" formulation in the continuous-time case. (Note that as T tends to zero, P_{k+1}^- tends to P_k; so that in the limit the a priori and a posteriori error covariance sequences become the same sequence.) We can say, however, that the term $-PH^T R^{-1} HP$ in the error covariance update represents the decrease in $P(t)$ due to the measurements. If this term is deleted, we recover the analog of Section 2.2 for the continuous case. The following example illustrates.

Example 3.1-1: *Linear Stochastic System*

If $H = 0$ so that there are no measurements, then

$$\dot{P} = AP + PA^T + GQG^T \tag{3.1-26}$$

represents the propagation of the error covariance for the linear stochastic system

$$\dot{x} = Ax + Bu + Gw. \tag{3.1-27}$$

It is a *continuous-time Lyapunov equation* which has similar behavior to the discrete version (2.2-4).

When $H = 0$ so that there are no measurements, (3.1-19) becomes

$$\dot{\hat{x}} = A\hat{x} + Bu; \tag{3.1-28}$$

so that the estimate propagates according to the deterministic version of the system. This is equivalent to (2.2-3).

In Part 2 of the book we shall need to know how the mean-square value of the state

$$X(t) = \overline{x(t)x^T(t)} \tag{3.1-29}$$

propagates. Let us derive a differential equation satisfied by $X(t)$ when the input $u(t)$ is equal to zero and there are no measurements.

Since there are no measurements, the optimal estimate is just the mean value of the unknown so that $\bar{x} = \hat{x}$. Then

$$P = \overline{(x - \bar{x})(x - \bar{x})^T} = X - \overline{x}\overline{x}^T,$$

so that by (3.1-28)

$$\dot{P} = \dot{X} - \dot{\bar{x}}\bar{x}^T - \bar{x}\dot{\bar{x}}^T = \dot{X} - A\bar{x}\bar{x}^T - \bar{x}\bar{x}^T A^T.$$

Equating this to (3.1-26) yields

$$\dot{X} = AX + XA^T + GQG^T. \tag{3.1-30}$$

The initial condition for this is

$$X(0) = P_0 + \bar{x}_0\bar{x}_0^T. \tag{3.1-31}$$

Thus, in the absence of measurements and a deterministic input, $X(t)$ and $P(t)$ satisfy the same Lyapunov equation.

Let us now assume the scalar case $x \in R$ and repeat Example 2.2-2 for continuous time. Suppose $x_0 \sim (0, p_0)$ and $u(t) = u_{-1}(t)$, the unit step. Let the control weighting $b = 1$, and suppose $g = 1$ and $w(t) \sim (0, 1)$. Then, by using the usual state equation solution there results from (3.1-28)

$$\hat{x}(s) = \frac{1}{s-a} \cdot \frac{1}{s} = \frac{1/a}{s-a} + \frac{-1/a}{s},$$

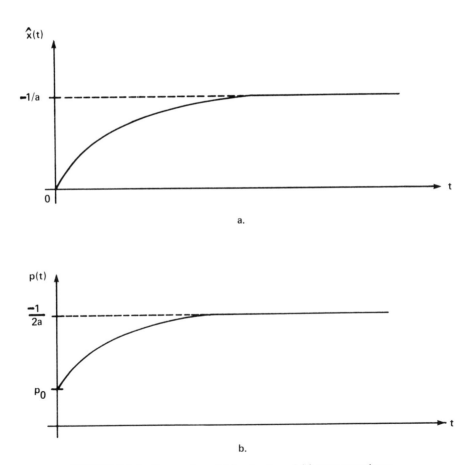

a.

b.

FIGURE 3.1-3 Propagation of (a) estimate and (b) error covariance.

so that

$$\hat{x}(t) = \frac{1}{a}(e^{at} - 1)u_{-1}(t). \tag{1}$$

This reaches a finite steady-state value only if $a < 0$; that is, if the system is asymptotically stable. In this case $\hat{x}(t)$ behaves as in Fig. 3.1-3a.

Equation (3.1-26) becomes

$$\dot{p} = 2ap + 1, \tag{2}$$

with $p(0) = p_0$. This may be solved by *separation of variables*. Thus

$$\int_{p_0}^{p(t)} \frac{dp}{2ap + 1} = \int_0^t dt,$$

or

$$p(t) = \left(p_0 + \frac{1}{2a}\right)e^{2at} - \frac{1}{2a}. \tag{3}$$

There is a bounded limiting solution if and only if $a < 0$. In this case $p(t)$ behaves as in Fig. 3.1-3b. ∎

3.2   SOME EXAMPLES

The best way to gain some insight into the continuous Kalman filter is to look at some examples.

Example 3.2-1:   *Estimation of a Constant Scalar Unknown*

This is the third example in the natural progression which includes Examples 2.3-2 and 2.4-2.

If x is a constant scalar unknown which is measured in the presence of additive noise then

$$\dot{x} = 0$$
$$z = x + v, \tag{1}$$

$v \sim (0, r^c)$. Let $x(0) \sim (x_0, p_0)$ be uncorrelated with $v(t)$.

Equation (3.1-21) becomes

$$\dot{p} = \frac{-p^2}{r^c}, \tag{2}$$

which can be solved by separation of variables:

$$\int_{p_0}^{p(t)} \frac{1}{p^2} dp = -\frac{1}{r^c}\int_0^t dt, \tag{3}$$

so that

$$p(t) = \frac{p_0}{1 + (p_0/r^c)t}. \tag{4}$$

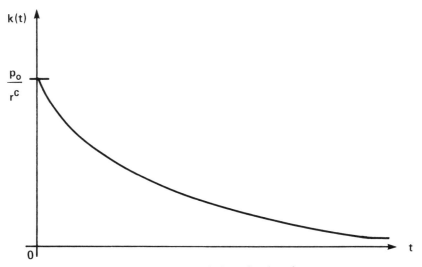

FIGURE 3.2-1   Kalman gain for estimation of a constant.

Comparing this to (5) in Example 2.4-2 we see that

$$p(kT) = p_k, \tag{5}$$

so that the discrete covariance sequence p_k simply corresponds to the samples of the continuous covariance $p(t)$.

The continuous Kalman gain is

$$K(t) = \frac{p_0/r^c}{1 + (p_0/r^c)t}, \tag{6}$$

which is shown in Fig. 3.2-1. Note that $K(t)$ vanishes as $t \to \infty$, meaning that for large t the measurements are disregarded as we become more and more certain of the estimate. As is seen from Example 2.4-2,

$$K(kT) = \frac{K_k}{T}. \tag{7}$$

In this example p_k and K_k correspond exactly to the samples of $p(t)$ and $TK(t)$ because the plant matrix and process noise in (1) are equal to zero (so that the discretized plant matrix is equal to 1). By (2.3-4), this means that $p_k = p^-_{k+1}$.   ■

Example 3.2-2:   *Continuous Filtering for a Scalar System*

a.   *Solution of Riccati Equation*

In the case of $x(t) \in R$ it is not difficult to solve the continuous Riccati equation analytically. We have, with $g = 1$ and $p(0) = p_0$ (lowercase letters emphasize these are scalars)

$$\dot{p} = -\frac{h^2}{r} p^2 + 2ap + q, \tag{3.2-1}$$

so that by separation of variables

$$\int_{p_0}^{p(t)} \frac{dp}{(h^2/r)p^2 - 2ap - q} = -\int_0^t dt. \tag{1}$$

Integrating, there results

$$\left| \frac{(h^2/r)p - a - \beta}{(h^2/r)p - a + \beta} \right| \left| \frac{(h^2/r)p_0 - a + \beta}{(h^2/r)p_0 - a - \beta} \right| = e^{-2\beta t} \tag{2}$$

where

$$\beta = \sqrt{a^2 + h^2 q/r}. \tag{3}$$

By defining (Bryson and Ho 1975)

$$p_1 = \frac{r}{h^2}(\beta - a), \qquad p_2 = \frac{r}{h^2}(\beta + a) \tag{4}$$

this yields

$$p(t) = p_2 + \frac{p_1 + p_2}{[(p_0 + p_1)/(p_0 - p_2)] e^{2\beta t} - 1}. \tag{3.2-2}$$

A sketch is provided in Fig. 3.2-2.

The steady-state value of $p(t)$ is given by p_2, or, if $a > 0$,

$$p_\infty = \frac{2q}{\lambda} \left(1 + \sqrt{1 + \frac{\lambda}{2a}} \right), \tag{3.2-3}$$

where

$$\lambda = \frac{2h^2 q}{ar} \tag{3.2-4}$$

is a signal-to-noise ratio. A similar expression holds if $a < 0$. (See Example 2.1-1.)

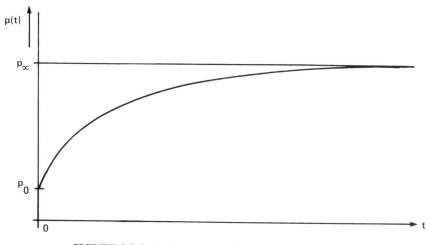

FIGURE 3.2-2 Error covariance for first-order system.

The steady-state value p_2 is bounded if $h \neq 0$, corresponding to observability. It is strictly positive if $q \neq 0$, which corresponds to reachability, and $h \neq 0$. The conditions for existence and uniqueness of a steady-state solution for the continuous Riccati equation are the same as those found in Section 2.5 for the discrete equation, as we shall discuss in Section 3.4.

The entire Kalman filter, including the Riccati equation, for this first-order system is shown in Fig. 3.2-3. Note that the Riccati equation manufacturing $p(t)$ is a sort of "doubled" or "squared" replica of the filter manufacturing $\hat{x}(t)$. This figure helps make the point that in the continuous Kalman filter, the data $z(t)$ are a continuous-time signal and all quantities vary continuously with time. To implement Fig. 3.2-3, we would use either an analog computer, or a digital computer with an integration routine such as Runge–Kutta (see program TRESP in Appendix B).

We now consider some special cases. The next two parts of this example should be compared to Example 2.3-3.

b. Completely Unreliable Measurements

Let $r \to \infty$ so that the measurements are totally unreliable. Then the Riccati equation is

$$\dot{p} = 2ap + q. \tag{3.2-5}$$

The solution is

$$p(t) = e^{2at}p_0 + \int_0^t e^{2a(t-\tau)}q \, d\tau, \tag{5}$$

or

$$p(t) = \left(p_0 + \frac{q}{2a} \right)e^{2at} - \frac{q}{2a}. \tag{3.2-6}$$

There is a steady-state solution of

$$p_\infty = -\frac{q}{2a} \tag{3.2-7}$$

if and only if the original plant is stable, $a < 0$.

In this case the Kalman gain $K(t)$ is equal to zero, so the estimate update is

$$\dot{\hat{x}} = a\hat{x} + bu, \tag{6}$$

with $\hat{x}(0) = \overline{x(0)}$. The estimate is a prediction based purely on the plant dynamics. This is Example 3.1-1.

c. No Process Noise

Let $q = 0$ so that the plant dynamics are purely deterministic. Then the solution of a applies with $\beta = |a|$.

If $a > 0$, the plant is unstable, then

$$\beta = a, \qquad p_1 = 0, \qquad p_2 = \frac{2ar}{h^2} \tag{7}$$

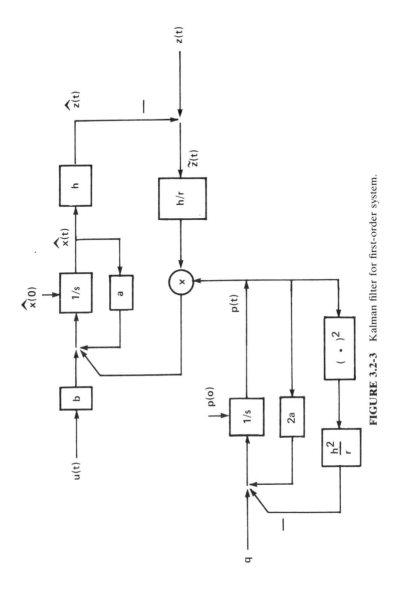

FIGURE 3.2-3 Kalman filter for first-order system.

155

and $p(t)$ is given by (3.2-2), which can be written as

$$p(t) = \frac{p_0}{p_0 h^2/2ar + (1 - p_0 h^2/2ar)\, e^{-2at}}.$$ (3.2-8)

The steady-state solution is p_2, or

$$p_\infty = \frac{2ar}{h^2},$$ (3.2-9)

which is nonzero, and is bounded if $h \neq 0$.

If $a < 0$, the plant is stable, then

$$\beta = -a, \qquad p_1 = \frac{-2ar}{h^2}, \qquad p_2 = 0.$$ (8)

In this case $p(t)$ is also given by (3.2-8), however, now

$$p_\infty = 0.$$ (3.2-10)

Figure 3.2-4 shows $p(t)$ for both unstable and stable plant. In either case, $p(t)$ is bounded if $h \neq 0$. If a is stable, then $K(t) \to 0$ and the measurements are discounted for large t. If a is unstable, on the other hand, then $K(t)$ approaches a nonzero steady-state value of

$$K_\infty = \frac{2a}{h},$$ (3.2-11)

so that the measurements are not discounted for large t.

d. Estimating a Scalar Wiener Process

Let $a = 0$ and $h = 1$ so that the plant is

$$\dot{x} = w$$ (3.2-12a)

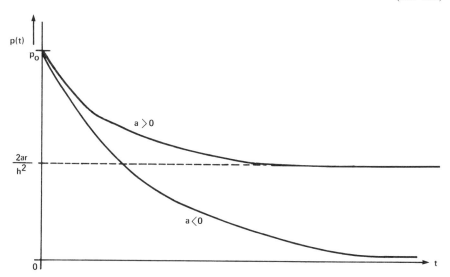

FIGURE 3.2-4 Error covariance with zero process noise for unstable plant and stable plant.

$$z = x + v, \qquad\qquad (3.2\text{-}12b)$$

$w \sim (0, q)$, $v \sim (0, r)$. Then

$$\beta = \sqrt{\frac{q}{r}}, \qquad p_1 = p_2 = \sqrt{qr}. \qquad\qquad (9)$$

The error covariance is

$$p(t) = \sqrt{qr} + \frac{2\sqrt{qr}}{[(p_0 + \sqrt{qr})/(p_0 - \sqrt{qr})]\, e^{2\sqrt{q/r}\,t} - 1}. \qquad\qquad (3.2\text{-}13)$$

Since $q \neq 0$ the process noise excites $x(t)$ and the error covariance reaches a nonzero limiting value. The steady-state error covariance and Kalman gain are

$$p_\infty = \sqrt{qr} \qquad\qquad (3.2\text{-}14)$$

$$K_\infty = \sqrt{\frac{q}{r}}. \qquad\qquad (3.2\text{-}15)$$

■

Example 3.2-3: *Continuous Filtering for Damped Harmonic Oscillator*

In Example 2.5-2 we applied the discrete Kalman filter to the discretized harmonic oscillator. Let us discuss here the continuous Kalman filter.

Suppose the plant is given by

$$\dot{x} = \begin{bmatrix} 0 & 1 \\ -\omega_n^2 & -2\alpha \end{bmatrix} x + \begin{bmatrix} 0 \\ 1 \end{bmatrix} w \qquad\qquad (1)$$

$$z = \begin{bmatrix} 0 & 1 \end{bmatrix} x + v, \qquad\qquad (2)$$

$w(t) \sim (0, q)$, $v(t) \sim (0, r)$, $x(0) \sim (\overline{x(0)}, P_0)$. This is a harmonic oscillator driven by white noise with a velocity measurement.

a. *Riccati Equation Solution*

To convert the Riccati equation into scalar equations, suppose that (P is symmetric)

$$P = \begin{bmatrix} p_1 & p_2 \\ p_2 & p_4 \end{bmatrix}. \qquad\qquad (3)$$

Then (3.1-21) yields the system of three coupled nonlinear differential equations

$$\dot{p}_1 = 2p_2 - \frac{1}{r} p_2^2 \qquad\qquad (3.2\text{-}16a)$$

$$\dot{p}_2 = p_4 - \omega_n^2 p_1 - 2\alpha p_2 - \frac{1}{r} p_2 p_4 \qquad\qquad (3.2\text{-}16b)$$

$$\dot{p}_4 = -2\omega_n^2 p_2 - 4\alpha p_4 - \frac{1}{r} p_4^2 + q. \qquad\qquad (3.2\text{-}16c)$$

These equations cannot easily be solved in closed form. A simulation was run on a digital computer using TRESP in Appendix B to obtain Fig. 3.2-5. The subroutine

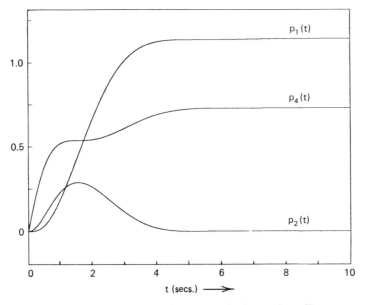

FIGURE 3.2-5 Error covariance terms for harmonic oscillator.

$f(t, P, \dot{P})$ which was used in the Runge–Kutta call is shown in Fig. 3.2-6. The Runge–Kutta integration period was 20 msec, and P_0 was 0. Note how easy it is to simulate the continuous filter using TRESP.

The Kalman gain is

$$K(t) = \frac{1}{r} \begin{bmatrix} p_2(t) \\ p_4(t) \end{bmatrix}. \tag{4}$$

```
      SUBROUTINE F(TIME,P,PP)
      REAL P(*), PP(*)
      DATA Q,R,OMSQ,AL/1.,1.,.64,.16/

C   COVARIANCE MATRIX ENTRIES ARE P(1), P(2), P(3)

      PP(1)= 2.*P(2) - P(2)**2/R
      PP(2)= P(3) - OMSQ*P(1) - 2.*AL*P(2) - P(2)*P(3)/R
      PP(3)= -2.*OMSQ*P(2) - 4.*AL*P(3) - P(3)**2/R + Q

      RETURN
      END
```

FIGURE 3.2-6 Harmonic oscillator error covariance subroutine for Runge–Kutta.

According to Fig. 3.2-5, the first component of $K(t)$ tends to zero, while its second component tends to a nonzero steady-state value.

The filter is

$$\dot{\hat{x}} = A\hat{x} + \frac{1}{r}\begin{bmatrix} p_2 \\ p_4 \end{bmatrix}(z - \hat{x}_2),$$

where x_2 is the second component of x. During the transient phase where $p_2(t) \neq 0$, the velocity measurement has a direct effect on the position estimate $\hat{x}_1$. At steady-state $p_2 = 0$ and the measurement has a direct effect only on the velocity estimate $\hat{x}_2$.

b. Steady-State Solution

It is easy to find the steady-state error covariance. At steady-state, $\dot{p}_1 = \dot{p}_2 = \dot{p}_3 = 0$, so that (3.2-16) become algebraic equations. The steady-state solution is given by

$$p_1(\infty) = \frac{2\alpha r}{\omega_n^2}\left(\sqrt{1 + \frac{q}{4\alpha^2 r}} - 1\right) \tag{3.2-17a}$$

$$p_2(\infty) = 0 \tag{3.2-17b}$$

$$p_4(\infty) = 2\alpha r\left(\sqrt{1 + \frac{q}{4\alpha^2 r}} - 1\right), \tag{3.2-17c}$$

where we have selected positive roots (assuming $\alpha > 0$) since $p(\infty) \geq 0$. Note that

$$\lambda = \frac{q}{\alpha r} \tag{3.2-18}$$

can be interpreted as a signal-to-noise ratio.

See Bryson and Ho (1975) and Parkus (1971). ∎

This is a good example of simplifying a filter by *preliminary analysis* prior to implementation. The scalar equations (3.2-16) are easier to program and require fewer operations to solve than the matrix Riccati equation. Even fairly complicated systems can be implemented by these techniques using TRESP in Appendix B if their order is not too high.

3.3 DERIVATION FROM WIENER–HOPF EQUATION

This section provides the complete connection between the Wiener filter and the Kalman filter.

In Section 1.5 we saw that the optimal linear filter $h(t, \tau)$ for estimating an unknown process $x(t)$ from a data process $z(t)$ given on (t_0, t) is the solution to the Weiner–Hopf equation

$$R_{XZ}(t, u) = \int_{t_0}^{t} h(t, \tau) R_Z(\tau, u)\, d\tau; \qquad t_0 < u < t. \tag{3.3-1}$$

We have selected the final time for which data are available as t to focus on the filtering problem. In terms of the time-varying filter $h(t, \tau)$, the optimal estimate for $x(t)$ is given by

$$\hat{x}(t) = \int_{t_0}^{t} h(t, \tau) z(\tau)\, d\tau. \tag{3.3-2}$$

The associated estimation error is defined as

$$\tilde{x}(t) = x(t) - \hat{x}(t), \tag{3.3-3}$$

with covariance given by

$$P(t) = R_X(t, t) - \int_{t_0}^{t} h(t, \tau) R_{ZX}(\tau, t)\, d\tau. \tag{3.3-4}$$

Wiener solved (3.3-1) for the steady-state case with $t_0 = -\infty$, and $\{z(t)\}$ and $\{x(t)\}$ stationary processes. We derived the solution for the Wiener filter in Section 1.5. In the steady-state case the optimal filter is time invariant, and Wiener's solution was in the frequency domain.

In 1961 Kalman and Bucy solved the Weiner–Hopf equation in the general nonstationary case. We follow here the solution procedure in Van Trees (1968).

The solution is somewhat roundabout, and the result is not an expression for $h(t, \tau)$. Instead, by working in the time domain a differential equation is found for $P(t)$. This turns out to be the Riccati equation. From $P(t)$, $h(t, t)$ can be found, which leads to a method of finding $\hat{x}(t)$. The full impulse response $h(t, \tau)$ is not explicitly constructed.

The solution is laid out in several subsections. They are:

1. Introduction of a shaping filter.
2. A differential equation for the optimal impulse response.
3. A differential equation for the estimate.
4. A differential equation for the error covariance.
5. Discussion.

Introduction of a Shaping Filter

The approach of Kalman and Bucy depends on assuming that the unknown $x(t)$ is generated by the *shaping filter*

$$\dot{x}(t) = Ax(t) + Gw(t), \tag{3.3-5a}$$

with $x(0) \sim (x_0, P_0)$. It is also assumed that the data $z(t)$ are related to $x(t)$ by the *linear measurement model*

$$z(t) = Hx(t) + v(t). \tag{3.3-5b}$$

The processes $w(t)$ and $v(t)$ are white with $w \sim (0, Q)$, $v \sim (0, R)$, and $Q \geqslant 0$, $R > 0$. Suppose $w(t)$, $v(t)$, and $x(0)$ are uncorrelated.

Under these circumstances, the required autocorrelations are given by

$$R_{XZ}(t, u) = \overline{x(t)z^T(u)} = \overline{x(t)[Hx(u) + v(u)]^T},$$

or

$$R_{XZ}(t, u) = R_X(t, u)H^T, \qquad (3.3\text{-}6)$$

and

$$R_Z(t, u) = \overline{z(t)z^T(u)}$$

$$= \overline{[Hx(t) + v(t)][Hx(u) + v(u)]^T},$$

or

$$R_Z(t, u) = HR_X(t, u)H^T + R\,\delta(t - u) \qquad (3.3\text{-}7)$$

The frequency domain solution of Wiener required a knowledge of $\Phi_{XZ}(\omega)$ and $\Phi_Z(\omega)$. By assuming (3.3-5) we have substituted for this requirement the alternative requirement of knowledge of the matrices A, G, H, P_0, Q, R.

By using the expression for $R_{XZ}(t, u)$ in the Wiener–Hopf equation, we obtain

$$R_X(t, u)H^T = \int_{t_0}^{t} h(t, \tau)R_Z(\tau, u)\,d\tau; \qquad t_0 < u < t. \qquad (3.3\text{-}8)$$

A Differential Equation for the Optimal Impulse Response

The solution does not provide an explicit expression for $h(t, \tau)$, but it depends on the fact that a differential equation can be found for the optimal impulse response.

Differentiate the Wiener–Hopf equation (3.3-8) using Leibniz's rule (Bartle 1976) to get

$$\frac{\partial R_X(t, u)}{\partial t} H^T = h(t, t)R_Z(t, u) + \int_{t_0}^{t} \frac{\partial h(t, \tau)}{\partial t} R_Z(\tau, u)\,d\tau; \qquad t_0 < u < t.$$

$$(3.3\text{-}9)$$

Now use (3.3-7) to write for $u < t$

$$h(t, t)R_Z(t, u) = h(t, t)HR_X(t, u)H^T.$$

By use of (3.3-8) this can be written

$$h(t, t)R_Z(t, u) = \int_{t_0}^{t} h(t, t)Hh(t, \tau)R_Z(\tau, u)\,d\tau; \qquad t_0 < u < t. \quad (3.3\text{-}10)$$

Now, use (3.3-5a) to say that

$$\frac{\partial R_X(t, u)}{\partial t} = E\left[\frac{dx(t)}{dt} x^T(u)\right]$$

$$= \overline{[Ax(t) + Gw(t)]x^T(u)}$$

$$= AR_X(t, u); \qquad u < t. \tag{3.3-11}$$

[We have $R_{wx}(t, u) = 0$ for $u < t$ since $w(t)$ is white.] Therefore, by (3.3-8)

$$\frac{\partial R_X(t, u)}{\partial t} H^T = \int_{t_0}^t Ah(t, \tau) R_Z(\tau, u) \, d\tau; \qquad t_0 < u < t. \tag{3.3-12}$$

Substituting into (3.3-9) using (3.3-10) and (3.3-12) yields

$$0 = \int_{t_0}^t \left[-Ah(t, \tau) + h(t, t)Hh(t, \tau) + \frac{\partial h(t, \tau)}{\partial t}\right] R_Z(\tau, u) \, d\tau; \qquad t_0 < u < t. \tag{3.3-13}$$

By (3.3-7)

$$0 = \int_{t_0}^t \left[-Ah(t, \tau) + h(t, t)Hh(t, \tau) + \frac{\partial h(t, \tau)}{\partial t}\right] HR_X(\tau, u)H^T \, d\tau$$

$$+ \left[-Ah(t, u) + h(t, t)Hh(t, u) + \frac{\partial h(t, u)}{\partial t}\right] R; \qquad t_0 < u < t.$$

Since $|R| \neq 0$ this is equivalent to

$$\frac{\partial h(t, \tau)}{\partial t} = Ah(t, \tau) - h(t, t)Hh(t, \tau); \qquad t_0 < \tau < t. \tag{3.3-14}$$

This is a differential equation satisfied by the optimal impulse response $h(t, \tau)$; it is the result we were seeking in this subsection. We will not find $h(t, \tau)$ explicitly, but will use (3.3-14) in the next subsection to find a differential equation for $\hat{x}(t)$.

A Differential Equation for the Estimate

To find a differential equation for the optimal estimate $\hat{x}(t)$, differentiate (3.3-2) using Leibniz's rule to get

$$\frac{d\hat{x}(t)}{dt} = h(t, t)z(t) + \int_{t_0}^t \frac{\partial h(t, \tau)}{\partial t} z(t) \, d\tau. \tag{3.3-15}$$

By using (3.3-14) this becomes

$$\dot{\hat{x}}(t) = h(t, t)z(t) + \int_{t_0}^t [Ah(t, \tau) - h(t, t)Hh(t, \tau)]z(\tau) \, d\tau.$$

Now substitute from (3.3-2) to obtain

$$\dot{\hat{x}}(t) = A\hat{x}(t) + h(t, t)[z(t) - H\hat{x}(t)]. \tag{3.3-16}$$

This is a differential equation for the estimate; now things are beginning to look familiar!

All that remains is to find an expression for $h(t, t)$, for then (3.3-16) provides a realization of the optimal filter $h(t, \tau)$. Unfortunately, we must tackle this is a roundabout manner. In the remainder of this subsection we shall note only that we can express $h(t, t)$ in terms of the error covariance $P(t)$.

Use (3.3-7) in the Wiener–Hopf equation (3.3-8) to obtain

$$R_X(t, u)H^T = \int_{t_0}^{t} h(t, \tau)[HR_X(\tau, u)H^T + R\,\delta(\tau - u)]\,d\tau,$$

or with $u = t$,

$$R_X(t, t)H^T - \int_{t_0}^{t} h(t, \tau)HR_X(\tau, t)H^T\,d\tau = h(t, t)R.$$

According to Equation (3.3-4) for the error covariance, and (3.3-6), this becomes

$$P(t)H^T = h(t, t)R;$$

so that

$$h(t, t) = P(t)H^T R^{-1}. \qquad (3.3\text{-}17)$$

At this point it is clear that $h(t, t)$ is nothing but the Kalman gain

$$K(t) = h(t, t) = P(t)H^T R^{-1}. \qquad (3.3\text{-}18)$$

By using $K(t)$ in (3.3-16) we can manufacture the estimate $\hat{x}(t)$ from the data $z(t)$. To compute $K(t)$ we need to find the error covariance $P(t)$, which we do in the next subsection.

A Differential Equation for the Error Covariance

To find a differential equation for the error covariance $P(t)$ we first need to find the *error system*. From (3.3-3), (3.3-5), and (3.3-16) we get

$$\dot{\tilde{x}}(t) = \dot{x}(t) - \dot{\hat{x}}(t)$$
$$= Ax(t) + Gw(t) - A\hat{x}(t) - K(t)[Hx(t) + v(t) - H\hat{x}(t)],$$

or

$$\dot{\tilde{x}}(t) = [A - K(t)H]\tilde{x}(t) - K(t)v(t) + Gw(t). \qquad (3.3\text{-}19)$$

This is the error system.

Let $\phi(t, t_0)$ be the state-transition matrix for $[A - K(t)H]$. Then the solution to (3.3-19) is

$$\tilde{x}(t) = \phi(t, t_0)\tilde{x}(t_0) - \int_{t_0}^{t} \phi(t, \tau)K(\tau)v(\tau)\,d\tau + \int_{t_0}^{t} \phi(t, \tau)Gw(\tau)\,d\tau. \qquad (3.3\text{-}20)$$

We shall soon need the cross-correlations $R_{v\tilde{x}}(t, t)$ and $R_{w\tilde{x}}(t, t)$. To find them use (3.3-20) and the uncorrelatedness assumptions. Thus

$$R_{v\tilde{x}}(t, t) = \overline{v(t)\tilde{x}^T(t)}$$

$$= -\int_{t_0}^{t} \overline{v(t)v^T(\tau)} K^T(\tau)\phi^T(t, \tau) \, d\tau. \qquad (3.3\text{-}21)$$

Note that

$$R_V(t, \tau) = R\,\delta(t - \tau), \qquad (3.3\text{-}22)$$

but the integral in (3.3-21) has an upper limit of t. Recall that the unit impulse can be expressed as

$$\delta(t) = \lim_{T \to 0} \frac{1}{T} \Pi\!\left(\frac{t}{T}\right), \qquad (3.3\text{-}23)$$

where the rectangle function is

$$\frac{1}{T} \Pi\!\left(\frac{t}{T}\right) = \begin{cases} \dfrac{1}{T}, & |t| < T \\ 0, & \text{otherwise.} \end{cases} \qquad (3.3\text{-}24)$$

Therefore, only half of the area of $\delta(t)$ should be considered as being to the left of $t = 0$. Hence, (3.3-21) is $[\phi(t, t) = I]$

$$R_{v\tilde{x}}(t, t) = -\frac{1}{2} RK^T(t) \qquad (3.3\text{-}25)$$

Similarly,

$$R_{w\tilde{x}}(t, t) = \overline{w(t)\tilde{x}^T(t)}$$

$$= \int_{t_0}^{t} \overline{w(t)w^T(\tau)} G^T\phi^T(t, \tau) \, d\tau,$$

or

$$R_{w\tilde{x}}(t, t) = \frac{1}{2} QG^T. \qquad (3.3\text{-}26)$$

To find a differential equation for $P(t)$, write

$$\dot{P}(t) = E\!\left[\frac{d\tilde{x}(t)}{dt} \tilde{x}^T(t)\right] + E\!\left[\tilde{x}(t) \frac{d\tilde{x}^T(t)}{dt}\right]. \qquad (3.3\text{-}27)$$

According to the error dynamics (3.3-19), the first term is

$$E\!\left[\frac{d\tilde{x}(t)}{dt} \tilde{x}^T(t)\right] = [A - K(t)H]P(t) + \frac{1}{2} K(t)RK^T(t) + \frac{1}{2}GQG^T, \qquad (3.3\text{-}28)$$

where we have used (3.3-25) and (3.3-26).

To (3.3-28) add its transpose to get

$$\dot{P} = (A - KH)P + P(A - KH)^T + KRK^T + GQG^T. \quad (3.3\text{-}29)$$

Finally, using (3.3-18) yields

$$\dot{P} = AP + PA^T - PH^T R^{-1} HP + GQG^T. \quad (3.3\text{-}30)$$

This is just the Riccati equation for the error covariance!

Discussion

The solution for $\hat{x}(t)$ provided by (3.3-30), (3.3-18), and (3.3-16) is the Kalman filter in Table 3.1-1. If the shaping filter (3.3-5) and covariances are time varying these equations still hold. If (3.3-5a) has a deterministic input it is easy to carry it through the derivation.

The Kalman filter provides the optimal estimate in the general non-stationary case. It is time varying even for a time-invariant shaping filter (3.3-5) since the Kalman gain $K(t) = h(t, t)$ in (3.3-16) is dependent on t. This accounts for the fact that the optimal filter cannot be found in the non-steady-state case using frequency domain methods.

The Kalman filter does not explicitly provide an expression for $h(t, \tau)$, however a realization for the optimal filter is given by (3.3-16).

In the limit as $t \to \infty$, the error covariance tends to a unique positive definite solution P_∞ under appropriate reachability and observability conditions (Section 3.4). Then the steady-state gain

$$K_\infty = P_\infty H^T R^{-1} \quad (3.3\text{-}31)$$

is a constant. The steady-state Kalman filter is

$$\dot{\hat{x}} = (A - K_\infty H)\hat{x} + K_\infty z$$

$$\hat{y} = H\hat{x}, \quad (3.3\text{-}32)$$

where $y = Hx$. Under these circumstances the stationarity assumptions hold if A is stable, and the transfer function

$$HF(s) = H[sI - (A - K_\infty H)]^{-1} K_\infty \quad (3.3\text{-}33)$$

is exactly the Wiener filter found by frequency domain methods in Section 1.5.

An analogous derivation of the discrete Kalman filter exists. In Section 2.6 we took a somewhat different (frequency domain) approach to show that the steady-state Kalman filter is the Wiener filter.

Example 3.3-1:   Comparison of Kalman Filter and Wiener Filter for Scalar System

Suppose the scalar unknown $x(t)$ has spectral density

$$\Phi_X(\omega) = \frac{2a\sigma^2}{\omega^2 + a^2} \quad (1)$$

and that data $z(t)$ are generated by

$$z(t) = x(t) + v(t) \tag{2}$$

with white noise $v(t) \sim (0, r)$ uncorrelated with $x(t)$.

a. Wiener Filter

In Example 1.5-4 we found that the Wiener filter is

$$H(s) = \frac{2\gamma}{1 + \sqrt{1 + \lambda}} \cdot \frac{1}{s + a\sqrt{1 + \lambda}} \tag{3}$$

with signal-to-noise ratios

$$\gamma = \frac{\sigma^2}{r} \tag{4}$$

and

$$\lambda = \frac{2\gamma}{a} = \frac{2\sigma^2}{ar}. \tag{5}$$

This can also be written as

$$H(s) = \frac{a(\sqrt{1 + \lambda} - 1)}{s + a\sqrt{1 + \lambda}}. \tag{6}$$

(It is easy to verify that $2\gamma/[a(\sqrt{1 + \lambda} + 1)(\sqrt{1 + \lambda} - 1)] = 1$.) The Wiener filter is a low-pass filter with a cutoff frequency of

$$\omega_c = a\sqrt{1 + \lambda}. \tag{7}$$

Recall that the Wiener filter is found by first factoring $\Phi_Z^{-1}(s)$ to find a whitening filter $W(s)$:

$$\Phi_Z^{-1}(\omega) = W(-j\omega) W(j\omega). \tag{8}$$

b. Kalman Filter

To find a shaping filter for the unknown $x(t)$, we can perform a spectral factorization on $\Phi_X(\omega)$ to obtain

$$\Phi_X(\omega) = \frac{1}{j\omega + a} (2a\sigma^2) \frac{1}{-j\omega + a}$$

$$= G(j\omega)(2a\sigma^2) G(-j\omega). \tag{9}$$

The shaping filter therefore has transfer function of

$$G(s) = \frac{1}{s + a}, \tag{10}$$

for which a state realization is

$$\dot{x} = -ax + w \tag{11}$$

with $w \sim (0, 2a\sigma^2)$. Let $x(0) \sim (x_0, p_0)$. The plant on which the Kalman filter should be run is (11), (2).

For this plant, the Riccati equation is

$$\dot{p} = -\frac{1}{r}p^2 - 2ap + 2a\sigma^2. \tag{12}$$

The solution to this is given in Example 3.2-2 (with a replaced by $-a$). The Kalman gain is

$$K(t) = \frac{1}{r}p(t) \tag{13}$$

and the Kalman filter is

$$\dot{\hat{x}} = -[a + K(t)]\hat{x} + K(t)z. \tag{14}$$

In the limit as $t \to \infty$ the steady-state filter is reached. It is

$$p_\infty = p_2 = ar(\sqrt{1+\lambda} - 1) \tag{15}$$

$$= \frac{2\sigma^2}{1 + \sqrt{1+\lambda}} \tag{16}$$

$$K_\infty = a(\sqrt{1+\lambda} - 1), \tag{17}$$

and

$$\dot{\hat{x}} = -a\sqrt{1+\lambda}\,\hat{x} + a(\sqrt{1+\lambda} - 1)z. \tag{18}$$

The transfer function of the steady-state Kalman filter (18) is

$$H(s) = \frac{a(\sqrt{1+\lambda} - 1)}{s + a\sqrt{1+\lambda}}, \tag{19}$$

which is exactly the Wiener filter (6). Moreover, the steady-state error covariance (16) agrees with the Wiener filter error covariance in Example 1.5-4.

Note that, since the measurement matrix $h = 1$, the numerator of the Wiener filter is the steady-state Kalman gain. ∎

Example 3.3-2:   *Estimating Motion Governed by Newton's Laws*

Newton's law

$$\ddot{y}(t) = \frac{u(t)}{m} \tag{1}$$

can be expressed, by defining the state as $x = [y \quad \dot{y}]^T$, in the reachable canonical form

$$\dot{x} = \begin{bmatrix} 0 & 1 \\ 0 & 0 \end{bmatrix} x + \begin{bmatrix} 0 \\ 1/m \end{bmatrix} u + \begin{bmatrix} 0 \\ 1 \end{bmatrix} w, \tag{2a}$$

where white process noise $w(t) \sim (0, \sigma^2)$ has been added to compensate for modeling errors. If position measurements are taken, then

$$z = [1 \quad 0]x + v. \tag{2b}$$

Let white noise $v(t) \sim (0, \rho^2)$ be uncorrelated with $w(t)$ and with the initial position $y(0)$ and velocity $\dot{y}(0)$.

If we represent the error covariance as

$$P = \begin{bmatrix} p_1 & p_2 \\ p_2 & p_4 \end{bmatrix} \tag{3}$$

then the Riccati equation yields after simplification

$$\dot{p}_1 = 2p_2 - \frac{p_1^2}{\rho^2} \tag{4}$$

$$\dot{p}_2 = p_4 - \frac{p_1 p_2}{\rho^2} \tag{5}$$

$$\dot{p}_4 = -\frac{p_2^2}{\rho^2} + \sigma^2. \tag{6}$$

A simulation could be run to plot $p_1(t)$, $p_2(t)$, $p_4(t)$ exactly as in Example 3.2-3, but we are concerned here with some rather interesting behavior at steady state which relates to some classical control concepts.

After the Kalman filter has been running for some time, a steady-state value of error-covariance P_∞ is reached for this system. (We investigate under what conditions this occurs in Section 3.4.) At steady state $\dot{P} = 0$; so let us set $\dot{p}_1$, $\dot{p}_2$, and $\dot{p}_4$ to zero. Then (4)–(6) becomes an algebraic set of equations which is easily solved to yield

$$p_2 = \sigma\rho \tag{7}$$

$$p_1 = \rho^2 \sqrt{\frac{2\sigma}{\rho}} \tag{8}$$

$$p_4 = \sigma\rho \sqrt{\frac{2\sigma}{\rho}}. \tag{9}$$

Defining signal-to-noise ratio

$$\gamma = \frac{\sigma}{\rho}, \tag{10}$$

the steady-state Kalman gain can be written as

$$K_\infty = P_\infty H^T R^{-1} = \frac{1}{\rho^2} \begin{bmatrix} p_1 \\ p_2 \end{bmatrix}$$

$$= \begin{bmatrix} \sqrt{2\gamma} \\ \gamma \end{bmatrix}. \tag{11}$$

The steady-state Kalman filter is therefore

$$\dot{\hat{x}} = (A - K_\infty H)\hat{x} + Bu + K_\infty z \tag{12}$$

or

$$\dot{\hat{x}} = \begin{bmatrix} -\sqrt{2\gamma} & 1 \\ -\gamma & 0 \end{bmatrix} \hat{x} + \begin{bmatrix} 0 \\ 1/m \end{bmatrix} u + \begin{bmatrix} \sqrt{2\gamma} \\ \gamma \end{bmatrix} z. \tag{13}$$

These equations describe in state-variable form the Wiener filter for reconstructing state $x(t)$ from the data.

The steady-state filter (13) is time invariant, so the transfer function from data $z(t)$

to estimate $\hat{x}(t)$ can be determined. It is

$$H(s) = (sI - A + K_\infty H)^{-1} K_\infty \tag{14}$$

$$= \frac{1}{2^2 + \sqrt{2}\gamma s + \gamma} \begin{bmatrix} \sqrt{2}\gamma s + \gamma \\ \gamma s \end{bmatrix}. \tag{15}$$

This is the conventional transfer-function form of the Wiener filter.

Now, notice something quite interesting. The characteristic polynomial of a second-order system is of the form

$$s^2 + 2\,\delta\omega_n s + \omega_n^2 \tag{16}$$

where δ is the damping ratio and ω_n the natural frequency. By comparing (15) and (16), we see that

$$\omega_n = \sqrt{\gamma} \tag{17}$$

$$\delta = \frac{1}{\sqrt{2}}, \tag{18}$$

so that the natural frequency of the observer system is dependent on the ratio of process to measurement noise variances. Furthermore, the damping ratio of the optimal steady-state filter is always $1/\sqrt{2}$ independent of the noise! This certainly provides some justification for the classical design criterion of placing poles so that $\delta = 1/\sqrt{2}$.

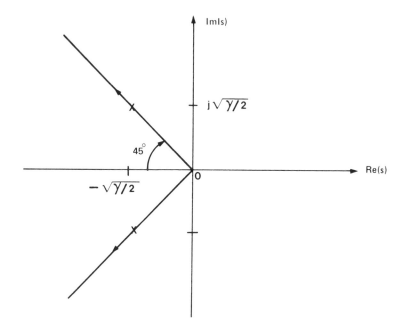

FIGURE 3.3-1 Root locus for Wiener filter as signal-to-noise ratio increases.

The roots of the characteristic equation are

$$s = -\sqrt{\frac{\gamma}{2}} \pm j\sqrt{\frac{\gamma}{2}}, \tag{19}$$

and a root-locus plot for the steady-state filter poles as γ increases from zero is shown in Fig. 3.3-1. Note that when the process noise is zero, the poles of the system (2a) and the Wiener filter (13) coincide.   ∎

3.4 ERROR DYNAMICS AND STATISTICAL STEADY STATE

All of our notions of the steady-state behavior of the discrete Kalman filter also hold for the continuous case. Some of the details are different. This section provides a summary of results; the derivations and proofs are very similar to those in the discrete development and are for the most part omitted. A time-invariant system is assumed here.

The Error System

The convergence and stability properties of the Kalman filter are defined in terms of the error system. The estimation error is

$$\tilde{x}(t) = x(t) - \hat{x}(t). \tag{3.4-1}$$

An expression for its dynamical behavior is given by (3.3-19);

$$\dot{\tilde{x}} = (A - KH)\tilde{x} + Gw - Kv. \tag{3.4-2}$$

This equation is driven by the process and the measurement noise. If $[A - K(t)H]$ tends to an asymptotically stable matrix as $t \to \infty$, then in the limit the estimate $\hat{x}(t)$ converges to the expected true plant state $x(t)$.

The residual is given by

$$\tilde{z} = H\tilde{x} + v. \tag{3.4-3}$$

The continuous error system is given by (3.4-2) and (3.4-3).

The error covariance is given by (3.3-29):

$$\dot{P}(t) = (A - KH)P + P(A - KH)^T + KRK^T + GQG^T. \tag{3.4-4}$$

Compare this to the discrete version (2.5-5). If we use $K(t) = P(t)H^T R^{-1}$ then (3.4-4) reduces to (3.1-21). Note that $\tilde{x}(t) = 0$ so that the estimate is unbiased.

If all statistics are Gaussian, then the Kalman filter manufactures the conditional mean $\hat{x}(t)$ and covariance $P(t)$ of $x(t)$ given all previous data. In this case it provides a scheme for updating the entire hyperstate [i.e., the conditional PDF of $x(t)$].

The Innovations Sequence

By (3.4-3) the residual has zero mean

$$\overline{\tilde{z}(t)} = 0. \tag{3.4-5}$$

The residual $\tilde{z}(t)$ is the new information in $z(t)$ which cannot be anticipated by using the data $z(\tau)$ for $\tau < t$. It is a *zero-mean white noise* process. It is also called the *innovations*.

The covariance of $\tilde{z}(t)$ has a form which makes the continuous Kalman filter simpler than the discrete filter. To find it, note that $\tilde{z}(t)$ is the limit as $T \to 0$ of the discrete innovations $\tilde{z}_k$. By (2.5-7), if R is the covariance of the *continuous* noise $v(t)$, then

$$P_{\tilde{z}_k} = HP_k^- H^T + \frac{R}{T}$$

$$= \frac{HP_k^- H^T T + R}{T}. \tag{3.4-6}$$

To make a discrete noise sequence equivalent in the limit to a continuous noise process, we must multiply its covariance by T (R is a spectral density matrix, but R/T is a covariance matrix in the limit). Therefore, letting $T \to 0$, we obtain

$$P_{\tilde{z}}(t) = R. \tag{3.4-7}$$

This accounts for the fact that in the continuous filter the factor R^{-1} appears, but in the discrete filter the corresponding factor is $(HP_k^- H^T + R)^{-1}$. [An alternative derivation of (3.4-7) is discussed in the problems.]

One way to test for the proper performance of an optimal filter is to monitor $\tilde{z}(t)$. If the innovations is not white noise, then the filter is performing suboptimality.

Technically, the innovations $v(t)$ are defined by

$$v(t) = R^{-1/2}\tilde{z}(t); \tag{3.4-8}$$

so that

$$P_v(t) = I. \tag{3.4-9}$$

The residual is, however, usually considered as the innovations; particularly because the measurement can always be prewhitened (Sections 1.4 and 4.4) to make $R = I$.

For an excellent discussion of these ideas see Kailath (1981).

The Algebraic Riccati Equation

The solution to the Riccati equation tends to a bounded steady-state value P if

$$\lim_{t \to \infty} P(t) = P \tag{3.4-10}$$

is bounded. In this case, for large t, $\dot{P}(t) = 0$ so that (3.1-21) tends to the continuous *algebraic Riccati equation (ARE)*

$$0 = AP + PA^T - PH^T R^{-1} HP + GQG^T. \qquad (3.4\text{-}11)$$

Any limiting solution to (3.1-21) is a solution to (3.4-11), although the converse is not true. The solutions to (3.4-11) are explored further in the problems.

The conditions under which:

1. bounded limiting solutions to (3.1-21) exist for all choices of $P(0)$,
2. the limiting solution to (3.1-21) is unique, independent of the choice of $P(0)$, and
3. the error system (3.4-2) is asymptotically stable in the limit

are the same as for the discrete filter. Now, however, detectability and stabilizability are defined with "stable" poles meaning poles in the open-left-half plane. Thus we have the next results.

Theorem 3.4-1

Let (A, H) be detectable. Then for every choice $P(0)$ there is a bounded limiting solution P to (3.1-21). Furthermore, P is a positive semidefinite solution to the algebraic Riccati equation (3.4-11).

By imposing a reachability condition we can make this result much stronger.

Theorem 3.4-2

Let $Q = \sqrt{Q}\sqrt{Q}^T \geq 0$, and $R > 0$. Suppose $(A, G\sqrt{Q})$ is reachable. Then (A, H) is detectable if and only if:

a. There is a unique positive definite limiting solution P to (3.1-21) which is independent of $P(0)$. Furthermore, P is the *unique* positive *definite* solution to the algebraic Riccati equation.
b. The steady-state error system defined by (3.4-2) with the *steady-state Kalman gain*

$$K = PH^T R^{-1} \qquad (3.4\text{-}12)$$

is asymptotically stable.

The proofs of these results are similar to the discrete-time case in Section 2.5.

These theorems of course explain the limiting behavior noted in the examples of Section 3.2.

If we relax the requirement in Theorem 3.4-2 to stabilizability of $(A, G\sqrt{Q})$ then the theorem still holds, but the unique solution P can then only be guaranteed to be positive semidefinite.

Time-Varying Plant

If the plant is time varying then we must redefine observability and reachability. Suppose the plant is

$$\dot{x} = A(t)x + B(t)u + G(t)w \qquad (3.4\text{-}13a)$$

$$z = H(t)x + v, \qquad (3.4\text{-}13b)$$

with $w \sim [0, Q(t)]$ and $v \sim [0, R(t)]$. Let $\phi(t, t_0)$ be the state transition matrix of A.

We say the plant is *uniformly completely observable* or *stochastically observable*, if for every final time T the *observability gramian* satisfies

$$\alpha_0 I \le \int_{t_0}^{T} \phi^T(\tau, t_0) H^T(\tau) R^{-1} H(\tau) \phi(\tau, t_0) \, d\tau \le \alpha_1 I \qquad (3.4\text{-}14)$$

for some $t_0 < T$, $\alpha_0 > 0$, and $\alpha_1 > 0$. We say the plant is *uniformly completely reachable*, or *stochastically reachable*, if for every initial time t_0 the *reachability gramian* satisfies

$$\alpha_0 I \le \int_{t_0}^{T} \phi(t, \tau) G(\tau) Q(\tau) G^T(\tau) \phi^T(t, \tau) \, d\tau \le \alpha_1 I \qquad (3.4\text{-}15)$$

for some $T > t_0$, $\alpha_0 > 0$, and $\alpha_1 > 0$.

Stochastic observability and reachability [and boundedness of $A(t)$, $Q(t)$, $R(t)$] guarantee that for large t the behavior of $P(t)$ is unique independent of $P(0)$. They also guarantee the uniform asymptotic stability of the error system, which means that the estimate $\hat{x}(t)$ converges to the expected value of $x(t)$. See Kalman and Bucy (1961).

3.5 FREQUENCY DOMAIN RESULTS

The presentation in this section parallels that in Section 2.6, and the proofs follow the same philosophy as in the discrete case. Because the continuous innovations process has a covariance equal simply to R, the measurement noise covariance, the results here are somewhat simpler than in Section 2.6.

We assume here a time-invariant plant, and begin with some frequency domain results for continuous linear stochastic systems corresponding to those at the end of Section 2.2. Then we derive a spectral factorization result and discuss further the relation between the Wiener and Kalman filters. Next we present the Chang–Letov design procedure for continuous systems, which gives a connection with the classical root-locus design method.

Spectral Densities for Linear Stochastic Systems

Consider the linear stochastic system

$$\dot{x} = Ax + Bu + Gw, \qquad (3.5\text{-}1a)$$

$$z = Hx + v, \qquad (3.5\text{-}1b)$$

with $w(t) \sim (0, Q)$ and $v(t) \sim (0, R)$ white; $x(0) \sim (x_0, P_0)$; $R > 0$; and $w(t)$, $v(t)$, and $x(0)$ uncorrelated.

If A is asymptotically stable then there is a positive semidefinite solution to the algebraic Lyapunov equation

$$0 = AP + PA^T + GQG^T. \qquad (3.5\text{-}2)$$

In this event, as $t \to \infty$ the effects of the initial condition $x(0)$ die away and if $w(t)$ is stationary [let us suppose for simplicity that $u(t) = 0$] then $x(t)$ also tends to a stationary process with covariance P.

Under these conditions of *stochastic steady state*, we can discuss spectral densities. According to (3.5-1), the spectral density of $x(t)$ is

$$\Phi_X(s) = (sI - A)^{-1} GQG^T (-sI - A)^{-T}. \qquad (3.5\text{-}3)$$

(See Section 1.5.) Likewise, the spectral density of the output $z(t)$ is

$$\Phi_Z(s) = H(sI - A)^{-1} GQG^T (-sI - A)^{-T} H^T + R. \qquad (3.5\text{-}4)$$

Let the white noise $w'(t)$ with unit spectral density be defined by $w(t) = \sqrt{Q} w'(t)$, and define the transfer function from $w'(t)$ to $z(t)$ as

$$H(s) = H(sI - A)^{-1} G\sqrt{Q}. \qquad (3.5\text{-}5)$$

Then we can say

$$\Phi_Z(s) = H(s)H^T(-s) + R. \qquad (3.5\text{-}6)$$

A Spectral Factorization Result

As in the discrete case, the Kalman filter manufactures a minimum-phase factorization for the spectral density of the data $z(t)$. Therefore, solution of the Riccati equation in the time domain is equivalent to a spectral factorization in the frequency domain. To show this, we proceed as follows.

The steady-state Kalman filter is given by

$$\dot{\hat{x}} = (A - KH)\hat{x} + Bu + Kz, \qquad (3.5\text{-}7)$$

where the steady-state Kalman gain satisfies $K = PH^T R^{-1}$ with P the positive semidefinite solution to the algebraic Riccati equation (3.4-11), which exists if (A, H) is detectable and $(A, G\sqrt{Q})$ is stabilizable.

Using a derivation like the one for (2.6-4) we have

$$\Delta^{cl}(s) = |I + H(sI - A)^{-1} K| \Delta(s), \qquad (3.5\text{-}8)$$

where the plant and filter characteristic polynomials are

$$\Delta(s) = |sI - A| \qquad (3.5\text{-}9)$$

and

$$\Delta^{cl}(s) = |sI - (A - KH)|. \qquad (3.5\text{-}10)$$

We can also show

$$\Phi_Z(s) = H(s)H^T(-s) + R$$
$$= [I + H(sI - A)^{-1}K]R[I + H(-sI - A)^{-1}K]^T, \qquad (3.5\text{-}11)$$

with plant transfer function $H(s)$ as in (3.5-5). By defining

$$W^{-1}(s) = [I + H(sI - A)^{-1}K]R^{1/2} \qquad (3.5\text{-}12)$$

this can be written as

$$\Phi_Z(s) = W^{-1}(s)W^{-T}(-s), \qquad (3.5\text{-}13)$$

which is a spectral factorization of $\Phi_Z(s)$.

The Kalman filter is guaranteed stable under the detectability and stability conditions. If in addition the plant is stable, then by (3.5-8), $W^{-1}(s)$ is a minimum-phase spectral factor of $\Phi_Z(s)$.

The *innovations representation* is given by

$$\dot{\hat{x}} = [A - K(t)H]\hat{x} + Bu + K(t)z \qquad (3.5\text{-}14a)$$
$$\tilde{z} = -H\hat{x} + z, \qquad (3.5\text{-}14b)$$

which at steady state has a transfer function from z to $\tilde{z}$ of

$$W'(s) = I - H[sI - (A - KH)]^{-1}K. \qquad (3.5\text{-}15a)$$

By the matrix inversion lemma

$$W'(s) = [I + H(sI - A)^{-1}K]^{-1}; \qquad (3.5\text{-}15b)$$

so that

$$W(s) = R^{-1/2}W'(s). \qquad (3.5\text{-}16)$$

The residual $\tilde{z}(t)$ is white noise with covariance

$$P_{\tilde{z}} = R, \qquad (3.5\text{-}17)$$

so that (3.5-14) is a *whitening filter* which turns the data $z(t)$ into the white noise process $\tilde{z}(t)$. It is clear then that $W(s)$ is a whitening filter which turns the data $z(t)$ into the innovations (3.4-8).

In terms of $W'(s)$ we can write

$$\Phi_Z(s) = [W'(s)]^{-1}R[W'(-s)]^{-T}. \qquad (3.5\text{-}18)$$

Note that

$$\lim_{s \to \infty} W'(s) = I, \qquad (3.5\text{-}19)$$

since $H(sI - A)^{-1}K$ has relative degree of at least 1. Hence (3.5-18) is the spectral factorization defined by (1.5-9).

These results show that the Kalman filter is a whitening filter which converts the data $z(t)$ to the residual $\tilde{z}(t)$, simultaneously providing the optimal estimate $\hat{x}(t)$. It can be viewed alternatively as a time-domain algorithm for computing a spectral factorization for the density of $z(t)$. For further discussion of these ideas see Shaked (1976).

We showed in the previous section that the Wiener filter and steady-state Kalman filter are identical, since they both provide the solution to the stationary Wiener–Hopf equation. Another demonstration of this follows.

In the case of linear measurements (3.5-1b) the Wiener filter that estimates $y(t) = Hx(t)$ is given by (see Section 1.5) the simple expression

$$HF(s) = I - R^{1/2} W(s), \tag{3.5-20}$$

where

$$\Phi_Z^{-1}(s) = W^T(-s) W(s). \tag{3.5-21}$$

In terms of $W'(s)$, therefore, the Wiener filter is

$$HF(s) = I - W'(s). \tag{3.5-22}$$

According to (3.5-15), this is just

$$HF(s) = H[sI - (A - KH)]^{-1} K, \tag{3.5-23}$$

which is the transfer function of the Kalman filter (3.5-7) from $z(t)$ to $\hat{z}(t) = H\hat{x}(t)$.

Recall from Section 2.6 that if (3.5-23) provides the optimal estimate of $y = Hx$ and (A, H) is observable, then the optimal estimate $\hat{x}$ of the state is given by the filter

$$F(s) = [sI - (A - KH)]^{-1} K. \tag{3.5-24}$$

Chang–Letov Design Procedure

We do not assume here that the plant matrix A is stable. By manipulation of our equations, we can derive a design procedure which provides a link with classical root locus techniques.

By using (3.5-11) and (3.5-8), we obtain the continuous *Chang–Letov* equation

$$\Delta^{cl}(s)\Delta^{cl}(-s) = |H(s)H^T(-s) + R| \cdot \Delta(s)\Delta(-s) \cdot |R|^{-1}, \tag{3.5-25}$$

where

$$H(s) = H(sI - A)^{-1} G\sqrt{Q}. \tag{3.5-26}$$

All matrices on the right-hand side are known, so that this represents a method of determining the closed-loop poles $\Delta^{cl}(s) = 0$ of the optimal filter (3.5-7). We know that the steady-state filter is stable if (A, H) is observable and $(A, G\sqrt{Q})$ is reachable. Thus, it is only necessary to compute the right-hand side of the Chang–Letov equation and find the roots of the resulting polynomial. The stable roots are the roots of $\Delta^{cl}(s)$, and the unstable roots are the roots of $\Delta^{cl}(-s)$. Once the desired closed-loop poles have been determined in this fashion, a technique such as Ackermann's formula can be used to find the required output injection K.

This procedure allows determination of the optimal filter by a method that does not involve either the solution of a Riccati equation or a spectral factorization.

In the single-output case we can represent $H(s)$ as

$$H(s) = \frac{N(s)}{\Delta(s)} \tag{3.5-27}$$

where $N(s) = H \operatorname{adj}(sI - A) G \sqrt{Q}$ is a row vector. Then the Chang–Letov equation is

$$\Delta^{\text{cl}}(s)\Delta^{\text{cl}}(-s) = \frac{1}{r} N(s) N^T(-s) + \Delta(s)\Delta(-s). \tag{3.5-28}$$

The right hand side is in exactly the canonical form for a root locus design: as r increases from 0 (perfect measurements) to ∞ (totally unreliable measurements), the roots of $\Delta^{\text{cl}}(s)\Delta^{\text{cl}}(-s)$ move from the roots of $N(s)N^T(-s)$ to the roots of $\Delta(s)\Delta(-s)$. This means the following.

If $r = 0$, then we examine the original plant *zeros* considering $w(t)$ as the input. If any of them are in the right-half of the s plane, we reflect them into the left-half plane by making them negative. The resulting n stable complex numbers are the required poles of $(A - KH)$.

If $r \to \infty$, then we examine the original plant *poles*. Unstable plant poles are reflected into the left-half plane to form a set of n stable poles for $(A - KH)$.

If $0 < r < \infty$, then the roots of $\Delta^{\text{cl}}(s)$ are on loci between the above two sets of stable numbers.

The following variation on the above procedure shows that the Chang–Letov method does not only apply to the state-variable formulation.

Example 3.5-1: Chang–Letov Design for Lateral Aircraft Dynamics

a. The Plant

The transfer function from rudder displacement $\delta_r(t)$ to sideslip angle $\beta(t)$ in a typical jet transport flying straight and level at 300 mph at sea level might be given by (Blakelock 1965)

$$H(s) = \frac{\beta}{\delta_r} = \frac{0.0364 s^2 + 1.3737 s - 0.0137}{s^3 + 0.376 s^2 + 1.8115 s - 0.00725} \tag{1}$$

$$= \frac{0.0364(s - 0.01)(s + 37.75)}{[(s + 0.19)^2 + 1.333^2](s - 0.004)}. \tag{2}$$

(Sideslip is the angle in the horizontal plane between the relative wind and the aircraft longitudinal axis. It should be zero in straight and level flight to conserve fuel.) The pole at $s = 0.004$ corresponds to the unstable spiral mode. The complex-conjugate pole pair corresponds to the Dutch roll mode, which causes the nose of

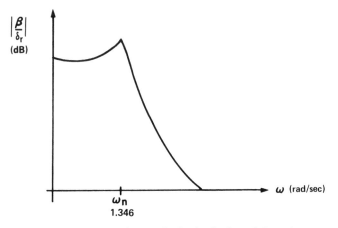

FIGURE 3.5-1   Bode magnitude plot for lateral dynamics.

the aircraft to move in a damped figure-of-eight pattern with a period of $0.75 = 1/1.333$ sec. The magnitude portion of the Bode plot for (1) is given in Fig. 3.5-1.

Wind gusts acting on the rudder are a source of perturbing noise. This noise is not white, but is low-frequency noise, so we should add a noise-shaping filter to our model (Section 3.6). Rather than complicate things in this manner, let us make the following approximation. The plant (1) is itself a low-pass filter, so let us assume the wind-gust noise is white. According to Fig. 3.5-1, only the low-frequency components of this white noise will have an appreciable effect on $\beta(t)$. If the wind-gust noise contains frequencies up to at least about $10\omega_n = 13.46$ rad/sec, this is a satisfactory approximation. Let the noise $\delta_r(t)$ have covariance of q.

Suppose sideslip angle $\beta(t)$ is measured in additive white noise so that

$$z(t) = \beta(t) + v(t), \tag{3}$$

with $v(t) \sim (0, r)$.

b. Chang–Letov Design of Optimal Steady-State Filter

To design the Wiener filter to estimate $\beta(t)$ from the data $z(t)$, let us use the Chang–Letov approach.

According to (3.5-28) we have

$$\Delta^{cl}(s)\Delta^{cl}(-s) = 0.0364^2 \frac{q}{r}(s-0.01)(s+37.75)(s+0.01)(s-37.75)$$

$$+[(s+0.19)^2+1.333^2](s-0.004)[(-s+0.19)^2+1.333^2](-s-0.004), \tag{4}$$

where we have included in the numerator of $H(s)$ the factor $\sqrt{q}$ required in (3.5-26). Symbolize (4) as

$$\Delta^{cl}(s)\Delta^{cl}(-s) = cn(s) + d(s) \tag{5}$$

where gain c is $0.0364^2 q/r$.

The root-locus plot for (5) as c varies from 0 to ∞ can be found by the

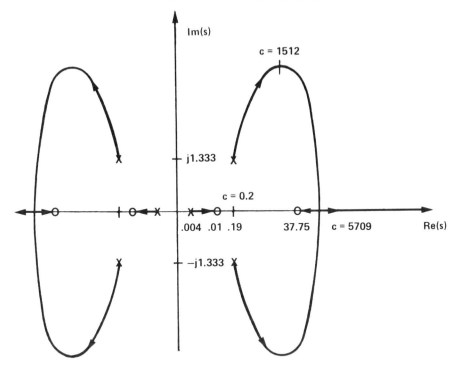

FIGURE 3.5-2 Chang–Letov root locus.

conventional rules. It is shown in Fig. 3.5-2, where 0's represent the roots of $n(s)$ and x's represent the roots of $d(s)$. Note that the roots of (5) move from the poles of $n(s)/d(s)$ to the zeros of $n(s)/d(s)$. [Observe from (4) that the last factor in $d(s)$ is negative, so that we must actually compute the conventional root locus with negative gain values $-c$.]

Now suppose that $q = 10$, $r = 0.1$. In this case, $c = 0.1325$ so that (4) yields

$$\Delta^{cl}(s)\Delta^{cl}(-s) = -s^6 - 3.3491s^4 - 192.11s^2 + 0.018935. \tag{6}$$

Note the even structure of this polynomial which has roots symmetric about the $j\omega$ axis. The roots of (6) are

$$s = 0.0099, \; -0.0099, \; 2.468 \pm j2.787, \; -2.468 \pm j2.787, \tag{7}$$

and assigning the stable roots to the optimal filter yields

$$\Delta^{cl}(s) = (s + 0.0099)[(s + 2.468)^2 + 2.787^2]$$
$$= s^2 + 4.946s^2 + 13.907s + 0.137. \tag{8}$$

This is the closed-loop characteristic polynomial of the desired Wiener filter.

c. Computation of Optimal Filter

The denominator of the filter transfer function is given by (8).

To find the numerator of the filter transfer function, realize (1) in observable canonical form (Kailath 1980). By inspection,

$$\dot{x} = \begin{bmatrix} -0.376 & 1 & 0 \\ -1.8115 & 0 & 1 \\ 0.00725 & 0 & 0 \end{bmatrix} x + \begin{bmatrix} 0.0364 \\ 1.3737 \\ -0.0137 \end{bmatrix} \delta_r$$

$$\beta = [1 \quad 0 \quad 0] x. \tag{9}$$

With Kalman gain $K = [K_1 \quad K_2 \quad K_3]^T$, we have

$$A - KH = \begin{bmatrix} -0.376 - K_1 & 1 & 0 \\ -1.8115 - K_2 & 0 & 1 \\ 0.00725 - K_3 & 0 & 0 \end{bmatrix}. \tag{10}$$

By comparison with the desired coefficients (8), therefore,

$$K = \begin{bmatrix} -0.376 \\ -1.8115 \\ 0.00725 \end{bmatrix} + \begin{bmatrix} 4.946 \\ 13.907 \\ 0.137 \end{bmatrix} = \begin{bmatrix} 4.57 \\ 12.0955 \\ 0.14425 \end{bmatrix}. \tag{11}$$

Now, by (3.5-23), the optimal filter for estimating the sideslip angle $\beta(t)$ from the measurements is

$$\frac{\hat{\beta}}{z} = \frac{4.57 s^2 + 12.0955 s + 0.14425}{s^3 + 4.946 s^2 + 13.907 s + 0.137}.$$

(Since we are using observable canonical form, the numerator coefficients are just the coefficients of the Kalman gain.) ■

3.6   CORRELATED NOISE AND SHAPING FILTERS

The philosophy in this section is the same as for the discrete situation. Where the details are different they are explicitly given.

Colored Process Noise

Let the plant be

$$\dot{x} = Ax + Bu + Gw \tag{3.6-1a}$$

$$z = Hx + v, \tag{3.6-1b}$$

with $v(t) \sim (0, R)$ white and uncorrelated with $x(0) \sim (x_0, P_0)$. If process noise $w(t)$ is uncorrelated with $x(0)$ and $v(t)$ but not white, then we should proceed as follows.

Factor the spectral density of $w(t)$ as

$$\Phi_w(s) = H(s) H^T(-s). \tag{3.6-2}$$

If $|\Phi_w(s)| \neq 0$ for $\text{Re}(s) = 0$, then $H(s)$ is minimum phase. Find a state representation for $H(s)$:

$$H(s) = H'(sI - A')^{-1} G' + D', \tag{3.6-3}$$

where primes denote additional variables. Then a *shaping filter* for $w(t)$ is given by

$$\dot{x}' = A'x' + G'w' \tag{3.6-4a}$$

$$w = H'x' + D'w'. \tag{3.6-4b}$$

In this shaping filter, $w'(t) \sim (0, I)$ is white noise.

To describe the dynamics of both the unknown $x(t)$ and the process noise $w(t)$, write the *augmented system*

$$\begin{bmatrix} \dot{x} \\ \dot{x}' \end{bmatrix} = \begin{bmatrix} A & GH' \\ 0 & A' \end{bmatrix}\begin{bmatrix} x \\ x' \end{bmatrix} + \begin{bmatrix} B \\ 0 \end{bmatrix}u + \begin{bmatrix} GD' \\ G' \end{bmatrix}w' \tag{3.6-5a}$$

$$z = \begin{bmatrix} H & 0 \end{bmatrix}\begin{bmatrix} z \\ x' \end{bmatrix} + v_k. \tag{3.6-5b}$$

The continuous Kalman filter is now applied to this augmented system.

Example 2.7-1 demonstrates the procedure, since the continuous Kalman filter can be applied to (6), (7) in that example. See Fig. 2.7-2 for some useful shaping filters.

Correlated Measurement and Process Noise

Now let the plant be described by (3.6-1) with $w(t) \sim (0, Q)$ and $v(t) \sim (0, R)$ both white, but with $v(t)$ and $w(t)$ correlated according to

$$\overline{v(t)w^T(\tau)} = S\,\delta(t - \tau). \tag{3.6-6}$$

Then

$$\overline{\begin{bmatrix} w(t) \\ v(t) \end{bmatrix}\begin{bmatrix} w(\tau) \\ v(\tau) \end{bmatrix}^t} = \begin{bmatrix} Q & S^T \\ S & R \end{bmatrix}\delta(t - \tau). \tag{3.6-7}$$

Rather than proceeding as in Section 2.7, let us use an interesting alternative (Gelb 1974).

We can convert this problem into an equivalent problem with uncorrelated process and measurement noises by adding zero to the right-hand side of (3.6-1a) to get

$$\dot{x} = Ax + Bu + Gw + D(z - Hx - v). \tag{3.6-8}$$

This results in the modified plant

$$\dot{x} = (A - DH)x + Dz + Bu + (Gw - Dv) \tag{3.6-9a}$$

$$z = Hx + v, \tag{3.6-9b}$$

which has process noise of $(Gw - Dv)$. This noise is white since it is zero mean and its covariance is given by

$$\overline{[Gw(t) - Dv(t)][Gw(\tau) - Dv(\tau)]^T}$$

$$= (GQG^T + DRD^T - DSG^T - GS^TD^T)\,\delta(t - \tau). \tag{3.6-10}$$

The motivation behind these manipulations is that D can be selected to make $(Gw - Dv)$ and v uncorrelated. To wit,

$$\overline{[Gw(t) - Dv(t)]v^T(\tau)} = (GS^T - DR)\,\delta(t - \tau), \qquad (3.6\text{-}11)$$

which is zero if we select

$$D = GS^T R^{-1}. \qquad (3.6\text{-}12)$$

With this choice, the modified plant is

$$\dot{x} = (A - GS^T R^{-1} H)x + GS^T R^{-1} z + Bu + (Gw - GS^T R^{-1} v) \quad (3.6\text{-}13\text{a})$$

$$z = Hx + v, \qquad (3.6\text{-}13\text{b})$$

where $GS^T R^{-1} z(t)$ is an additional deterministic input. The new process noise autocorrelation (3.6-10) is

$$\overline{[Gw(t) - Dv(t)][Gw(\tau) - Dv(\tau)]^T} = G(Q - S^T R^{-1} S)G^T \,\delta(t - \tau), \quad (3.6\text{-}14)$$

which is less than the GQG^T term arising from the original process noise $w(t)$.

Applying the continuous Kalman filter to (3.6-13), there results the estimate update

$$\dot{\hat{x}} = (A - GS^T R^{-1} H)\hat{x} + Bu + GS^T R^{-1} z + PH^T R^{-1}(z - H\hat{x}),$$

or

$$\dot{\hat{x}} = A\hat{x} + Bu + (PH^T + GS^T)R^{-1}(z - H\hat{x}). \qquad (3.6\text{-}15)$$

Defining a *modified Kalman gain* by

$$K = (PH^T + GS^T)R^{-1}, \qquad (3.6\text{-}16)$$

this becomes

$$\dot{\hat{x}} = A\hat{x} + Bu + K(z - H\hat{x}) \qquad (3.6\text{-}17)$$

which is identical in form to the estimate update equation in Table 3.1-1.

The error covariance update equation for (3.6-13) with process noise autocorrelation (3.6-14) is

$$\dot{P} = (A - GS^T R^{-1} H)P + P(A - GS^T R^{-1} H)^T$$
$$+ G(Q - S^T R^{-1} S)G^T - PH^T R^{-1} HP.$$

This can be written in terms of the modified gain as

$$\dot{P} = AP + PA^T + GQG^T - KRK^T, \qquad (3.6\text{-}18)$$

which is identical to (3.1-24). Nonzero S results in a smaller error covariance than the uncorrelated noise case due to the additional information provided by the cross-correlation term S.

We have found that if the measurement and process noises are correlated then it is only necessary to define a modified Kalman gain and use the formulation (3.6-18) for the Riccati equation; all other equations are the same as for the uncorrelated case. This is the same result as for the discrete Kalman filter.

Colored Measurement Noise

If the plant (3.6-1) has $w(t) \sim (0, Q)$ white with $w(t)$ and $v(t)$ uncorrelated, but nonwhite $v(t)$ of the form (2.7-25) then a continuous analog to (2.7-30) can be written. However, if the term v' in (2.7-25) is zero then this approach cannot be used unless $|R| = |D'(D')^T| \neq 0$.

An alternative to state augmentation is to define a *derived measurement* that has white measurement noise.

Suppose that, by spectral factorization or some other means, a shaping filter for $v(t)$ has been found to be

$$\dot{x}' = A'x' + G'x' \tag{3.6-19a}$$

$$v = x' \tag{3.6-19b}$$

with $w'(t) \sim (0, I)$ white and uncorrelated with plant process noise $w(t)$. Define the derived measurement by

$$z' = \dot{z} - A'z - HBu. \tag{3.6-20}$$

Then, if H is time invariant

$$z' = H(Ax + Bu + Gw) + \dot{v} - A'(Hx + v) - HBu$$

$$= (HA - A'H)x + G'w' + HGw.$$

Defining a new measurement matrix by

$$H' = HA - A'H, \tag{3.6-21}$$

we can write

$$z' = H'x + (G'w' + HGw). \tag{3.6-22}$$

This is a new output equation in terms of the derived measurement.

The reason for these manipulations is that the new measurement noise is white with

$$[G'w'(t) + HGw(t)] \sim \{0, [G'(G')^T + HGQG^TH^T]\}, \tag{3.6-23}$$

so that (3.6-1a) and (3.6-22) define a new plant driven by white noises. The order of this plant is still n, which is a reason for preferring the derived measurement approach over the augmented state approach.

In this new plant, the process and measurement noises are correlated with

$$E\{[G'w'(t) + HGw(t)]w^T(\tau)\} = HGQ, \tag{3.6-24}$$

so that for optimality the modified Kalman gain of the previous subsection should be used. The optimal filter is therefore given by

$$\dot{\hat{x}} = A\hat{x} + Bu + K(\dot{z} - A'z - HBu - H'\hat{x}), \qquad (3.6\text{-}25)$$

$$\dot{P} = AP + PA^T + GQG^T - K[G'(G')^T + HGQG^TH^T]K^T \qquad (3.6\text{-}26)$$

$$K = [P(H')^T + GQG^TH^T][G'(G')^T + HGQG^TH^T]^{-1}. \qquad (3.6\text{-}27)$$

A disadvantage of (3.6-25) is that the data must be differentiated. To avoid this (Gelb 1974), note that

$$\frac{d}{dt}(Kz) = \dot{K}z + K\dot{z}, \qquad (3.6\text{-}28)$$

so we can write (3.6-25) in the alternate form

$$\frac{d}{dt}(\hat{x} - Kz) = A\hat{x} + Bu - \dot{K}z - K(A'z + HBu + H'\hat{x}). \qquad (3.6\text{-}29)$$

Figure 3.6-1 gives a block diagram of this version of the optimal filter.

To initialize the derived filter we must use

$$\hat{x}(0^+) = x_0 + P_0H^T(HP_0H^T + R)^{-1}[z(0) - Hx_0] \qquad (3.6\text{-}30)$$

and

$$P(0^+) = P_0 - P_0H^T(HP_0H^T + R)^{-1}HP_0, \qquad (3.6\text{-}31)$$

Since the instant after data are available ($t = 0^+$), discontinuities in $\hat{x}(t)$ and $P(t)$ occur. These equations amount to a discrete Kalman filter measurement update. For details see Bucy (1967), Bryson and Johansen (1965), and Stear and Stubberud (1968).

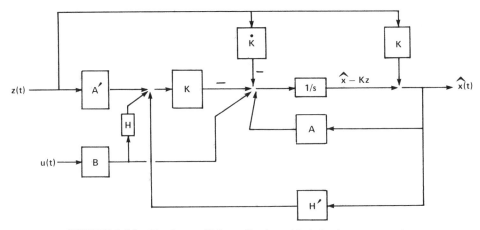

FIGURE 3.6-1 Continuous Kalman filtering with derived measurements.

Example 3.6-1:   *Continuous Filtering with No Measurement Noise*

In the continuous Kalman filter it is required that the measurement noise covariance R be nonsingular. Reexamining the last few pages it becomes apparent that the derived measurement approach allows us to derive a filter that does not require $|R| \neq 0$.

Let the plant be

$$\dot{x} = Ax + Bu + Gw \qquad (3.6\text{-}32a)$$

$$z = Hx, \qquad (3.6\text{-}32b)$$

$w(t) \sim (0, Q)$ with the usual assumptions. Then in the "measurement-noise shaping filter" (3.6-19) we evidently have $A' = G' = 0$, and the derived measurement is given by (3.6-20), or

$$z' = \dot{z} - HBu. \qquad (3.6\text{-}33)$$

According to (3.6-25)–(3.6-27) the optimal filter is now

$$\dot{\hat{x}} = A\hat{x} + Bu + K(\dot{z} - HBu - HA\hat{x}) \qquad (3.6\text{-}34a)$$

$$= (I - KH)A\hat{x} + (I - KH)Bu + K\dot{z} \qquad (3.6\text{-}34b)$$

$$\dot{P} = AP + PA^T + GQG^T - K(HGQG^TH^T)K^T \qquad (3.6\text{-}35)$$

$$K = (PA^T + GQG^T)H^T(HGQG^TH^T)^{-1}. \qquad (3.6\text{-}36)$$

If $|HGQG^TH^T| = 0$ then we must differentiate the data once more to define a new derived measurement, and obtain the resulting filter.

This is basically the *Deyst* filter. See Deyst (1969) and Maybeck (1979).   ■

3.7   DISCRETE MEASUREMENTS OF CONTINUOUS-TIME SYSTEMS

In this section, we discuss an alternative to the method of Section 2.4 for discrete filtering of continuous-time systems. In that section we proceeded by first discretizing the continuous system. Here, we show how to avoid system discretization by propagating the estimate and error covariance between measurements *in continuous time* using an integration routine such as Runge–Kutta. This approach yields the optimal estimate continuously at all times, including times *between* the data arrival instants.

Let the plant be given by

$$\dot{x} = Ax + Bu + Gw, \qquad (3.7\text{-}1a)$$

with $w(t) \sim (0, Q)$. Suppose measurements are given at discrete times t_k according to

$$z_k = Hx_k + v_k, \qquad (3.7\text{-}1b)$$

where we represent $x(t_k)$ by x_k. Let white noises $w(t)$ and $v_k \sim (0, R)$ be uncorrelated with each other and with $x(0) \sim (\bar{x}_0, P_0)$. Note that R is a covariance matrix while Q is a spectral density matrix.

The approach we are about to describe applies to time-varying systems, though for ease of notation all plant and noise matrices are written as constants here.

Applying the Kalman filter directly to (3.7-1) with no preliminary discretization is very easy since the estimate at each time is the conditional mean given all previous data.

The time update between measurements for the continuous linear stochastic system (3.7-1a) is found from Table 3.1-1 by setting $H = 0$;

$$\dot{\hat{x}} = A\hat{x} + Bu \qquad\qquad (3.7\text{-}2)$$

$$\dot{P} = AP + PA^T + GQG^T. \qquad\qquad (3.7\text{-}3)$$

See Example 3.1-1. The measurement update to incorporate a discrete data sample z_k is given in Table 2.3-2.

The optimal filter for the plant (3.7-1) is given by combining these two updates. The resulting *continuous-discrete Kalman filter* is summarized in Table 3.7-1. In the Gaussian case, this is *the* optimal filter. In the case of arbitrary statistics, it is the best *linear* filter.

Note that we do not require $|R| > 0$ as is the case for continuous measurements. All we require is $R \geq 0$ and $|HP^-(t_k)H^T + R| \neq 0$ for each k.

TABLE 3.7-1   Continuous-Discrete Kalman Filter

System and measurement models:

$$\dot{x} = Ax + Bu + Gw$$

$$z_k = Hx_k + v_k$$

$$x(0) \sim (\bar{x}_0, P_0), \; w(t) \sim (0, Q), \; v_k \sim (0, R)$$

Assumptions:

$\{w(t)\}$ and $\{v_k\}$ are white noise processes uncorrelated with each other and with $x(0)$.

Initialization:

$$P(0) = P_0, \; \hat{x}(0) = \bar{x}_0$$

Time update between measurements:

$$\dot{P} = AP + PA^T + GQG^T$$

$$\dot{\hat{x}} = A\hat{x} + Bu$$

Measurement update at times t_k:

$$K_k = P^-(t_k)H^T[HP^-(t_k)H^T + R]^{-1}$$

$$P(t_k) = (I - K_k H)P^-(t_k)$$

$$\hat{x}_k = \hat{x}_k^- + K_k(z_k - H\hat{x}_k^-)$$

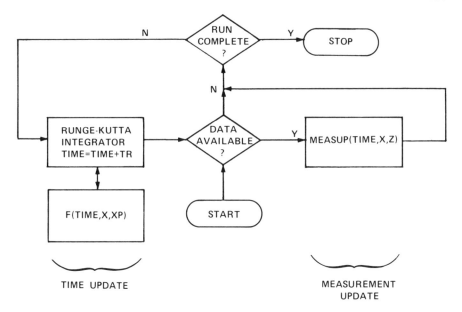

FIGURE 3.7-1 Software implementation of the continuous-discrete Kalman filter.

One advantage of the continuous-discrete filter over the alternative approach using system discretization is that in the former, it is not necessary for the sample times t_k to be equally spaced. This means that the cases of irregular and intermittent measurements are easy to handle. In the absence of data the optimal *prediction* is given by performing only the time update portion of the algorithm.

A block diagram/flowchart of a continuous-discrete Kalman filter using a Runge–Kutta integrator with step size TR is shown in Fig. 3.7-1. A driver program can be written which is the same for every problem. Such a program is included in Appendix B (CDKAL). Subroutines F(TIME, X, XP), containing the continuous dynamics, and MEASUP(TIME, X, Z), containing the measurement details, depend on the particular problem. (XP means $\dot{X}$.) Subroutine MEASUP can be based on one of the measurement update routines (stabilized Kalman, UDU^T) in Bierman (1977). Figure 3.7-1 should be compared to Fig. 2.4-5.

As an illustration of the continuous-discrete Kalman filter let us apply it to one of the examples we did by discretization in Section 2.4.

Example 3.7-1: *Continuous-Discrete α-β Tracker*

Consider the range subsystem in the radar tracking problem of Example 2.4-3:

$$\dot{x} = \begin{bmatrix} 0 & 1 \\ 0 & 0 \end{bmatrix} x + \begin{bmatrix} 0 \\ 1 \end{bmatrix} w, \tag{1}$$

where $x \triangleq [r \quad \dot{r}]^T$. Let $w \sim (0, q)$ and range r be measured every T sec so that $t_k = kT$ and

$$z_k = [1 \quad 0]x_k + v_k, \tag{2}$$

$v_k \sim (0, R)$. We use the notation $x_k \triangleq x(kT)$.

In Examples 2.4-4 and 2.4-5 we derived simple scalar updates implementing the optimal filter by first discretizing (1). Here we intend to avoid this discretization by using the continuous-discrete optimal filter formulation. In this simple example, discretization is quite easy since $e^{AT} = I + AT$. In general, however, it can be quite complicated to find e^{AT}, and if a good integration routine is available, this is not required. An additional advantage we will have in the solution we are about to present is that up-to-date estimates for range and range rate will be available at all times, including times *between* the appearance of the target on the radar screen. With slowly scanning radars this can be quite a valuable benefit!

a.   Time Update

Between measurements the optimal estimate propagates according to

$$\dot{\hat{x}} = A\hat{x} \tag{3}$$

or

$$\frac{d\hat{r}}{dt} = \hat{\dot{r}} \tag{4a}$$

$$\frac{d}{dt}\hat{\dot{r}} = 0. \tag{4b}$$

The error covariance obeys the Lyapunov equation $\dot{P} = AP + PA^T + GQG^T$. Let

$$P \triangleq \begin{bmatrix} p_1 & p_2 \\ p_2 & p_4 \end{bmatrix} \tag{5}$$

and substitute to obtain the scalar equations

$$\dot{p}_1 = 2p_2 \tag{6a}$$

$$\dot{p}_2 = p_4 \tag{6b}$$

$$\dot{p}_4 = q. \tag{6c}$$

We could easily solve these analytically, but shall deliberately refrain from doing so to make the point that this is not required to implement the optimal filter.

b.   Measurement Update

At a data point the Kalman gain is found as follows. Let

$$\delta \triangleq HP^-(t_k)H^T + R. \tag{7}$$

Write

$$P^-(t_k) \triangleq \begin{bmatrix} p_1^- & p_2^- \\ p_2^- & p_4^- \end{bmatrix} \tag{8}$$

```
            SUBROUTINE F(TIME,X,XP)

C   ALPHA-BETA TRACKER CONTINUOUS DYNAMICS FOR TIME UPDATE
C   CALLED BY SUBROUTINE RUNKUT

C   X(1)-X(2) STATE ESTIMATE
C   X(3)-X(5) ERROR COVARIANCE ENTRIES P(1),P(2),P(4)
C   XP        TIME DERIVATIVES OF X

            REAL X(*), XP(*)
            DATA Q/1./

C   ERROR COVARIANCE UPDATE EQUATION

            XP(3)= 2.*X(4)
            XP(4)= X(5)
            XP(5)= Q

C   ESTIMATE EQUATION

            XP(1)= X(2)
            XP(2)= 0
            RETURN
            END




            SUBROUTINE MEASUP(TIME,X,Z)

C   ALPHA-BETA TRACKER DISCRETE MEASUREMENT UPDATE

C   X(1)-X(2) STATE ESTIMATE
C   X(3)-X(5) ERROR COVARIANCE ENTRIES P(1),P(2),P(4)
C   Z         DATA SAMPLE

            REAL X(*)
            DATA R/1./

C   KALMAN GAIN

            DEL= X(3) + R
            AK1= X(3)/DEL
            AK2= X(4)/DEL

C   ERROR COVARIANCE

            X(5)= X(5) - AK2*X(4)
            X(3)= (1.-AK1)*X(3)
            X(4)= (1.-AK1)*X(4)

C   ESTIMATE

            RES= Z - X(1)
            X(1)= X(1) + AK1*RES
            X(2)= X(2) + AK2*RES

            RETURN
            END
```

FIGURE 3.7-2 Continuous-discrete-tracker time and measurement update subroutines.

189

(the time dependence of the scalar components is not shown since in a software implementation they will be updated by FORTRAN replacement) so that

$$\delta = p_1^- + R. \tag{9}$$

Then

$$K \overset{\Delta}{=} \begin{bmatrix} k_1 \\ k_2 \end{bmatrix} = \begin{bmatrix} \dfrac{p_1^-}{\delta} \\ \dfrac{p_2^-}{\delta} \end{bmatrix}. \tag{10}$$

The error covariance update is

$$P(t_k) = \begin{bmatrix} 1 - k_1 & 0 \\ -k_2 & 1 \end{bmatrix} P^-(t_k), \tag{11}$$

or

$$p_1 = (1 - k_1)p_1^- = \frac{Rp_1^-}{\delta} \tag{12a}$$

$$p_2 = (1 - k_1)p_2^- = \frac{Rp_2^-}{\delta} \tag{12b}$$

$$p_4 = p_4^- - k_2 p_2^- \tag{12c}$$

The estimate update is (omitting the time subscript k)

$$\hat{r} = \hat{r}^- + k_1(z - \hat{r}^-) \tag{13a}$$

$$\hat{\dot{r}} = \hat{\dot{r}}^- + k_2(z - \hat{r}^-). \tag{13b}$$

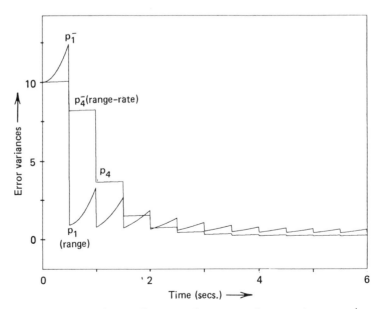

FIGURE 3.7-3   Continuous-discrete-tracker range and range-rate error variances.

c. Filter Implementation

Our preliminary analysis has yielded scalar updates which are easy to program and faster to run than matrix updates. FORTRAN implementations of the time update between measurements (subroutine F in Fig. 3.7-1) and the measurement update (MEASUP) are given in Fig. 3.7-2. The argument X is a vector containing both the estimate and the error covariance entries. We have been able to avoid using two sets of covariance entries, one for $P(t_k)$ and one for $P^-(t_k)$, by paying careful attention to the order of the operations. In the program, (12c) is done before (12b) since it requires the term p_2^- before that term is updated.

Using this software and the driver program CDKAL in Appendix B, the error variances for range and range rate shown in Fig. 3.7-3 were obtained. Measurements were taken every 0.5 sec, and each time a measurement update was made, the error variances decreased. Steady state has nearly been attained after 6 sec. The Runge–Kutta integration step size was 10 msec. This figure is similar in concept to Fig. 2.4-6, though different parameters were used. ∎

3.8 OPTIMAL SMOOTHING

We have discussed the *filtering* problem at some length. Given a stochastic system, the solution to the *prediction* problem is straightforward. All we must do to find the optimal prediction is use only the time update portion of the Kalman filter, which simply involves setting $H = 0$ in Table 3.1-1. We discuss here optimal *smoothing*. This is the continuous version of Section 2.8. We shall basically follow the presentation in Gelb (1974).

Suppose data are available over a time interval $[0, T]$, and we desire optimal estimates $\hat{x}(t)$ for every t within the interval. All smoothed estimates are based somehow on filtered estimates, so the first step is to run the usual Kalman filter on the data to generate a filtered estimate $\hat{x}_f(t)$. We proceed as follows.

Let the unknown $x(t)$ be described by

$$\dot{x} = Ax + Bu + Gw \qquad (3.8\text{-}1a)$$

with measurements

$$z = Hx + v. \qquad (3.8\text{-}1b)$$

White noises $w(t) \sim (0, Q)$ and $v(t) \sim (0, R)$ are uncorrelated with each other and with $x(0) \sim (\bar{x}_0, P_0)$. Let $Q \geq 0$ and $R > 0$. (In contrast to the discrete case, we will not require $|Q| \neq 0$.)

To find the optimal smoothed estimate $\hat{x}(t)$ for $t \in [0, T]$ we must take into account the data $z(\tau)$ for $0 \leq \tau \leq t$ and also for $t < \tau \leq T$. Suppose that the data for the subinterval preceding time t have been used to construct via the Kalman filter an estimate $\hat{x}_f(t)$. This estimate is given by

$$\dot{P}_f = AP_f + P_f A^T + GQG^T - P_f H^T R^{-1} HP_f \qquad (3.8\text{-}2)$$

$$K_f = P_f H^T R^{-1} \qquad (3.8\text{-}3)$$

$$\dot{\hat{x}}_f = A\hat{x}_f + Bu + K_f(z - H\hat{x}_f). \qquad (3.8\text{-}4)$$

Then we can use the filtered estimate $\hat{x}_f(t)$ to manufacture the smoothed estimate $\hat{x}(t)$, which includes the effects of future data.

To do this, we first construct a filter that runs *backward* in time to generate an estimate $\hat{x}_b(t)$ which includes the data $z(\tau)$ for $t < \tau \leq T$. This is known as the *information filter*. Then we use a smoother algorithm to combine $\hat{x}_f(t)$ and $\hat{x}_b(t)$ into the smoothed estimate $\hat{x}(t)$.

The smoother scheme we will present gives the best linear estimate if the statistics are arbitrary, and the optimal estimate if all statistics are normal. It extends directly to the time-varying case.

The Information Filter

All $\hat{x}$ and P in this subsection should be considered as having a subscript "b" (i.e., "backward"), which we suppress for convenience.

To reverse time, define $\tau = T - t$. Then

$$\frac{d}{d\tau} x = -\frac{d}{dt} x, \tag{3.8-5a}$$

so that

$$\frac{d}{d\tau} x = -Ax - Bu - Gw \tag{3.8-5b}$$

replaces plant equation (3.8-1a). The measurement equation remains the same. Evidently, then, the filter equations are simply

$$\frac{d}{d\tau} P = -AP - PA^T + GQG^T - PH^TR^{-1}HP \tag{3.8-6}$$

$$\frac{d}{d\tau} \hat{x} = -A\hat{x} - Bu + PH^TR^{-1}(z - H\hat{x}). \tag{3.8-7}$$

It is more convenient to express these equations in terms of the information matrix

$$S(\tau) = P^{-1}(\tau), \tag{3.8-8}$$

and to define an intermediate variable

$$\hat{y}(\tau) = P^{-1}(\tau)\hat{x}(\tau) = S(\tau)\hat{x}(\tau). \tag{3.8-9}$$

By using

$$\frac{d}{d\tau} P^{-1} = -P^{-1}\left(\frac{d}{d\tau} P\right)P^{-1} \tag{3.8-10}$$

equation (3.8-6) yields

$$\frac{d}{d\tau} P^{-1} = P^{-1}A + A^TP^{-1} - P^{-1}GQG^TP^{-1} + H^TR^{-1}H \tag{3.8-11}$$

or

$$\frac{d}{d\tau}S = SA + A^{T}S - SGQG^{T}S + H^{T}R^{-1}H. \qquad (3.8\text{-}12)$$

Differentiating (3.8-9) with respect to τ and using (3.8-7) and (3.8-12) yields

$$\frac{d}{d\tau}\hat{y} = (A^{T} - SGQG^{T})\hat{y} - SBu + H^{T}R^{-1}z. \qquad (3.8\text{-}13)$$

By defining the *backward filter gain*

$$K_{b} = SGQ \qquad (3.8\text{-}14)$$

we can write

$$\frac{d}{dt}\hat{y} = (A^{T} - K_{b}G^{T})\hat{y} - SBu + H^{T}R^{-1}z. \qquad (3.8\text{-}15)$$

Equations (3.8-12) and (3.8-13) are the information formulation of the continuous Kalman filter. The information filter runs backward in time beginning at time $t = T$, or $\tau = 0$.

Optimal Smoothed Estimate

Suppose now that a Kalman filter has been run forward on the interval $[0, T]$ to compute for each t in $[0, T]$, a filtered estimate $\hat{x}_{f}(t)$ based on data prior to t. Suppose also that a separate information filter has been run backward on $[0, T]$ to find for each t an estimate $\hat{x}_{b}(t)$ based on data subsequent to t. To combine $\hat{x}_{f}(t)$ and $\hat{x}_{b}(t)$ into an overall smoothed estimate $\hat{x}(t)$ which takes into account all the available data is very simple.

Since $w(t)$ and $v(t)$ are white, the estimates $\hat{x}_{f}(t)$ and $\hat{x}_{b}(t)$ are independent. We can therefore use the "Millman's theorem" result from the problems of Chapter 1 to write for the smoothed error covariance

$$P = (P_{f}^{-1} + P_{b}^{-1})^{-1}, \qquad (3.8\text{-}16)$$

or

$$P = (P_{f}^{-1} + S)^{-1}. \qquad (3.8\text{-}17)$$

[We shall suppress subscripts "b" on $S(t)$ and $\hat{y}(t)$.] By the matrix inversion lemma

$$P = P_{f} - P_{f}S(I + P_{f}S)^{-1}P_{f}, \qquad (3.8\text{-}18)$$

or

$$P = (I - K)P_{f} \qquad (3.8\text{-}19)$$

where the *smoother gain* is

$$K = P_{f}S(I + P_{f}S)^{-1}. \qquad (3.8\text{-}20)$$

TABLE 3.8-1 Optimal Smoothing Scheme

System and measurement model;

$$\dot{x} = Ax + Bu + Gw$$

$$z = Hx + v,$$

$$x(0) \sim (\bar{x}_0, P_0),\ w(t) \sim (0, Q),\ v(t) \sim (0, R).$$

$x(0),\ w(t),\ v(t)$ uncorrelated.

Forward filter:

  initialization:

$$\hat{x}_f(0) = \bar{x}_0,\ P_f(0) = P_0$$

  update:

$$\dot{P}_f = AP_f + P_f A^T + GQG^T - P_f H^T R^{-1} H P_f$$

$$K_f = P_f H^T R^{-1}$$

$$\dot{\hat{x}}_f = A\hat{x}_f + Bu + K_f(z - H\hat{x}_f)$$

Backward filter: $(\tau = T - t)$

  initialization:

$$\hat{y}(0) = 0,\ S(0) = 0$$

  update:

$$\frac{d}{d\tau} S(T - \tau) = S(T - \tau)A + A^T S(T - \tau)$$

$$- S(T - \tau)GQG^T S(T - \tau) + H^T R^{-1} H$$

$$K_b(T - \tau) = S(T - \tau)GQ$$

$$\frac{d}{d\tau} \hat{y}(T - \tau) = [A^T - K_b(T - \tau)G^T]\hat{y}(T - \tau) - S(T - \tau)Bu(T - \tau)$$

$$+ H^T R^{-1} z(T - \tau)$$

Smoother:

$$K(t) = P_f(t)S(t)[I + P_f(t)S(t)]^{-1}$$

$$P(t) = [I - K(t)]P_f(t)$$

$$\hat{x}(t) = [I - K(t)]\hat{x}_f(t) + P(t)\hat{y}(t)$$

The smoothed estimate is

$$\hat{x} = P(P_f^{-1}\hat{x}_f + P_b^{-1}\hat{x}_b) \tag{3.8-21}$$

or, using (3.8-19) and (3.8-9),

$$\hat{x} = (I - K)\hat{x}_f + P\hat{y}, \tag{3.8-22}$$

$$\hat{x} = \hat{x}_f + (P\hat{y} - K\hat{x}_f). \tag{3.8-23}$$

The smoothing scheme contains a Kalman filter, an information filter, and a smoother; it is shown in its entirety in Table 3.8-1.

Since for time T, the filtered and smoothed estimates are the same, we have the boundary condition $S(\tau = 0) = 0$. According to (3.8-9) then, $\hat{y}(\tau = 0) = 0$. [The boundary condition $\hat{x}_b(T)$ is unknown, and this is the reason for using the intermediate variable $\hat{y}(\tau)$ in the backward filter instead of $\hat{x}_b(\tau)$.]

In some cases P may be equal to P_f, so that the smoothed and filtered estimates are the same. In this case, smoothing does not offer an improvement over filtering alone. States whose estimates are improved by smoothing are said to be *smoothable*. It can be shown that if the plant is $(A, G\sqrt{Q})$ reachable so that all states are disturbed by the process noise, then all states are smoothable. See Gelb (1974) and Fraser (1967).

Figure 2.8-3 shows the relation between forward, backward, and smoothed error covariances.

These are other types of smoothers ("fixed-point," "fixed-lag"). The type we are discussing in which the final time T is fixed is known as the *fixed-interval* smoother.

Rauch–Tung–Striebel Smoother

The smoothing scheme in Table 3.8-1 requires a forward filter, a backward filter, and a smoother. We can quickly derive a more convenient scheme which combines the backward filter and the smoother into a single backward smoother recursion (Rauch, Tung, and Striebel 1965).

To this end, it is necessary to eliminate S and $\hat{y}$ from the smoother equations. In order to accomplish this, use (3.8-10) to obtain from (3.8-16)

$$\frac{d}{dt}P^{-1} = -P_f^{-1}\dot{P}_f P_f^{-1} + \frac{d}{dt}(P_b^{-1}). \tag{3.8-24}$$

As in (3.8-5a) we have

$$\frac{d}{d\tau}P^{-1} = P^{-1}\dot{P}P^{-1} = P_f^{-1}\dot{P}_f P_f^{-1} + \frac{d}{d\tau}(P_b^{-1}), \tag{3.8-25}$$

so that (3.8-2), (3.8-11), and (3.8-16) imply

$$P^{-1}\dot{P}P^{-1} = P_f^{-1}A + A^T P_f^{-1} + P_f^{-1} GQG^T P_f^{-1} - H^T R^{-1} H$$
$$+ P_b^{-1}A + A^T P_b^{-1} - P_b^{-1} GQG^T P_b^{-1} + H^T R^{-1} H$$
$$= P^{-1}A + A^T P^{-1} + P_f^{-1} GQG^T P_f^{-1}$$
$$- (P^{-1} - P_f^{-1}) GQG^T (P^{-1} - P_f^{-1}),$$

or

$$\dot{P} = AP + PA^T + PP_f^{-1} GQG^T + GQG^T P_f^{-1} P - GQG^T.$$

Finally,

$$\dot{P} = (A + GQG^T P_f^{-1})P + P(A + GQG^T P_f^{-1})^T - GQG^T. \quad (3.8\text{-}26)$$

Defining a smoother gain as

$$F = GQG^T P_f^{-1}. \quad (3.8\text{-}27)$$

this becomes

$$\dot{P} = (A + F)P + P(A + F)^T - GQG^T. \quad (3.8\text{-}28)$$

In a similar fashion we obtain the recursion for the smoothed estimate

$$\dot{\hat{x}} = A\hat{x} + F(\hat{x} - \hat{x}_f). \quad (3.8\text{-}29)$$

The Rauch–Tung–Striebel smoother algorithm is then as follows. First run the usual Kalman filter on the given data. Then run the recursive smoother defined by (3.8-27)–(3.8-29) backward for $t \leq T$ to generate the smoothed estimate $\hat{x}(t)$ and error covariance $P(t)$. Since at $t = T$, the filtered and smoothed estimates are identical, the initial conditions for the backward recursion are

$$P(T) = P_f(T), \qquad \hat{x}(T) = \hat{x}_f(T). \quad (3.8\text{-}30)$$

PROBLEMS

Section 3.1

3.1-1: The Lyapunov Equation Written as a Vector Equation. Let $\otimes$ denote the Kronecker product (Appendix A) and $s(\cdot)$ the stacking operator. Show that the continuous matrix Lyapunov equation $\dot{P} = AP + PA^T + CC^T$ is equivalent to the *vector* equation

$$\frac{d}{dt} s(P) = [(A \otimes I) + (I \otimes A)]s(P) + s(CC^T).$$

3.1-2: Solutions to Algebraic Lyapunov Equations. Use the results of the previous problem to show that $0 = AP + PA^T + CC^T$ has a unique solution if and only if $\lambda_i + \lambda_j \neq 0$ for all i and j, where λ_i is an eigenvalue of A. Show that the stability of A is sufficient to guarantee this.

3.1-3:

a. Find all solutions to $0 = AP + PA^T + CC^T$ if

$$A = \begin{bmatrix} 0 & 0 \\ 1 & -1 \end{bmatrix}, \qquad C = \begin{bmatrix} 0 \\ 1 \end{bmatrix}.$$

Interpret this in light of the previous problem.

b. Find all symmetric solutions.

c. Find the positive definite solution(s).

d. Which eigenvalues of A are (A, C) reachable?

3.1-4: For the A and C of the previous problem, solve $\dot{P} = AP + PA^T + CC^T$ for $t \geq 0$ if $P(0) = 0$ and if $P(0) = I$. Now let t go to infinity. To which of the solutions found in the previous problem do the solutions for $P(0) = 0$ and $P(0) = I$ converge?

3.1-5: System Described by Newton's Laws. A system is described by $\ddot{x}(t) = w(t)$, with white noise $w(t) \sim (0, q)$ independent of the initial state

$$\begin{bmatrix} x(0) \\ \dot{x}(0) \end{bmatrix} \sim \left(\begin{bmatrix} \bar{x}_0 \\ \bar{v}_0 \end{bmatrix}, \begin{bmatrix} p_x(0) & 0 \\ 0 & p_{\dot{x}}(0) \end{bmatrix} \right).$$

(v_0 means initial velocity.)

a. Find and sketch the means $\bar{x}(t)$, $\bar{\dot{x}}(t)$.

b. Find and sketch the variances $p_x(t)$, $p_{\dot{x}}(t)$, and $p_{x\dot{x}}(t)$.

3.1-6: Computer Simulation of Stochastic Systems. Let

$$\dot{x} = \begin{bmatrix} 0 & 1 \\ -\omega_n^2 & -2\alpha \end{bmatrix} x + \begin{bmatrix} 0 \\ 1 \end{bmatrix} w$$

$$z = [1 \quad 0]x + v,$$

with $w(t) \sim (0, 1)$, $v(t) \sim (0, 1/2)$ independent white noise processes with *uniform* distribution.

Write subroutine F(TIME, X, XP) required for program TRESP in Appendix B to simulate the system. Use the FORTRAN supplied function to generate the uniform noises.

3.1-7: Stochastic Harmonic Oscillator. Solve analytically for the mean and covariance of $x(t)$ in Problem 3.1-6 if $x(0) \sim (1, I)$, $\omega_n = 1$, $\alpha = 0.1$.

3.1-8: Periodic Solutions of Riccati Equations

a. Find the general solution to $\dot{p} = ap^2 + bp + c$ for the case $q = 4ca - b^2 > 0$ if the initial condition is $p(0) = p_0$.

b. Show that $p(t)$ exhibits periodic behavior and never converges to a steady-state value. Sketch $p(t)$.

3.1-9: (Friedland 1969). Let $\bar{P}$ be any solution to (3.1-21). Then show that any other solution can be expressed as

$$P = \bar{P} + VMV^T, \tag{1}$$

where

$$\dot{V} = (A - \bar{P}H^T R^{-1} H) V \tag{2}$$

and

$$\dot{M} = -MV^T H^T R^{-1} HVM. \tag{3}$$

Section 3.2

3.2-1: Kalman Filter Simulation. Consider the plant in Example 3.2-3. Write subroutine F(TIME,X,XP) for TRESP in Appendix B to simulate the entire Kalman filter. You must integrate the Riccati equation *and* the actual filter dynamics (3.1-23). To run your program, use $\omega_n^2 = 2$, $\alpha = 1.5$, $q = 0.1$, $r = 0.1$, $\bar{x}(0) = [1 \;\; 1]^T$, $P_0 = I$. For data, use $z(t) = -29e^{-1.1t} + 41e^{-1.9t}$; $t \geq 0$.

3.2-2: Harmonic Oscillator with Position Measurements. In Example 3.2-3, suppose measurements of position instead of velocity are taken so that

$$z = [1 \;\; 0]x + v.$$

a. Write Riccati equation as three coupled scalar equations.

b. Find steady-state error covariance.

c. Find steady-state Kalman filter (3.1-23).

d. Write subroutine F(TIME,X,XP) to simulate the entire time-varying filter.

3.2-3: Filtering with No Process Noise. Show that if there is no process noise so that $Q = 0$, the Riccati equation can be written in terms of the information matrix $P^{-1}(t)$ as the Lyapunov equation

$$\frac{d}{dt}(P^{-1}) = -P^{-1}A - A^T P^{-1} + H^T R^{-1} H.$$

[Hint: $d/dt(P^{-1}) = -P^{-1}\dot{P}P^{-1}$.]

3.2-4: Filtering with No Process Noise. Consider Newton's system

$$\ddot{x} = u$$

$$z = x + v,$$

with $v(t) \sim (0, r)$ white and $x(0) \sim (\overline{x(0)}, P_0)$.

a. Find the error covariance $P(t)$ for the maximum-likelihood estimator. Note that no a priori information about $x(t)$ means $\bar{x}(0) = 0$, $P(0) = \infty$, which can more conveniently be expressed as $P^{-1}(0) = 0$.

b. Find the Kalman gain and write down the maximum-likelihood Kalman filter.

A filter to estimate the state of this system is called a second-order polynomial tracking filter.

3.2-5: Repeat Problem 3.2-4 for the harmonic oscillator in Example 3.2-3 with no process noise.

3.2-6: Discrete Polynomial Tracking Filter. Repeat Problem 3.2-4, first discretizing the plant using a sampling period of T sec.

Section 3.3

3.3-1: Steady-State Filter for Harmonic Oscillator. Consider the harmonic oscillator in Example 3.2-3, where velocity measurements were made.

a. Find steady-state Kalman filter. Show it is always stable (consider the two cases $\alpha > 0$ and $\alpha < 0$ separately).
b. Find transfer function of Wiener filter.
c. Plot filter root locus as q/r goes from 0 to ∞. Show that the damping ratio of the filter is $\sqrt{\alpha^2 + q/4r}/\omega_n$, and that the natural frequency is the same as that of the original plant.

3.3-2: Harmonic Oscillator with Position Measurements. Redo Problem 3.3-1 if position measurements are taken instead of velocity measurements. Note that now the filter natural frequency changes with q/r, while the real part of the poles stays constant.

3.3-3: Newton's System with Velocity Measurements. Repeat Problem 3.3-1 for the system in Example 3.3-2 if velocity measurements are taken so that

$$z = [0 \quad 1]x + v.$$

Section 3.4

3.4-1: Algebraic Riccati Equation Analytic Solution
a. Show that a solution to the equation

$$0 = AP^T + PA^T + PR^{-1}P^T + Q \tag{1}$$

for $R > 0$ is given by

$$P = \sqrt{ARA^T - Q}L\sqrt{R} - AR, \tag{2}$$

where L is any orthogonal matrix, $LL^T = I$. Show that this reduces to the well-known quadratic equation solution in the scalar case.
b. Show that if we want $P = P^T$ in (2), then L must be selected as the solution to a Lyapunov equation.

3.4-2:   Covariance of Residual.   Use (3.3-20) [with $\tilde{x}(t_0) = 0$] to show that $P_{\tilde{z}}(t, u) = \overline{\tilde{z}(t)\tilde{z}^T(u)} = R\,\delta(t - u)$. [Hint: let the error system have state-transition matrix $\phi(t, t_0)$. Then show $\overline{\tilde{x}(t)\tilde{x}^T(u)} = \phi(t, u)P(u)$ for $t \geq u$ (Bryson and Ho 1975).]

3.4-3:   Estimation of a Voltage Using Continuous and Discrete Filters.   Consider the circuit of Fig. P3.4-1 where $u(t)$ is an input voltage and $w(t) \sim (0, q)$ is a noise source. The initial state is $x(0) \sim (0, p_0)$, uncorrelated with $w(t)$.

a.   Write state equation.

b.   Find and sketch $x(t)$ if $u(t) = u_{-1}(t)$ the unit step, for the deterministic case $p_0 = 0$ and $w(t) = 0$. Now let $RC = 1$ and measurements be taken according to

$$z(t) = x(t) + v(t), \tag{1}$$

with noise $v(t) \sim (0, r)$ uncorrelated with $w(t)$ and $x(0)$.

c.   Solve the algebraic Riccati equation for p_∞, the steady-state error variance. When is p_∞ greater than zero? Find a differential equation for the estimate $\hat{x}(t)$ in the steady-state filter. Find the transfer function of the Wiener filter.

d.   Let $q = 3$, $r = 1$. Find $\hat{x}(t)$ if $u(t) = u_{-1}(t)$ and the data are $z(t) = 0.75(1 - e^{-1.2t})u_{-1}(t)$.

e.   Find the differential equation for $\hat{x}(t)$ in the steady-state filter if r tends to infinity. Comment. What happens as r goes to zero so that the measurements are perfect?

f.   Let $q = 3$, $r = 1$. Use separation of variables to solve the Riccati equation for $p(t)$ if $p_0 = 2$. Sketch $p(t)$. Verify that the limiting behavior agrees with p_∞ in part c.

g.   Now, convert to the discretized system with $T = 0.1$ sec. Let $q = 3$, $r = 1$, $p_0 = 2$.

h.   For the discrete system, find p_k for $k = 0–5$. Do these values agree with $p(kT)$ in part f?

i.   For the discrete system, use data $z_k = z(kT)$ with $z(t)$ as in part d.

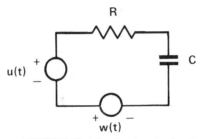

FIGURE P3.4-1   An electric circuit.

(Disregard z_0.) Let $u_k = u_{-1}(k)$. Find optimal estimates $\hat{x}_k$ for $k = 0\text{–}3$. Compare to the (suboptimal) estimates $\hat{x}(kT)$ in part d.

3.4-4: An Alternate Design Method for Steady-State Filters. An observer for the plant (3.3-5) is given by

$$\dot{\hat{x}} = A\hat{x} + K(z - H\hat{x}).$$

The observer gain K can be chosen, for example, to place the observer poles so that $\hat{x}(t)$ converges to $x(t)$ as quickly as desired. The error system using this observer is given by (3.3-19).

Write an equation that may be solved for P_∞, the steady-state error covariance. Now select K to minimize P_∞. One way to do this is to differentiate the error covariance equation with respect to K, using $\partial P_\infty/\partial K = 0$ for a minimum (Schultz and Melsa 1967).

3.4-5: ARE Solutions

a. Find all symmetric solutions to (3.4-11) if

$$A = \begin{bmatrix} 0 & 8 \\ 1 & -2 \end{bmatrix}, \qquad G = \begin{bmatrix} \sqrt{2} & 0 \\ 0 & 1 \end{bmatrix}, \qquad H = [0 \;\; 1], \qquad Q = I, \qquad R = 1.$$

b. Check the stability of closed-loop system $(A - KC)$, where $K = PH^T R^{-1}$, for all your solutions.

Section 3.5

3.5-1: Prove (3.5-11).

3.5-2: Steady-State Filters. Design an optimal steady-state filter to separate a signal $x(t)$ with spectral density

$$\Phi_X(\omega) = \frac{\omega^2 + 4}{\omega^4 + 6\omega^2 + 25}$$

from additive white noise $v(t) \sim (0, r)$:

a. Using Kalman theory.
b. Using Wiener theory.
c. Sketch filter poles as a function of r.
d. Sketch $\Phi_X(\omega)$, noise spectral density, and filter frequency response for $r = 1$.

3.5-3: Filter Design Using Hamiltonian Matrix. The Hamiltonian matrix

$$M = \begin{bmatrix} A & -GQG^T \\ -H^T R^{-1} H & -A^T \end{bmatrix}.$$

provides an alternative design procedure for the optimal steady-state filter.

a. Perform a state-space transformation

$$T = \begin{bmatrix} I & P \\ 0 & I \end{bmatrix}.$$

Show that if P satisfies the ARE, then

$$TMT^{-1} = \begin{bmatrix} A - KH & 0 \\ -H^T R^{-1} H & -(A - KH)^T \end{bmatrix},$$

where $K = PH^T R^{-1}$.

b. Hence, show that

$$|sI - M| = \Delta^{cl}(s)\Delta^{cl}(-s),$$

where $\Delta^{cl}(s) = |sI - (A - KH)|$ is the characteristic polynomial of the optimal steady-state filter. Therefore, the filter poles are the stable eigenvalues of M. Knowing these, a technique like Ackermann's formula can be used to determine the optimal filter gain K without solving the ARE for P (Kailath 1980).

3.5-4: Filtering with No Process Noise. Use Problem 3.5-3 to show that if $GQG^T = 0$, then the optimal steady-state filter poles are the "stabilized" poles of the plant (i.e., unstable plant poles s_i are replaced by $-s_i$).

3.5-5: Steady-State Filter Design. In Example 3.2-3, let $\omega_n^2 = 5$, $\alpha = 1$, $q = 1$, $r = 1$. Design the steady-state filter using:

a. Chang–Letov.

b. The results of Problem 3.5-3.

c. Wiener theory.

3.5-6: Redo Problem 3.5-2 using:

a. The Hamiltonian matrix.

b. The Chang–Letov procedure.

3.5-7: Steady-State Filter Poles. Let

$$\dot{x}_1 = -x_1 + 0.5x_2$$
$$\dot{x}_2 = 8.5x_1 - 2x_2 + w$$
$$z = x_1 + v,$$

where $w(t) \sim (0, q)$, $v \sim (0, r)$ are uncorrelated white noises. Sketch the poles of the optimal steady-state filter as q/r goes from 0 to ∞. Find the optimal filter if $q/r = 2$.

Section 3.6

3.6-1: Derived Measurement for Time-Varying Measurement Model. Redo the derived measurement analysis for the case of time-varying measurement matrix $H(t)$.

3.6-2: Colored Measurement Noise. Design an optimal time-invariant filter to estimate a signal with spectral density

$$\Phi_X(\omega) = \frac{2}{\omega^2 + 4}$$

which is measured in the presence of additive noise with spectral density

$$\Phi_V(\omega) = \frac{3\omega^2}{\omega^4 + 4}$$

a. Using Kalman filter theory with state augmentation.
b. Using Wiener filter theory.
c. Using the Kalman filter with derived measurements.
d. Sketch the signal and noise spectral densities.

3.6-3: Aircraft Longitudinal Dynamics with Gust Noise. Consider the situation in Example 2.7-1.
a. Sketch $\Phi_w(\omega)$ and find and sketch the autocorrelation function $R_w(t)$.
b. Find a shaping filter for $w(t)$. Augment the state equation (1).
c. If $\omega_n = 3$ rad/sec, $\delta = 0.1$, $\beta = 1$, $\sigma^2 = 2$, use Chang–Letov to plot the poles of the optimal steady-state filter as r goes from ∞ to 0.
d. Find the steady-state filter if $r = 2$.
e. Find the optimal estimate for pitch $\theta(t)$ if the data are $z(t) = 0.1 \cos 4t u_{-1}(t)$. Use the steady-state filter.
f. Find steady-state error covariance.

3.6-4: Correlated Process and Measurement Noise. Let

$$\dot{x} = x + u + w$$

$$z = x + v$$

where $w(t) \sim (0, 1)$, $v(t) \sim (0, r)$, and $\overline{vw} = \mu$.
a. Find optimal filter gain $K(t)$ and error covariance $P(t)$.
b. Find optimal steady-state filter and error covariance.
c. Sketch poles of steady-state filter as μ/r goes from 0 to ∞.
d. Let $\mu = 1$, $r = 2$. Find optimal estimate for $x(t)$ if the data are $z(t) = e^{0.9t}u_{-1}(t)$ and the input $u(t) = 0$. Use steady-state filter.

Section 3.7

3.7-1: Newton's System. In Example 3.7-1, let $x(0) \sim (1, I)$, $q = 1$, $r = 0.5$, $T = 0.5$ sec. Solve analytically for and sketch the components of $P(t)$ for $t = 0$ to 2 sec if the continuous-discrete filter is used.

3.7-2: Harmonic Oscillator
a. Repeat Problem 3.7-1 for the harmonic oscillator in Example 3.2-3 if $x(0) \sim (1, I)$, $q = 1$, $r = 0.5$, $T = 0.5$ sec, $\omega_n = \sqrt{2}$ rad/sec, $\alpha = 1$.

b. Write subroutines for this example to implement the continuous-discrete Kalman filter using CDKAL in Appendix B.

3.7-3: Comparison of Continuous, Discrete, and Continuous-Discrete Filters.   Let

$$\dot{x} = 0.2x + w$$

$$z = x + v$$

with $w(t) \sim (0, 1)$ and $v(t) \sim (0, 0.5)$ uncorrelated white noises. Let $x(0) \sim (1, 1)$.

a. Write the continuous filter equations. Solve for the error covariance $p(t)$.
b. Measurements are now taken every second. Discretize and write the discrete Kalman filter equations. Solve for error covariance p_k for $t = 0$ to 5 sec.
c. Write the continuous-discrete filter equations. Solve for $p(t)$ for $t = 0$ to 5 sec.
d. Plot all three error covariances on the same graph.

3.7-4: Convergence for Continuous-Discrete Filter. Show that if $(A, G\sqrt{Q})$ is reachable and (A, H) is detectable, then in the continuous-discrete filter, $p(t)$ displays a bounded steady-state behavior independent of $p(0)$. Furthermore, the Kalman filter is stable. (What does stability mean in this case?)

Section 3.8

3.8-1: Smoothing for Scalar System

$$\dot{x} = ax + bu + w$$

$$z = x + v$$

with $w(t) \sim (0, q)$, $v(t) \sim (0, r)$.

a. Formulate the two-filter smoother in Table 3.8-1. Now, if $a = 0$, $q = 1$, $r = 0.5$, final time $T = 5$ sec, $x(0) \sim (1, 1)$ solve analytically for smoothed error covariance $p(t)$. Sketch $p_f(t)$, $s(t)$, and $p(t)$.
b. Repeat using the Rauch–Tung–Striebel formulation.

3.8-2: Smoothing for Newton's System.   Let

$$\ddot{x} = w$$

$$z = x + v$$

$w(t) \sim (0, q)$, $v(t) \sim (0, r)$. Repeat Problem 3.8-1. For the numerical work, let $x(0) \sim (1, I)$, $q = 0$, $r = 1$.

4

KALMAN FILTER DESIGN AND IMPLEMENTATION

In this chapter we discuss some practical aspects of Kalman filter design and implementation. There are two general topics. In Section 4.1 we analyze the effects of assuming a model which does not accurately describe the physical system whose state is to be estimated. In the remaining sections we present some techniques for designing filters of simplified structure.

4.1 MODELING ERRORS, DIVERGENCE, AND EXPONENTIAL DATA WEIGHTING

We have discussed Kalman filter design for the ideal situation when all plant dynamics and noise statistics are exactly known. If this is not the case serious divergence problems can occur where the theoretical behavior of the filter and its actual behavior do not agree. The Kalman filter has no mechanism built into its theory for dealing with inaccurate dynamics A, B, G, H, or statistics P_0, Q, R. If the plant is inaccurately known we should, to be theoretically rigorous, apply not Kalman filter theory but adaptive control theory; this can result in more complicated designs.

Fortunately, by understanding the properties of the Kalman filter we can often use its robustness in the face of process noise to compensate for inaccurately known plant dynamics. Alternatively, we can modify the filter to get better performance. These topics are the subject of this section.

Modeling Errors

For this discussion we need to define two concepts. They must be clearly distinguished from each other. The *plant* is the actual physical system we are interested in. The *model* is the mathematical description of the physical system which we use for the filter design stage. (The plant is also called the "truth model.")

Let the model be given by

$$x_{k+1} = Ax_k + Bu_k + Gw_k \qquad (4.1\text{-}1a)$$

$$z_k = Hx_k + v_k, \qquad (4.1\text{-}1b)$$

with w_k and v_k white and uncorrelated with each other and with x_0. Let $w_k \sim (0, Q)$, $v_k \sim (0, R)$, $x_0 \sim [E(x_0), P_0]$. The Kalman filter designed on the basis of this model is given by:

Time Update:

$$P_{k+1}^- = AP_kA^T + GQG^T \qquad (4.1\text{-}2)$$

$$\hat{x}_{k+1}^- = A\hat{x}_k + Bu_k \qquad (4.1\text{-}3)$$

Measurement Update:

$$K_k = P_k^-H^T(HP_k^-H^T + R)^{-1} \qquad (4.1\text{-}4)$$

$$P_k = (I - K_kH)P_k^-(I - K_kH)^T + K_kRK_k^T \qquad (4.1\text{-}5)$$

$$\hat{x}_k = \hat{x}_k^- + K_k(z_k - H\hat{x}_k^-). \qquad (4.1\text{-}6)$$

The a priori error covariance is defined as

$$P_k^- = E[(x_k - \hat{x}_k^-)(x_k - \hat{x}_k^-)^T]. \qquad (4.1\text{-}7)$$

Now suppose that in reality the plant is described not by (4.1-1) but by the state x_k' with dynamics

$$x_{k+1}' = \bar{A}x_k' + \bar{B}u_k' + \bar{G}w_k' \qquad (4.1\text{-}8a)$$

$$z_k' = \bar{H}x_k' + v_k'. \qquad (4.1\text{-}8b)$$

Assume w_k' and v_k' are white and uncorrelated with each other and with x_0', and let $w_k' \sim (0, \bar{Q})$, $v_k' \sim (0, \bar{R})$, $x_0' \sim [E(x_0'), \bar{P}_0]$. We assume here that the model and the plant have the same order n. Matrices $\bar{A}, \bar{B}, \bar{G}, \bar{H}, \bar{P}_0, \bar{Q}, \bar{R}$ are in general all different from the values we assumed in our model (4.1-1) which we used to design the "optimal" filter. (In this section, overbars denote not expected values, but plant quantities.)

The word "optimal" is in inverted commas because, although it is indeed optimal for the assumed model, it is not necessarily optimal for the actual plant. In fact, if the filter (4.1-2)–(4.1-6) is applied to the plant (4.1-8) some very serious problems may result.

Now we need to develop a theoretical background for analyzing *mismatch* between the model and the plant.

Define the a priori and a posteriori errors between the estimate generated by the filter (4.1-2)–(4.1-6) and the state of the *actual plant* (4.1-8) as

$$\tilde{x}_k^- = x_k' - \hat{x}_k^-$$ (4.1-9)

and

$$\tilde{x}_k = x_k' - \hat{x}_k.$$ (4.1-10)

These are *not* the same as the *predicted* error $(x_k - \hat{x}_k^-)$ between the estimate and the *model* state whose covariance P_k^- in (4.1-7) is minimized by the filter (4.1-2)–(4.1-6)! They are, however, the errors we are interested in. If the filter is designed and implemented correctly then the *actual* errors $\tilde{x}_k^-$ and $\tilde{x}_k$ should be acceptably small during the identification run.

To find expressions for the *actual error covariances* $E[\tilde{x}_k^-(\tilde{x}_k^-)^T]$ and $E(\tilde{x}_k\tilde{x}_k^T)$ we proceed as follows. First consider the time update. We have

$$\tilde{x}_{k+1}^- = x_{k+1}' - \hat{x}_{k+1}^-$$
$$= \bar{A}x_k' + \bar{B}u_k' + \bar{G}w_k' - A\hat{x}_k - Bu_k.$$

By defining the *model mismatch system matrix*

$$\Delta A = \bar{A} - A$$ (4.1-11)

this can be written as

$$\tilde{x}_{k+1}^- = A\tilde{x}_k + \Delta A x_k' + \bar{B}u_k' - Bu_k + \bar{G}w_k'.$$ (4.1-12)

Now append the error onto the plant state to define an augmented state, and write (4.1-8a) and (4.1-12) in one equation (Gelb 1974),

$$\begin{bmatrix} x_{k+1}' \\ \tilde{x}_{k+1}^- \end{bmatrix} = \begin{bmatrix} \bar{A} & 0 \\ \Delta A & A \end{bmatrix}\begin{bmatrix} x_k' \\ \tilde{x}_k \end{bmatrix} + \begin{bmatrix} \bar{B} & 0 \\ \bar{B} & -B \end{bmatrix}\begin{bmatrix} u_k' \\ u_k \end{bmatrix} + \begin{bmatrix} \bar{G} \\ \bar{G} \end{bmatrix}w_k'.$$ (4.1-13)

Define correlation matrices by

$$\begin{bmatrix} U_k & (V_k^-)^T \\ V_k^- & S_k^- \end{bmatrix} = E\begin{bmatrix} x_k' \\ \tilde{x}_k^- \end{bmatrix}\begin{bmatrix} x_k' \\ \tilde{x}_k^- \end{bmatrix}^T$$ (4.1-14)

and

$$\begin{bmatrix} U_k & V_k^T \\ V_k & S_k \end{bmatrix} = E\begin{bmatrix} x_k' \\ \tilde{x}_k \end{bmatrix}\begin{bmatrix} x_k' \\ \tilde{x}_k \end{bmatrix}^T.$$ (4.1-15)

Then from (4.1-13)

$$\begin{bmatrix} U_{k+1} & (V_{k+1}^-)^T \\ V_{k+1}^- & S_{k+1}^- \end{bmatrix} = \begin{bmatrix} \bar{A} & 0 \\ \Delta A & A \end{bmatrix}\begin{bmatrix} U_k & V_k^T \\ V_k & S_k \end{bmatrix}\begin{bmatrix} \bar{A}^T & \Delta A^T \\ 0 & A^T \end{bmatrix}$$

$$+ \begin{bmatrix} \bar{G} \\ \bar{G} \end{bmatrix}\bar{Q}[\bar{G}^T \quad \bar{G}^T],$$

where we have set $u_k' = 0$ and $u_k = 0$ for simplicity. Multiplying this out there result the time-update equations

$$U_{k+1} = \bar{A} U_k \bar{A}^T + \bar{G} \bar{Q} \bar{G}^T \tag{4.1-16}$$

$$V_{k+1}^- = \Delta A U_k \bar{A}^T + A V_k \bar{A}^T + \bar{G} \bar{Q} \bar{G}^T \tag{4.1-17}$$

$$S_{k+1}^- = A S_k A^T + \bar{G} \bar{Q} \bar{G}^T + \Delta A U_k \Delta A^T$$
$$+ (A V_k \Delta A^T + \Delta A V_k^T A^T). \tag{4.1-18}$$

We should make a digression to ensure we have a feel for these equations before proceeding.

Note that U_k is the plant state correlation matrix $E[x_k'(x_k')^T]$. Note also that if $\Delta A = 0$, $G = \bar{G}$, and $Q = \bar{Q}$ so that there is no model mismatch, then (4.1-18) is identical to (4.1-2). The extra terms in (4.1-18) represent the increase in the error covariance due to modeling inaccuracies. We are not directly interested in the plant state correlation U_k or the cross-correlation matrix V_k^-; they are required only to update the quantity of interest S_k^-, which is the correlation matrix of the actual error $\tilde{x}_k^- = x_k' - \hat{x}_k^-$. In the presence of modeling errors, the actual error correlation depends on the plant state correlation.

If S_k^- is small then our estimate $\hat{x}_k^-$ matches well the *actual* plant state x_k', which is what we really want.

Now let us derive the measurement update equations. We have

$$\tilde{x}_k = x_k' - \hat{x}_k = x_k' - \hat{x}_k^- - K_k(z_k' - H\hat{x}_k^-). \tag{4.1-19}$$

A comment needs to be made at this point. What we are doing is running the designed filter (4.1-2)–(4.1-6) on the plant (4.1-8). The measurements are therefore the actual plant output. For this reason, z_k' and not the model output z_k is used in (4.1-19). The model is used only to *design* the filter.

Proceeding, (4.1-19) becomes

$$\tilde{x}_k = \tilde{x}_k^- - K_k \bar{H} x_k' - K_k v_k' + K_k H \hat{x}_k^-.$$

If we define the *model mismatch output matrix*

$$\Delta H = \bar{H} - H \tag{4.1-20}$$

this can be written as

$$\tilde{x}_k = (I - K_k H) \tilde{x}_k^- - K_k \Delta H x_k' - K_k v_k'. \tag{4.1-21}$$

By defining an augmented state vector we can write

$$\begin{bmatrix} x_k' \\ \tilde{x}_k \end{bmatrix} = \begin{bmatrix} I & 0 \\ -K_k \Delta H & I - K_k H \end{bmatrix} \begin{bmatrix} x_k' \\ \tilde{x}_k^- \end{bmatrix} + \begin{bmatrix} 0 \\ -K_k \end{bmatrix} v_k'. \tag{4.1-22}$$

Now take correlations of both sides to obtain

$$\begin{bmatrix} U_k & V_k^T \\ V_k & S_k \end{bmatrix} = \begin{bmatrix} I & 0 \\ -K_k \Delta H & I - K_k H \end{bmatrix} \begin{bmatrix} U_k & (V_k^-)^T \\ V_k^- & S_k^- \end{bmatrix} \begin{bmatrix} I & -\Delta H^T K_k^T \\ 0 & (I - K_k H)^T \end{bmatrix}$$
$$+ \begin{bmatrix} 0 \\ -K_k \end{bmatrix} \bar{R} [0 \quad -K_k^T], \tag{4.1-23}$$

which yields the measurement update

$$U_k = U_k \qquad (4.1\text{-}24)$$

$$V_k = (I - K_k H) V_k^- - K_k \Delta H U_k \qquad (4.1\text{-}25)$$

$$S_k = (I - K_k H) S_k^- (I - K_k H)^T + K_k \bar{R} K_k^T + K_k \Delta H U_k \Delta H^T K_k^T$$
$$- [(I - K_k H) V_k^- \Delta H^T K_k^T + K_k \Delta H (V_k^-)^T (I - K_k H)^T]. \qquad (4.1\text{-}26)$$

Note that if $\Delta H = 0$ and $R = \bar{R}$ so that there are no modeling errors, then update (4.1-26) for the actual error correlation is equal to update (4.1-5) for the *predicted covariance*. Again, we are not interested in U_k and V_k, but we are forced to carry them along as "excess baggage" in order to find the quantity of interest S_k.

Since initially the uncertainty in the estimate is the same as the uncertainty in the plant state, we would initialize the recursion by

$$S_0 = - V_0 = U_0 = \bar{P}_0. \qquad (4.1\text{-}27)$$

By combining (4.1-12) and (4.1-21) we obtain the *error system* for the actual error

$$\tilde{x}_{k+1}^- = A(I - K_k H)\tilde{x}_k^- + (\Delta A - A K_k \Delta H)x_k'$$
$$+ \bar{B}u_k' - Bu_k + \bar{G}w_k' - A K_k v_k'. \qquad (4.1\text{-}28)$$

Several things are worth mentioning about this equation. First, it should be compared to (2.5-2), to which it reduces if the model is an exact description of the plant. Second, it is driven by the *plant* noise processes. Third, if the model mismatch matrices ΔA and ΔH are not both zero, the error system is driven by the plant state x_k'. Unless x_k' is bounded, the error $\tilde{x}_k^-$ will grow. We are thus faced with the realization that in the presence of modeling inaccuracies, the plant must in most cases be stable to avoid causing an unbounded estimation error.

The effect of an inaccurately known deterministic input is also clearly seen in (4.1-28), for the error system is driven by the term $(\bar{B}u_k' - Bu_k)$. If this is not zero, unbounded errors can again occur in $\tilde{x}_k^-$.

It is evident from this brief discussion that, even if $A(I - K_k H)$ is stable for large k, which is guaranteed by the steady-state properties of the Kalman filter, modeling inaccuracies can lead to nonconvergence and unbounded actual estimation errors.

A final point is worth making. The expected error is given by

$$E(\tilde{x}_{k+1}^-) = A(I - K_k H)E(\tilde{x}_k^-) + (\Delta A - A K_k \Delta H)E(x_k') + \bar{B}u_k' - Bu_k. \qquad (4.1\text{-}29)$$

Even if $E(\tilde{x}_0^-) = 0$, it is not guaranteed that $E(\tilde{x}_k^-) = 0$ in the presence of modeling inaccuracies. Therefore U_k, V_k, and S_k should not be considered as covariances, but as correlation matrices. Modeling inaccuracies can result in biased estimates.

We shall call P_k given by (4.1-2)–(4.1-6) the *predicted* error covariance, and S_k given by (4.1-16)–(4.1-18) and (4.1-24)–(4.1-26) the *actual* error correlation.

Example 4.1-1: *Inaccurate Modeling of Wiener Process as Constant Unknown*

To illustrate the use of (4.1-16)–(4.1-18) and (4.1-24)–(4.1-26) in the analysis of model mismatch, suppose the model is assumed to be the scalar system

$$x_{k+1} = x_k \tag{1}$$

$$z_k = x_k + v_k, \tag{2}$$

$v_k \sim (0, r)$, $x_0 \sim (m_0, p_0)$.

Then according to Example 2.3-2, using (4.1-2)–(4.1-6) to design a filter for this model results in an error covariance given by

$$p_k = \frac{p_0}{1 + k(p_0/r)}, \tag{3}$$

and in a Kalman gain given by

$$K_k = \frac{p_0/r}{1 + k(p_0/r)}. \tag{4}$$

The predicted error covariance (3) is sketched in Fig. 4.1-1 for $p_0 = 1$, $r = 10$; it shows the anticipated decrease to zero of the error as more and more measurements are made. The Kalman gain also goes to zero with k; this means that as more

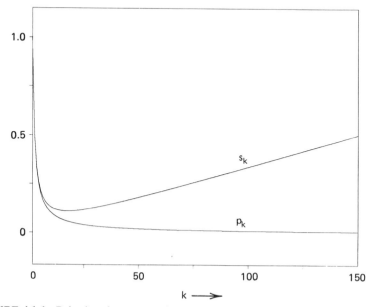

FIGURE 4.1-1   Behavior of error covariances with model mismatch. Predicted error covariance p_k and actual error covariance s_k are shown.

measurements are made, the estimate theoretically becomes more reliable so that each subsequent measurement has less and less effect.

Now let the actual plant, which is supposedly modeled by (1) and (2), be given by

$$x'_{k+1} = x'_k + w'_k \tag{5}$$

$$z_k' = x'_k + v'_k, \tag{6}$$

$w'_k \sim (0, q)$, $v'_k \sim (0, r)$, $x'_0 \sim (m_0, p_0)$. Thus, the actual unknown is a Wiener process, which we have incorrectly modeled as a random constant.

Applying the Kalman filter (4) to the actual plant results in

$$\hat{x}_{k+1} = \hat{x}_k + \frac{p_0/r}{1 + (k+1)(p_0/r)} (z'_{k+1} - \hat{x}_k). \tag{7}$$

To find the resulting *actual* error correlation for $\tilde{x}_k = x'_k - \hat{x}_k$, we need to use (4.1-16)–(4.1-18) and (4.1-24)–(4.1-26). Since $\Delta A = 0$ and $\Delta H = 0$, we do not need to compute U_k, V_k, V_k^-, but can simply use the recursion for the actual error correlation s_k:

$$s_0 = p_0. \tag{8}$$

Time Update:

$$s_{k+1}^- = s_k + q. \tag{9}$$

Measurement Update:

$$s_k = (1 - K_k)^2 s_k^- + K_k^2 r \tag{10}$$

$$= \left(\frac{1 + (k-1)p_0/r}{1 + kp_0/r} \right)^2 s_k^- + \frac{p_0^2/r}{(1 + kp_0/r)^2}. \tag{11}$$

We get

$k = 0$

$$s_0 = p_0. \tag{12}$$

$k = 1$

Time Update:

$$s_1^- = p_0 + q. \tag{13}$$

Measurement Update:

$$s_1 = \frac{p_0 + q + p_0^2/r}{(1 + p_0/r)^2} = \frac{p_0(1 + p_0/r) + q}{(1 + p_0/r)^2}. \tag{14}$$

$k = 2$

Time Update:

$$s_2^- = \frac{p_0(1 + p_0/r) + q[1 + (1 + p_0/r)^2]}{(1 + p_0/r)^2}. \tag{15}$$

Measurement Update:

$$s_2 = \frac{p_0(1 + p_0/r) + q[1 + (1 + p_0/r)^2] + p_0^2/r}{(1 + 2p_0/r)^2}$$

$$= \frac{p_0(1 + 2p_0/r) + q[1 + (1 + p_0/r)^2]}{(1 + 2p_0/r)^2}. \qquad (16)$$

And, in general

$$s_k = \frac{p_0(1 + kp_0/r) + q \sum_{i=0}^{k-1} (1 + ip_0/r)^2}{(1 + kp_0/r)^2}$$

$$= p_k + \frac{q \sum_{i=0}^{k-1} (1 + ip_0/r)^2}{(1 + kp_0/r)^2}. \qquad (17)$$

This actual error correlation is also shown in Fig. 4.1-1 for $p_0 = 1$, $r = 10$, $q = 0.01$. The behavior of the actual correlation s_k is quite different from the behavior of the predicted covariance p_k!

To find the limiting behavior of s_k, use the identities

$$\sum_{i=0}^{k-1} i = \frac{(k-1)k}{2} \qquad (18)$$

$$\sum_{i=0}^{k-1} i^2 = \frac{(k-1)k(2k-1)}{6} \qquad (19)$$

to write

$$s_k \to \frac{q \sum_{i=0}^{k-1} [1 + (2ip_0/r) + i^2(p_0^2/r^2)]}{1 + 2k(p_0/r) + k^2(p_0^2/r^2)}$$

$$\to \frac{q[k^2(p_0/r) + (k^3/3 - k^2/2)(p_0^2/r^2)]}{k^2 p_0^2/r^2}$$

$$= q \left[\left(\frac{r}{p_0} - \frac{1}{2} \right) + \frac{k}{3} \right]. \qquad (20)$$

While we think that all is well for large k because of (3), the actual error correlation s_k is increasing linearly with k!

The program used to make Fig. 4.1-1 is shown in Fig. 4.1-2. It implements the predicted error covariance recursion

$$p_{k+1} = p_k - \frac{p_k^2}{p_k + r} \qquad (21)$$

and the actual error covariance recursion (9), (10). Note that carrying out a *performance analysis* often requires only simple programs. The analysis we have done subsequent to (10) is not required in a practical performance study! ∎

```
      PROGRAM MODMIS
      DATA Q,R,P,S/.01,1.,1.,1./
      REWIND 7
      REWIND 8

      DO 10 I= 1,500
      WRITE(7,*) P
      WRITE(8,*) S
      P= P - P**2/(P+R)
      AK= P/R
      S= S + Q
      S= S*(1.-AK)**2 + R*AK**2
10    CONTINUE

      STOP
      END
```

FIGURE 4.1-2 Program for Kalman filter performance study.

Exercise 4.1-2: *Model Mismatch in Reduced-Order Filters*

Let the actual plant be given by (4.1-8) and the assumed model by (4.1-1). It can happen that the order n_m of the model is less than the order n_p of the plant. This can occur in two ways. Either we can fail to account for all the states through ignorance of the plant, or we can deliberately select a reduced-order model to describe the plant by omitting some of the plant states. If done properly, this reduced-order modeling can result in a (nonoptimal!) reduced-order filter which has a simplified structure and yet gives us the essential information we require about the plant state (Gelb 1974, D'Appolito 1971, Nash et al. 1972). The design of reduced-order filters is discussed further later on in the chapter. Here we discuss the actual error correlation equations.

Let $T \in R^{n_m \times n_p}$ be the state-space transformation which takes the actual plant state x'_k to the model state x_k (i.e., $x_k = Tx'_k$). Let T have full row rank n_m. Define modified model mismatch matrices by

$$\Delta A = \bar{A} - T^+ A T \qquad (4.1\text{-}30)$$

$$\Delta H = \bar{H} - HT \qquad (4.1\text{-}31)$$

where T^+ is the Moore–Penrose matrix inverse of T. Note that $TT^+ = I$ since T has full row rank, so T^+ is a right inverse here. [Since T has full row rank, its right inverse is given by $T^+ = T^T(TT^T)^{-1}$.] Then the actual errors are defined as the n_p vectors

$$\bar{x}_k^- = x'_k - T^+ \hat{x}_k^- \qquad (4.1\text{-}32)$$

$$\bar{x}_k = x'_k - T^+ \hat{x}_k. \qquad (4.1\text{-}33)$$

a. Show that the actual error correlation S_k^- for a reduced-order filter that does not estimate the full plant state is given by (4.1-16)–(4.1-18) with A replaced everywhere by $T^+ A T$, and (4.1-24)–(4.1-26) with H and K_k replaced everywhere by HT and $T^+ K_k$, respectively. Let $u'_k = 0$, $u_k = 0$ for simplicity.

b. Show that the actual error for the reduced-order filter develops according to (4.1-28) with the above replacements, and with B also replaced by $T^+ B$. ∎

Exercise 4.1-3: Actual Error Correlation for Continuous Kalman Filter

In the continuous-time case, define the actual error as

$$\tilde{x}(t) = x'(t) - \hat{x}(t) \tag{4.1-34}$$

where $x'(t)$ is the actual plant state and $\hat{x}(t)$ is the estimate produced by the filter designed based on the model. Define correlation matrices by

$$\begin{bmatrix} U & V^T \\ V & S \end{bmatrix} = E \begin{bmatrix} x' \\ \tilde{x} \end{bmatrix} \begin{bmatrix} x' \\ \tilde{x} \end{bmatrix}^T. \tag{4.1-35}$$

a. Show that the actual error correlation $S(t)$ is given by

$$\dot{S} = (A - KH)S + S(A - KH)^T + \bar{G}\bar{Q}\bar{G}^T$$
$$\qquad + K\bar{R}K^T + (\Delta A - K\Delta H)V^T + V(\Delta A - K\Delta H)^T \tag{4.1-36a}$$

$$\dot{V} = V\bar{A}^T + (A - KH)V + (\Delta A - K\Delta H)U + \bar{G}\bar{Q}\bar{G}^T \tag{4.1-36b}$$

$$\dot{U} = \bar{A}U + U\bar{A}^T + \bar{G}\bar{Q}\bar{G}^T, \tag{4.1-36c}$$

where overbars denote actual plant quantities, and ΔA and ΔH are the same as for the discrete case. The initial conditions are given by (4.1-27).

b. Show that the actual error system is

$$\dot{\tilde{x}} = (A - KH)\tilde{x} + (\Delta A - K\Delta H)x' + \bar{B}u' - Bu + \bar{G}w' - Kv'. \tag{4.1-37}$$

■

Exercise 4.1-4: Model Mismatch in Continuous Reduced-Order Filters

For the continuous Kalman filter the results of Exercise 4.1-2 are also valid.

a. Defining the actual error as

$$\tilde{x} = x' - T^+\hat{x}, \tag{4.1-38}$$

show that the actual error correlation $S(t)$ for a filter that does not estimate the entire plant state is given for the special case $T^+ = T^T$ by

$$\dot{S} = T^T(A - KH)TS + ST^T(A - KH)^TT + \bar{G}\bar{Q}\bar{G}^T$$
$$\qquad + T^TK\bar{R}K^TT + (\Delta A - T^TK\Delta H)V^T + V(\Delta A - T^TK\Delta H)^T \tag{4.1-39}$$

$$\dot{V} = V\bar{A}^T + T^T(A - KH)TV + (\Delta A - T^TK\Delta H)U + \bar{G}\bar{Q}\bar{G}^T \tag{4.1-40}$$

$$\dot{U} = \bar{A}U + U\bar{A}^T + \bar{G}\bar{Q}\bar{G}^T. \tag{4.1-41}$$

The model mismatch matrices ΔA and ΔH are as defined in Exercise 4.1-2.

b. Show that the actual error evolves according to

$$\dot{\tilde{x}} = T^T(A - KH)T\tilde{x} + (\Delta A - T^TK\Delta H)x'$$
$$\qquad + \bar{B}u' - T^TBu + \bar{G}w' - T^TKv'. \tag{4.1-42}$$

■

Kalman Filter Divergence

Filter divergence occurs when the actual error correlation S_k is not bounded as expected from the predicted error covariance P_k (Gelb 1974, Anderson

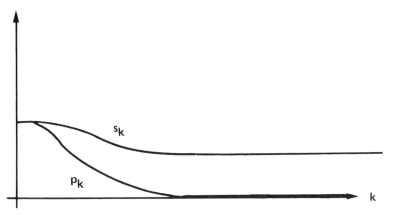

FIGURE 4.1-3 Apparent filter divergence.

and Moore 1979). In *apparent divergence*, S_k remains bounded, but it approaches a larger bound than does P_k. See Fig. 4.1-3. This occurs when there are modeling errors that cause the implemented filter to be suboptimal. In *true divergence*, P_k remains bounded while S_k tends to infinity. See Fig. 4.1-1. This occurs when the filter is not stable (Section 2.5), or when there are unmodeled unstable plant states. An illustration of an unstable filter is provided by Example 4.1-1 [$(A, G\sqrt{Q})$ is not reachable in that example]. An illustration of unmodeled unstable states is provided in the next example.

Example 4.1-5:   *Unmodeled Unstable States*

Suppose the assumed model is the scalar random bias

$$x_{k+1} = x_k \tag{1}$$

$$z_k = x_k + v_k, \tag{2}$$

$v_k \sim (0, r)$, $x_0 \sim (\bar{x}_0, p_0)$. Then by the results of Example 2.3-2, the predicted error covariance [for $(x_k - \hat{x}_k)$] is

$$p_k = \frac{p_0}{1 + k(p_0/r)} \tag{3}$$

and the Kalman gain is

$$K_k = \frac{p_0/r}{1 + k(p_0/r)}. \tag{4}$$

As k increases P_k and K_k go to zero.

Now suppose the actual plant is given by

$$x'_{k+1} = \begin{bmatrix} 1 & \tau \\ 0 & 1 \end{bmatrix} x'_k \tag{5}$$

$$z'_k = [1 \quad 0] x'_k + v'_k, \tag{6}$$

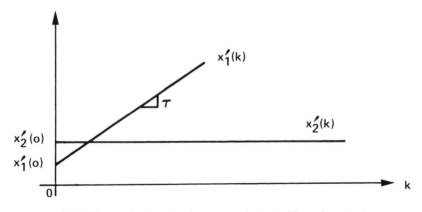

FIGURE 4.1-4   Zero-input response of plant with random ramp.

$v'_k \sim (0, r)$, $x'_0 \sim (\bar{x}'_0, \bar{P}_0)$, where τ is the sampling period. Thus, the plant has an unmodeled random ramp behavior. The zero-input response of (5) is shown in Fig. 4.1-4, where $x'_k \triangleq [x'_1(k) \quad x'_2(k)]^T$.

Applying the Kalman filter designed based on model (1)–(2) to the plant (5)–(6) there results

$$\hat{x}_{k+1} = \hat{x}_k + \frac{p_0/r}{1+(k+1)(p_0/r)}(z'_{k+1} - \hat{x}_k) \tag{7}$$

as the estimate for $x'_1(k)$.

We can use Exercise 4.1-2 to investigate the behavior of the actual error in this situation. What we have inadvertently done is assume a reduced-order model related to the plant by $x_k = Tx'_k$ with

$$T = [1 \quad 0]. \tag{8}$$

Hence $T^+ = [1 \quad 0]^T$ is a right inverse for T and

$$T^+ AT = \begin{bmatrix} 1 & 0 \\ 0 & 0 \end{bmatrix}, \tag{9}$$

$$HT = [1 \quad 0]. \tag{10}$$

The model-mismatch matrices are

$$\Delta A = \bar{A} - T^+ AT = \begin{bmatrix} 0 & \tau \\ 0 & 1 \end{bmatrix} \tag{11}$$

and

$$\Delta H = \bar{H} - HT = 0, \tag{12}$$

and

$$T^+ K_k = \begin{bmatrix} \dfrac{p_0/r}{1+k(p_0/r)} \\ 0 \end{bmatrix}. \tag{13}$$

The actual error system therefore becomes [see (4.1-28) and Exercise 4.1-2]

$$\bar{x}^-_{k+1} = \begin{bmatrix} \dfrac{1+(k-1)(p_0/r)}{1+k(p_0/r)} & 0 \\ 0 & 0 \end{bmatrix} \bar{x}^-_k + \begin{bmatrix} 0 & \tau \\ 0 & 1 \end{bmatrix} x'_k + \begin{bmatrix} \dfrac{p_0/r}{1+k(p_0/r)} \\ 0 \end{bmatrix} v'_k. \tag{14}$$

If the plant were indeed described by (1)–(2), then the error system would be only the scalar system obtained by keeping the $(1, 1)$ entries of the coefficient matrices. In this example, however, we are not so fortunate. Error system (14) is driven by the plant state x'_k. According to Fig. 4.1-4, $x'_2(k)$ is constant at $x'_2(0)$, so this term tends to make the first component of $\bar{x}^-_k$ increase linearly with k at the rate τ. In the limit of large k, the Kalman gain tends to zero and (14) becomes

$$\bar{x}^-_{k+1} = \begin{bmatrix} 1 & 0 \\ 0 & 0 \end{bmatrix} \bar{x}^-_k + \begin{bmatrix} 0 & \tau \\ 0 & 1 \end{bmatrix} x'_k, \tag{15}$$

so that the error increases without bound because of the unmodeled unstable plant dynamics!

It is also easy to see this behavior in terms of an unbounded actual error correlation matrix S_k by using (4.1-16)–(4.1-18) and (4.1-24)–(4.1-26). Let us suppose $\bar{P}_0 = I$. Then $S_0 = -V_0 = U_0 = I$.

According to (4.1-16) and (4.1-24), the plant state correlation matrix develops by the recursion

$$U_{k+1} = \bar{A} U_k \bar{A}^T, \tag{16}$$

which has solution

$$U_k = \bar{A}^k U_0 (\bar{A}^T)^k = \begin{bmatrix} 1 & k\tau \\ 0 & 1 \end{bmatrix} \begin{bmatrix} 1 & 0 \\ k\tau & 1 \end{bmatrix} = \begin{bmatrix} 1+k^2\tau^2 & k\tau \\ k\tau & 1 \end{bmatrix}. \tag{17}$$

The correlation of $x'_1(k)$ increases quadratically with k. Define the components of V_k and S_k by

$$V_k \triangleq \begin{bmatrix} V_1(k) & V_2(k) \\ V_3(k) & V_4(k) \end{bmatrix} \tag{18}$$

and

$$S_k \triangleq \begin{bmatrix} S_1(k) & S_2(k) \\ S_2(k) & S_4(k) \end{bmatrix}. \tag{19}$$

Now use (4.1-17) and (4.1-25) (and Exercise 4.1-2!) to see that

$$V_{k+1} = \begin{bmatrix} \dfrac{1+k(p_0/r)}{1+(k+1)(p_0/r)} & 0 \\ 0 & 1 \end{bmatrix} \begin{bmatrix} (k+1)\tau^2 + V_1(k) + V_2(k)\tau & V_2(k)+\tau \\ (k+1)\tau & 1 \end{bmatrix}. \tag{20}$$

Thus, for large k the first factor is I and then $V_1(k)$, $V_2(k)$, and $V_3(k)$ increase with k.

Finally, by (4.1-18) and (4.1-26)

$$S_k^- = \begin{bmatrix} 1 & 0 \\ 0 & 0 \end{bmatrix} S_k \begin{bmatrix} 1 & 0 \\ 0 & 0 \end{bmatrix} + \begin{bmatrix} 0 & \tau \\ 0 & 1 \end{bmatrix} U_k \begin{bmatrix} 0 & 0 \\ \tau & 1 \end{bmatrix} + \begin{bmatrix} 1 & 0 \\ 0 & 0 \end{bmatrix} V_k \begin{bmatrix} 0 & 0 \\ \tau & 1 \end{bmatrix} + \begin{bmatrix} 0 & \tau \\ 0 & 1 \end{bmatrix} V_k^T \begin{bmatrix} 1 & 0 \\ 0 & 0 \end{bmatrix}$$

$$= \begin{bmatrix} S_1(k) + 2V_2(k)\tau + \tau^2 & V_2(k)+\tau \\ V_2(k)+\tau & 1 \end{bmatrix}, \tag{21}$$

$$S_{k+1} = \begin{bmatrix} \dfrac{1 + k(p_0/r)}{1 + (k+1)(p_0/r)} & 0 \\[2ex] 0 & 1 \end{bmatrix} \begin{bmatrix} S_1(k) + 2V_2(k)\tau + \tau^2 & V_2(k) + \tau \\ V_2(k) + \tau & 1 \end{bmatrix} \begin{bmatrix} \dfrac{1 + k(p_0/r)}{1 + (k+1)(p_0/r)} & 0 \\[2ex] 0 & 1 \end{bmatrix}$$

$$+ \begin{bmatrix} \dfrac{p_0^2/r}{[1 + (k+1)(p_0/r)]^2} & 0 \\[2ex] 0 & 0 \end{bmatrix}. \tag{22}$$

Again, for large k the first and third factors of the first term in (22) tend to I, and $S_1(k)$ then increases with k. Note that $S_4(k)$ does not increase but remains at $S_4(0) = 1$. The actual error associated with the unmodeled state $x_2'(k)$ does not increase; the effect here of the unmodeled unstable dynamics is to provide an unaccounted for "input" to error component one which makes it increase. ■

It turns out that it is not very difficult to correct the divergence in this example. The problem is that the Kalman gain K_k goes to zero with k, and so the filter disregards the measurements for large k. All we need to do is prevent this from happening! We give two methods for doing this in the next subsections.

Fictitious Process Noise Injection

We discuss two methods for preventing divergence: *fictitious process noise injection* and *exponential data weighting*. Both methods work by preventing the predicted error covariance P_k, and hence the Kalman gain K_k, from going to zero with k.

To prevent divergence, we can ensure that *in the model* all states are sufficiently excited by the process noise. Thus, we must ensure that $(A, G\sqrt{Q})$ is reachable, even though in the plant there may be no process noise $(w_k' = 0)$. If $(A, G\sqrt{Q})$ is reachable, then as was seen in Section 2.5, the detectability of (A, H) guarantees that P_k approaches a strictly positive definite (i.e., nonzero) steady-state value P.

At this point, it is worthwhile to look back at the radar tracking problem in Example 2.4-3. In that example the actual plant was nonlinear. To apply the Kalman filter in Table 2.3-1, we assumed a model of simplified linear structure. To prevent divergence, although we were not yet speaking in those terms, we added fictitious process noise (maneuver noise) in the model to compensate for the unmodeled nonlinear dynamics.

It is easy to see that adding process noise to the model in Example 4.1-1 will correct the divergence problem, for this will make the model the same as the plant!

The next example shows how to use fictitious process noise injection to correct the divergence in Example 4.1-5.

Example 4.1-6: Prevention of Divergence by Process Noise Injection

The plant in Example 4.1-5 is a random ramp with no process noise. The model assumed for filter design was a random bias with no process noise. It was found that while the predicted error covariance p_k tended to zero with k, the actual error correlation S_k increased with k.

We can use the simplified random bias model to design a first-order nondivergent filter for this second-order plant by adding fictitious process noise w_k to the model. Thus, assume a model of the form

$$x_{k+1} = x_k + w_k \tag{1}$$

$$z_k = x_k + v_k, \tag{2}$$

$w_k \sim (0, q)$, $v_k \sim (0, r)$, $x_0 \sim (\bar{x}_0, p_0)$.

According to Table 2.3-1, the predicted error covariance is given by

$$p_{k+1}^- = p_k + q \tag{3}$$

$$p_k = \frac{p_k^-}{1 + (p_k^-/r)}. \tag{4}$$

With nonzero q it is not easy to find a closed-form expression for the error covariance like the one we gave in Example 4.1-5; however, p_k is sketched in Fig. 4.1-5 for $p_0 = 10$, $r = 10$, $q = 0.1$. This figure was made using a simple program for (3), (4), like the one in Fig. 4.1-2. Note that p_k does not go to zero. We can find its steady-state value (cf., Example 2.1-1) by noting that at steady-state (3) and (4) can be combined with $p_{k+1}^- = p_k^- \triangleq p^-$ to yield

$$p^- = \frac{rp^-}{r + p^-} + q$$

or

$$(p^-)^2 - qp^- - qr = 0, \tag{5}$$

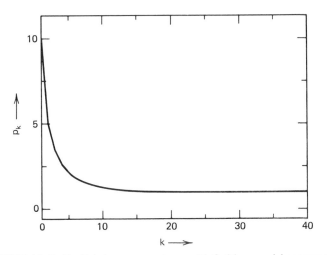

FIGURE 4.1-5 Predicted error covariance with fictitious model process noise.

whose positive solution is

$$p^- = \frac{q}{2}\left(1 + \sqrt{1 + \frac{4r}{q}}\right).$$ (6)

Hence the steady-state value of p_k is

$$p = \frac{q}{2}\left(\sqrt{1 + \frac{4r}{q}} - 1\right).$$ (7)

The Kalman gain is

$$K_k = \frac{p_k}{r},$$ (8)

which reaches a nonzero steady-state value of

$$K = \frac{q}{2r}\left(\sqrt{1 + \frac{4r}{q}} - 1\right).$$ (9)

The steady-state Kalman gain is plotted as a function of q/r in Fig. 4.1-6. Note that it varies from zero to 1 as q/r increases.

Now suppose the actual plant is the random ramp

$$x'_{k+1} = \begin{bmatrix} 1 & \tau \\ 0 & 1 \end{bmatrix} x'_k$$ (10)

$$z'_k = [1 \quad 0]x'_k + v'_k,$$ (11)

$v'_k \sim (0, r) x'_0 = (\bar{x}'_0, \bar{P}_0)$.

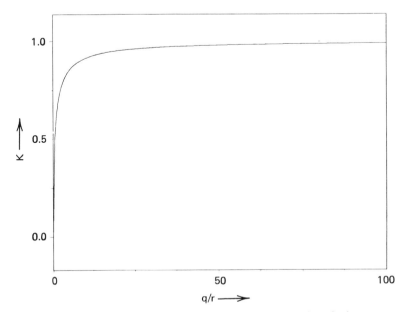

FIGURE 4.1-6   Steady-state Kalman gain as a function of q/r.

If we apply the above Kalman filter to this plant, the model mismatch matrices are as in Example 4.1-5 and the actual error system becomes

$$\tilde{x}_{k+1}^- = \begin{bmatrix} 1 - K_k & 0 \\ 0 & 0 \end{bmatrix} \tilde{x}_k^- + \begin{bmatrix} 0 & \tau \\ 0 & 1 \end{bmatrix} x_k' + \begin{bmatrix} K_k \\ 0 \end{bmatrix} v_k' \tag{12}$$

with gain K_k given by (8). The difference between this and equation (14) in Example 4.1-5 is that, now that fictitious process noise has been added to the model, K_k does not go to zero. This means that as k becomes large the measurements are not discounted. In fact, (12) approaches the steady-state error system

$$\tilde{x}_{k+1}^- = \begin{bmatrix} 1 - K & 0 \\ 0 & 0 \end{bmatrix} \tilde{x}_k^- + \begin{bmatrix} 0 & \tau \\ 0 & 1 \end{bmatrix} x_k' + \begin{bmatrix} K \\ 0 \end{bmatrix} v_k'. \tag{13}$$

Two effects are of interest here. First, if $q \neq 0$, then the plant measurement noise v_k' continues to excite the error as $k \to \infty$, so that the limiting value of the first component $\tilde{x}_1^-(k)$ is nonzero even if plant state component $x_2(k)$ is zero. Fictitious process noise injection into the model has resulted in suboptimal behavior of the filter. This degradation is compensated for by the second effect: if q is correctly chosen, the error remains bounded even in the face of the unmodeled plant dynamics. To see this, note that since in the model (A, H) is detectable and $(A, G\sqrt{Q})$ is reachable, $A(I - KH) = (1 - K)$ is always stable (i.e., $|1 - K| < 1$) for all q, so the error in (13) remains bounded. As q increases, $(1 - K)$ tends to zero (see Fig. 4.1-6). In the extreme, if we select q so that $(1 - K) \approx 0$, then the error is given in steady state by

$$\tilde{x}_1^-(k) = \tau x_2'(k) + v_k' \tag{14}$$

$$\tilde{x}_2^-(k) = x_2'(k), \tag{15}$$

which is bounded since, according to Fig. 4.1-4, $x_2'(k)$ has a constant value of $x_2'(0)$.

This bounded behavior of $\tilde{x}_k^-$ induced by the fictitious process noise can also be seen by examining the actual error correlation S_k. From (4.1-16)–(4.1-18) and (4.1-24)–(4.1-26) it is apparent that we still have (if $\bar{P}_0 = I$)

$$U_k = \begin{bmatrix} 1 + k^2\tau^2 & k\tau \\ k\tau & 1 \end{bmatrix}. \tag{16}$$

Now, however,

$$V_{k+1} = \begin{bmatrix} 1 - K_{k+1} & 0 \\ 0 & 1 \end{bmatrix} \begin{bmatrix} (k+1)\tau^2 + V_1(k) + V_2(k)\tau & V_2(k) + \tau \\ (k+1)\tau & 1 \end{bmatrix}, \tag{17}$$

and

$$S_{k+1} = \begin{bmatrix} 1 - K_{k+1} & 0 \\ 0 & 1 \end{bmatrix} \begin{bmatrix} S_1(k) + 2V_2(k)\tau + \tau^2 & V_2(k) + \tau \\ V_2(k) + \tau & 1 \end{bmatrix} \begin{bmatrix} 1 - K_{k+1} & 0 \\ 0 & 1 \end{bmatrix}$$
$$+ \begin{bmatrix} K_{k+1}^2 r & 0 \\ 0 & 0 \end{bmatrix}. \tag{18}$$

To make the point, if we choose q very large so that $K \approx 1$, then in steady state there results

$$V_k = \begin{bmatrix} 0 & 0 \\ k\tau & 1 \end{bmatrix} \tag{19}$$

$$S_k = \begin{bmatrix} r & 0 \\ 0 & 1 \end{bmatrix}, \tag{20}$$

so that S_k is bounded.   ∎

Exponential Data Weighting

Another way to prevent divergence by keeping the filter from discounting measurements for large k is to use *exponential data weighting*. Additional references for this material are Anderson (1973), Anderson and Moore (1979), and Gelb (1974).

Let us set the model covariance matrices equal to

$$R_k = R\alpha^{-2(k+1)} \tag{4.1-43}$$

$$Q_k = Q\alpha^{-2(k+1)} \tag{4.1-44}$$

for some $\alpha \geq 1$ and constant matrices Q and R. Since $\alpha \geq 1$, this means that as time k increases, the process and measurement noise covariances decrease, so that we are giving more credibility to the recent data by decreasing the noise covariance exponentially with k. See Fig. 4.1-7.

According to Table 2.3-3 the Kalman filter is now given as follows. The error covariance develops by

$$P_{k+1}^- = A[P_k^- - P_k^- H^T(HP_k^- H^T + R\alpha^{-2k-2})^{-1}HP_k^-]A^T + GQG^T\alpha^{-2k-2} \tag{4.1-45}$$

or

$$P_{k+1}^- \alpha^{2(k+1)} = \alpha^2 A\left[P_k^-\alpha^{2k} - P_k^-\alpha^{2k}H^T\left(HP_k^-\alpha^{2k}H^T + \frac{R}{\alpha^2}\right)^{-1}HP_k^-\alpha^{2k}\right]A^T + GQG^T.$$

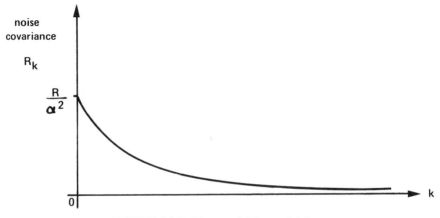

FIGURE 4.1-7   Exponential data weighting.

By defining the *weighted covariance*

$$P_k^\alpha \triangleq P_k^- \alpha^{2k} \tag{4.1-46}$$

this becomes

$$P_{k+1}^\alpha = \alpha^2 A \left[P_k^\alpha - P_k^\alpha H^T \left(H P_k^\alpha H^T + \frac{R}{\alpha^2} \right)^{-1} H P_k^\alpha \right] A^T + GQG^T. \tag{4.1-47}$$

The initial condition is $P_0^\alpha = P_0$.
   The Kalman gain is

$$K_k = P_k^- H^T (H P_k^- H^T + R\alpha^{-2k-2})^{-1},$$

which can be written in terms of P_k^α as

$$K_k = P_k^\alpha H^T \left(H P_k^\alpha H^T + \frac{R}{\alpha^2} \right)^{-1}. \tag{4.1-48}$$

Finally, the estimate update equation is

$$\hat{x}_{k+1}^- = A\hat{x}_k^- + Bu_k + AK_k(z_k - H\hat{x}_k^-) \tag{4.1-49}$$

or

$$\hat{x}_{k+1}^- = A(I - K_k H)\hat{x}_k^- + Bu_k + AK_k z_k. \tag{4.1-50}$$

This filter has time-varying noise covariance Q_k and R_k, however, since they are exponentially varying we see from (4.1-47) and (4.1-48) that we can compute P_k^α and K_k by considering the equivalent *time-invariant* model

$$x_{k+1} = \alpha A x_k + Bu_k + Gw_k \tag{4.1-51a}$$

$$z_k = Hx_k + v_k, \tag{4.1-51b}$$

$w_k \sim (0, Q)$, $v_k \sim (0, R/\alpha^2)$. The error covariance update equation for this model which is given in Table 2.3-3 is identical to 4.1-47.
   While P_k^α is the error covariance for the time-invariant model (4.1-51), it is *not* the error covariance for the original model (4.1-1) with time-varying covariances (4.1-43), (4.1-44). By using the sensitivity relations (4.1-16)–(4.1-18), (4.1-24)–(4.1-26) it can be shown (Anderson 1973) that P_k^α provides an upper bound for the actual error covariance.
   The error covariance P_k^- with no data weighting reaches a bounded steady-state value if (A, H) is detectable. The covariance P_k^α reaches such a value if $(\alpha A, H)$ is detectable. But detectability of (A, H) is equivalent to detectability of $(\alpha A, H)$, so that if the plant (4.1-1) is detectable then P_k^α is bounded.
   To investigate the stability of the filter (4.1-47), (4.1-48), (4.1-50), note the following. If $(\alpha A, G\sqrt{Q})$ is stabilizable, then the filter designed from Table 2.3-3 for the equivalent model (4.1-51) is asymptotically stable for

large k. Although the covariance for this filter is given by (4.1-47) and the Kalman gain by (4.1-48), the estimate update is not given by (4.1-50). It is given, from Table 2.3-3, by

$$\hat{x}^-_{k+1} = \alpha A(I - K_k H)\hat{x}^-_k + Bu_k + \alpha A K_k z_k. \qquad (4.1-52)$$

The stabilizability property for (4.1-51) guarantees that for large k the poles of $\alpha A(I - K_k H)$ are inside $|z| = 1$. This means that the poles of $A(I - K_k H)$ are inside

$$|z| = \frac{1}{\alpha} < 1, \qquad (4.1-53)$$

so that in the limit, (4.1-50) has at least exponential stability with terms decaying like α^{-k}.

Reachability of $(A, G\sqrt{Q})$ [i.e., of the model (4.1-1)] is equivalent to reachability of $(\alpha A, G\sqrt{Q})$ [i.e., of the equivalent model (4.1-51)]. Thus, by using exponential data weighting, reachability of $(A, G\sqrt{Q})$ guarantees a *stability margin* of α^{-1} for the filter (4.1-50).

We can show that exponential data weighting is equivalent to fictitious process noise injection. To wit, write (4.1-47) as

$$P^\alpha_{k+1} = A \left[P^\alpha_k - P^\alpha_k H^T \left(H P^\alpha_k H^T + \frac{R}{\alpha^2} \right)^{-1} H P^\alpha_k \right] A^T + Q'_k \quad (4.1-54)$$

where

$$Q'_k = GQG^T + (\alpha^2 - 1)A \left[P^\alpha_k - P^\alpha_k H^T \left(H P^\alpha_k H^T + \frac{R}{\alpha^2} \right)^{-1} H P^\alpha_k \right] A^T. \qquad (4.1-55)$$

Evidently, Q'_k corresponds to a process noise whose covariance is increased over the original Q.

If $\alpha \neq 1$, then the actual error covariance is increased since the filter is suboptimal. Compare with the discussion of (13) in Example 4.1-6. This means that P^1_k (i.e., P^α_k for $\alpha = 1$) provides a lower bound for the actual error covariance, while P^α_k (for the α actually used) provides an upper bound.

The next example shows that if $\alpha > 1$, then P^α_k reaches a nonzero steady-state value. Thus, exponential data weighting can be used to prevent divergence. It works by preventing the discounting of data for large k, much as fictitious process noise injection prevented divergence in Example 4.1-6.

Example 4.1-7:   *Steady-State Kalman Gain with Exponential Data Weighting*

Let us consider again the scalar system of Example 4.1-1,

$$x_{k+1} = x_k \qquad (1)$$

$$z_k = x_k + v_k,$$
(2)

$v_k \sim (0, r)$, $x_0 \sim (\bar{x}_0, p_0)$. In Examples 4.1-1 and 4.1-5, we showed that the predicted error covariance p_k went to zero with k, which resulted in divergence problems. We showed in Example 4.1-6 how to prevent divergence by keeping p_k from going to zero with fictitious process noise. In this example, we show how to prevent p_k from going to zero with exponential data weighting.

If we use the Kalman filter with exponential data weighting, then (4.1-47) becomes

$$p_{k+1}^\alpha = \alpha^2 \left[p_k^\alpha - \frac{(p_k^\alpha)^2}{p_k^\alpha + r/\alpha^2} \right]$$

$$= \left(\frac{p_k^\alpha r/\alpha^2}{p_k^\alpha + r/\alpha^2} \right) \alpha^2.$$
(3)

This can be written (cf. (2.3-21)) as

$$\frac{\alpha^2}{p_{k+1}^\alpha} = \frac{1}{p_k^\alpha} + \frac{\alpha^2}{r}.$$
(4)

We can more easily solve this for the information matrix

$$l_k \triangleq (p_k^\alpha)^{-1}.$$
(5)

Write (4) as the linear difference equation in l_k,

$$l_{k+1} = \alpha^{-2} l_k + \frac{1}{r}.$$
(6)

The solution to this (for $\alpha > 1$) is

$$l_k = \alpha^{-2k} l_0 + \sum_{i=0}^{k-1} \alpha^{-2(k-i-1)} \left(\frac{1}{r} \right),$$

or,

$$l_k = \alpha^{-2k} l_0 + \frac{\alpha^{-2(k-1)}}{r} \sum_{i=0}^{k-1} \alpha^{2i}$$

$$= \alpha^{-2k} l_0 + \frac{\alpha^{-2(k-1)}}{r} \left(\frac{1 - \alpha^{2k}}{1 - \alpha^2} \right);$$

so

$$l_k = \alpha^{-2k} l_0 + \frac{1}{r} \left(\frac{1 - \alpha^{-2k}}{1 - \alpha^{-2}} \right).$$
(7)

In terms of p_k^α,

$$p_k^\alpha = \left[\alpha^{-2k} p_0^{-1} + \frac{1}{r} \left(\frac{1 - \alpha^{-2k}}{1 - \alpha^{-2}} \right) \right]^{-1},$$
(8)

or

$$p_k^\alpha = \frac{p_0 \alpha^{2k}}{1 + (p_0/r) \alpha^{2k} [(1 - \alpha^{-2k})/(1 - \alpha^{-2})]}.$$
(9)

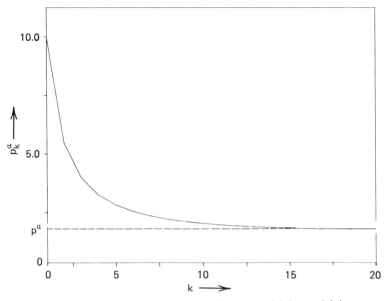

FIGURE 4.1-8   Covariance matrix with exponential data weighting.

This can be profitably compared to (3) in Example 4.1-1. Figure 4.1-8 shows p_k^α for $p_0 = 10$, $r = 10$, and $\alpha = 1.1$.

The steady-state value of p_k^α is easily seen by (8) to be

$$p^\alpha = r\left(1 - \frac{1}{\alpha^2}\right), \tag{10}$$

so that the steady-state Kalman gain is

$$K = \frac{p^\alpha}{p^\alpha + r/\alpha^2} = \left(1 - \frac{1}{\alpha^2}\right). \tag{11}$$

This is nonzero if $\alpha > 1$, which shows that if exponential data weighting is used, data are not discounted as k becomes large. ∎

Exercise 4.1-8:   *Expression for Covariance with Exponential Data Weighting*

a.   Suppose that in (4.1-1), $A = I$, $G = 0$, and H is time varying so that $z_k = H_k x_k + v_k$. Then show that the solution to (4.1-47) can be written explicitly as

$$(P_k^\alpha)^{-1} = \alpha^{-2k}(P_0)^{-1} + \sum_{i=0}^{k-1} H_{i+1}^T R^{-1} H_{i+1} \alpha^{-2(k-i-1)}. \tag{4.1-56}$$

b.   Suppose that H_k does not decrease with k and that $\alpha = 1$, corresponding to no exponential data weighting. Show that in this case P_k^α approaches a limiting value of zero for all P_0 as k increases.

c.   Suppose that $H_k = H \neq 0$ is constant and $\alpha > 1$. Then show that P_k^α approaches a bounded but nonzero positive semidefinite limit which is independent of P_0. ∎

4.2 REDUCED-ORDER FILTERS AND DECOUPLING

The dominant factor in determining the computational complexity of the Kalman filter is the dimension n of the model state vector. The number of computations per iteration is on the order of n^3. Any reduction in the number of states will therefore pay off directly in terms of increased computation time. In this section we discuss methods of reducing the dimension of the state vector in the Kalman filter.

Decoupling and Parallel Processing

If the model is block diagonal of the form

$$\begin{bmatrix} \dot{x}_1 \\ \dot{x}_2 \end{bmatrix} = \begin{bmatrix} A_1 & 0 \\ 0 & A_2 \end{bmatrix} \begin{bmatrix} x_1 \\ x_2 \end{bmatrix} + \begin{bmatrix} w_1 \\ w_2 \end{bmatrix} \tag{4.2-1a}$$

$$\begin{bmatrix} z_1 \\ z_2 \end{bmatrix} = \begin{bmatrix} H_1 & 0 \\ 0 & H_2 \end{bmatrix} \begin{bmatrix} x_1 \\ x_2 \end{bmatrix} + \begin{bmatrix} v_1 \\ v_2 \end{bmatrix}, \tag{4.2-1b}$$

$w_1 \sim (0, Q_1)$, $w_2 \sim (0, Q_2)$, $v_1 \sim (0, R_1)$, $v_2 \sim (0, R_2)$, $P(0) = \text{diag}[P_1(0), P_2(0)]$, where $x_1 \in R^{n_1}$ and $x_2 \in R^{n_2}$, then the filtering problem decouples into two separate and unrelated problems. A Kalman filter can be designed for each subsystem $(A_i, G_i, H_i, Q_i, R_i)$. Each filter has the form

$$\dot{\hat{x}}_i = A_i \hat{x}_i + K_i(z_i - H_i \hat{x}_i), \tag{4.2-2}$$

and they can be implemented in a parallel fashion on separate microprocessors.

We have seen an example of this parallel filter implementation in the α-β radar tracker of Example 2.4-4. In that case, the range and azimuth filters are independent.

The effect of this decoupling into independent filters is to reduce the number of operations from the order of $(n_1 + n_2)^3$ to the order of $(n_1^3 + n_2^3)$. If $n = 8$, for example, and if we have two 4×4 subsystems, then the number of operations is reduced from about 512 to about 128.

If model matrices A, H, Q, and R are block diagonal, then using decoupled filters, each of which is optimal for its subsystem, results in an *optimal* overall filter. (See the problems.) If A is not block diagonal it may still be desirable in some instances to implement suboptimal decoupled filters by ignoring the off-diagonal terms.

Example 4.2-1: Suboptimal Filters for Weakly Coupled Systems

Let the second-order plant be

$$\begin{bmatrix} \dot{x}_1 \\ \dot{x}_2 \end{bmatrix} = \begin{bmatrix} a_1 & \varepsilon \\ 0 & a_2 \end{bmatrix} \begin{bmatrix} x_1 \\ x_2 \end{bmatrix} + \begin{bmatrix} w_1 \\ w_2 \end{bmatrix} \tag{1}$$

$$\begin{bmatrix} z_1 \\ z_2 \end{bmatrix} = \begin{bmatrix} 1 & 0 \\ 0 & 1 \end{bmatrix} \begin{bmatrix} x_1 \\ x_2 \end{bmatrix} + \begin{bmatrix} v_1 \\ v_2 \end{bmatrix}, \tag{2}$$

with $w_i(t) \sim (0, q_i)$, $v_i(t) \sim (0, r_i)$, $x_i(0) \sim \overline{(x_i(0), 0)}$ uncorrelated (Gelb 1974).

a. Optimal Filter

To find the optimal second-order filter, we must solve the second-order Riccati equation. If we let

$$P = \begin{bmatrix} p_1 & p_{12} \\ p_{12} & p_2 \end{bmatrix}, \tag{3}$$

then the Riccati equation can be written as the three scalar equations

$$\dot{p}_1 = -\frac{p_1^2}{r_1} - \frac{p_{12}^2}{r_2} + 2a_1 p_1 + 2\varepsilon p_{12} + q_1 \tag{4a}$$

$$\dot{p}_{12} = -\frac{p_1 p_{12}}{r_1} - \frac{p_{12} p_2}{r_2} + (a_1 + a_2) p_{12} + \varepsilon p_2 \tag{4b}$$

$$\dot{p}_2 = -\frac{p_{12}^2}{r_1} - \frac{p_2^2}{r_2} + 2a_2 p_2 + q_2. \tag{4c}$$

The Kalman gain is then $PH^T R^{-1}$ or

$$K = \begin{bmatrix} p_1/r_1 & p_{12}/r_2 \\ p_{12}/r_1 & p_2/r_2 \end{bmatrix}, \tag{5}$$

and the (second-order) optimal filter is

$$\begin{bmatrix} \dot{\hat{x}}_1 \\ \dot{\hat{x}}_2 \end{bmatrix} = \begin{bmatrix} a_1 & \varepsilon \\ 0 & a_2 \end{bmatrix} \begin{bmatrix} \hat{x}_1 \\ \hat{x}_2 \end{bmatrix} + \begin{bmatrix} p_1/r_1 & p_{12}/r_2 \\ p_{12}/r_1 & p_2/r_2 \end{bmatrix} \begin{bmatrix} z_1 - \hat{x}_1 \\ z_2 - \hat{x}_2 \end{bmatrix}. \tag{6}$$

b. Optimal Filter for $\varepsilon = 0$

If $\varepsilon = 0$, then (1) decouples into two separate systems. Writing a Riccati equation for each of these first-order systems we get

$$\dot{p}_1 = -\frac{p_1^2}{r_1} + 2a_1 p_1 + q_1 \tag{7a}$$

$$\dot{p}_2 = -\frac{p_2^2}{r_2} + 2a_2 p_2 + q_2. \tag{7b}$$

The Kalman gains are

$$K_1 = \frac{p_1}{r_1} \tag{8a}$$

$$K_2 = \frac{p_2}{r_2}, \tag{8b}$$

and the optimal filter in case $\varepsilon = 0$ is composed of the two separate filters

$$\dot{\hat{x}}_1 = a_1 \hat{x}_1 + \frac{p_1}{r_1}(z_1 - \hat{x}_1) \tag{9a}$$

$$\dot{\hat{x}}_2 = a_2 \hat{x}_2 + \frac{p_2}{r_2}(z_2 - \hat{x}_2). \tag{9b}$$

Note that, since $p_{12}(0) = 0$, if $\varepsilon \to 0$, then the filter equations (4) and the filter (6) tend to (7) and (9).

c. Suboptimal Filter for $\varepsilon \neq 0$

The filter in a is optimal for any ε. If $\varepsilon = 0$, the filter in b is optimal.

If ε is small but nonzero, it is worth asking whether a state estimate based on the simplified filter (7), (9) would be accurate enough for our purposes. To answer this question, the optimal error covariances (4) must first be computed.

A subroutine $f(t, P, \dot{P})$ to implement (4), for use with the driver program TRESP in Appendix B, is shown in Fig. 4.2-1. An integration period of 20 msec was used to obtain Fig. 4.2-2, which shows the results for several values of ε. The steady-state variances are indicated.

Now suppose that we use the decoupled filter (7), (9) to estimate the plant state in (1). This is suboptimal when $\varepsilon \neq 0$. For the sake of discussion, let $\varepsilon = 0.7$ so that the curves in Fig. 4.2-2c represent the *optimal* error variances, which are obtained by using the filter (4), (6). The curves for $\varepsilon = 0$ in the figure are the *predicted* error variances using the suboptimal filter (7), (9). It is clear that we do not yet have enough information to judge the performance of our suboptimal filter on the plant (1) when $\varepsilon = 0.7$. If $\hat{x} = [\hat{x}_1 \quad \hat{x}_2]^T$ is the estimate provided by the suboptimal filter (7), (9) and $x'(t)$ is the state of the plant (1), then the *actual* error is

$$\tilde{x}(t) = x'(t) - \hat{x}(t). \tag{10}$$

The *actual* error correlation $S(t) = E[\tilde{x}(t)\tilde{x}^T(t)]$ must be examined to determine the adequacy of our suboptimal filter. Note that we must deal with *three* error covariances to do a complete performance analysis.

According to Exercise 4.1-3, $S(t)$ is found as follows. The model mismatch matrices are

$$\Delta A = \begin{bmatrix} a_1 & \varepsilon \\ 0 & a_2 \end{bmatrix} - \begin{bmatrix} a_1 & 0 \\ 0 & a_2 \end{bmatrix} = \begin{bmatrix} 0 & \varepsilon \\ 0 & 0 \end{bmatrix} \tag{11}$$

and

$$\Delta H = I - I = 0.$$

```
SUBROUTINE F(TIME,P,PP)
REAL P(*), PP(*)
DATA A1,A2,Q1,Q2,R1,R2/-2.,-1.,1.,1.,1.,1./
DATA EPS/3./

PP(1)= -P(1)**2/R1 - P(2)**2/R2 + 2.*A1*P(1) + 2.*EPS*P(2) + Q1
PP(2)= -P(1)*P(2)/R1 - P(2)*P(3)/R2 + (A1+A2)*P(2) + EPS*P(3)
PP(3)= -P(2)**2/R1 - P(3)**2/R2 + 2.*A2*P(3) + Q2

RETURN
END
```

FIGURE 4.2-1 Subroutine for Runge–Kutta to find optimal error covariance.

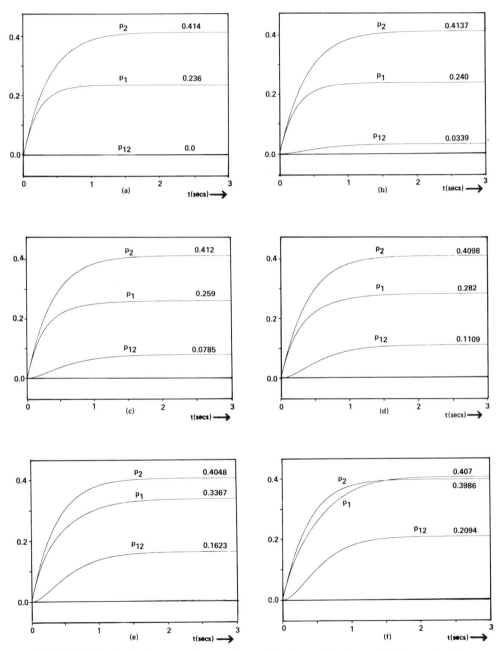

FIGURE 4.2-2 Optimal error variances. (a) $\varepsilon = 0.0$. (b) $\varepsilon = 0.3$. (c) $\varepsilon = 0.7$. (d) $\varepsilon = 1.0$. (e) $\varepsilon = 1.5$. (f) $\varepsilon = 2.0$.

```
C   SUBROUTINE TO IMPLEMENT ACTUAL ERROR COVARIANCE

        SUBROUTINE F(TIME,S,SP)
        REAL S(*), SP(*)
        DATA A1,A2,Q1,Q2,R1,R2/-2.,-1.,1.,1.,1.,1./
        DATA EPS/1.5/


C   SUBOPTIMAL (PREDICTED) ERROR COVARIANCE

        SP(1)= -S(1)**2/R1 + 2.*A1*S(1) + Q1
        SP(2)= -S(2)**2/R2 + 2.*A2*S(2) + Q2


C   SUBOPTIMAL GAINS

        AK1= S(1)/R1
        AK2= S(2)/R2


C   PLANT STATE CORRELATIONS

        SP(3)= 2.*A1*S(3) + 2.*EPS*S(4) + Q1
        SP(4)= (A1+A2)*S(4) + EPS*S(5)
        SP(5)= 2.*A2*S(5) + Q2


C   ACTUAL ERROR/PLANT STATE CROSSCORRELATIONS

        SP(6)= (2.*A1-AK1)*S(7) + EPS*S(4)
        SP(7)= (A1+A2-AK1)*S(7) + EPS*S(5)
        SP(8)= (A1+A2-AK2)*S(8) + EPS*S(9)
        SP(9)= (2.*A2-AK2)*S(9) + Q2


C   ACTUAL ERROR CORRELATIONS

        SP(10)= 2.*(A1-AK1)*S(10) + R1*AK1**2 + 2.*EPS*S(7) + Q1
        SP(11)= (A1+A2-AK1-AK2)*S(11) + EPS*S(9)
        SP(12)= 2.*(A2-AK2)*S(12) + R2*AK2**2 + Q2

        RETURN
        END
```

FIGURE 4.2-3 Subroutine to find actual error covariance.

$S(t)$ is found by using (4.1-36). Let

$$U = \begin{bmatrix} U_1 & U_{12} \\ U_{12} & U_2 \end{bmatrix}, \qquad V = \begin{bmatrix} V_1 & V_{12} \\ V_{21} & V_2 \end{bmatrix}, \qquad S = \begin{bmatrix} s_1 & s_{12} \\ s_{12} & s_2 \end{bmatrix}. \tag{12}$$

[Note that correlations $U(t)$ and $S(t)$ are symmetric while cross-correlation $V(t)$ is not.] Then we obtain the scalar equations

$$\dot{U}_1 = 2a_1 U_1 + 2\varepsilon U_{12} + q_1 \tag{13a}$$

$$\dot{U}_{12} = (a_1 + a_2) U_{12} + \varepsilon U_2 \tag{13b}$$

$$\dot{U}_2 = 2 a_2 U_2 + q_2 \tag{13c}$$

$$\dot{V}_1 = (2 a_1 - K_1) V_{12} + \varepsilon U_{12} \tag{14a}$$

$$\dot{V}_{12} = (a_1 + a_2 - K_1) V_{12} + \varepsilon U_2 \tag{14b}$$

$$\dot{V}_{21} = (a_1 + a_2 - K_2) V_{21} + \varepsilon V_2 \tag{14c}$$

$$\dot{V}_2 = (2 a_2 - K_2) V_2 + q_2 \tag{14d}$$

$$\dot{s}_1 = 2(a_1 - K_1) s_1 + r_1 K_1^2 + 2 \varepsilon V_{12} + q_1 \tag{15a}$$

$$\dot{s}_{12} = (a_1 + a_2 - K_1 - K_2) s_{12} + \varepsilon V_2 \tag{15b}$$

$$\dot{s}_2 = 2(a_2 - K_2) s_2 + r_2 K_2^2 + q_2. \tag{15c}$$

In these equations, K_1 and K_2 are the gains actually used to find $\hat{x}$. In this case they are the suboptimal values in (8). Note that if $\varepsilon = 0$ then (15) reduces to (4), which is equivalent to (7) since $p_{12}(0) = 0$.

These equations are rather tedious, but it is very easy to integrate them by using a Runge–Kutta integrator (i.e., TRESP in Appendix B) to call a subroutine that implements them. This subroutine is shown in Fig. 4.2-3. Note that we have been

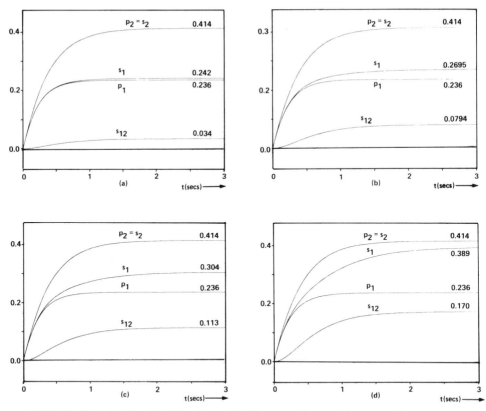

FIGURE 4.2-4   Predicted $[p(t)]$ and actual $[s(t)]$ error variances. (a) $\varepsilon = 0.3$. (b) $\varepsilon = 0.7$. (c) $\varepsilon = 1.0$. (d) $\varepsilon = 1.5$.

forced to include (7) in order to find the suboptimal gains K_1 and K_2. Figure 4.2-4 shows the predicted error variances $P(t)$ from (7) and the actual error variances $S(t)$ from (15) for several values of ε. The actual variances in this figure should be compared with the optimal variances in Fig. 4.2-2 to decide for a given ε if the suboptimal filter has a satisfactory performance.

Note that (7b) and (15c) with gains as in (8) are identical. Thus the predicted and actual error variances for the second plant state are the same. They are independent of ε, and they are both greater than the optimal error variance (4c) which decreases with ε.

The predicted error variance (7a) and actual error correlation (15a) for the first plant state are not the same. As ε increases, the actual correlation becomes significantly larger than the predicted variance. The actual variance is again larger than the optimal error variance given by (4a), though our prediction says it should be smaller. ∎

Reduced-Order Filters

We discuss here two topics: elimination of a portion of the state vector by substitution, and ignoring of a portion of the state vector.

Suppose first that the system is of the form

$$\begin{bmatrix} x_{k+1} \\ \xi_{k+1} \end{bmatrix} = \begin{bmatrix} A_1 & 0 \\ A_{21} & A_2 \end{bmatrix} \begin{bmatrix} x_k \\ \xi_k \end{bmatrix} + \begin{bmatrix} w_k \\ \eta_k \end{bmatrix} \tag{4.2-3a}$$

$$z_k = \begin{bmatrix} H_1 & H_2 \end{bmatrix} \begin{bmatrix} x_k \\ \xi_k \end{bmatrix}, \tag{4.2-3b}$$

with $w_k \sim (0, Q_1)$, $\eta_k \sim (0, Q_2)$ white and uncorrelated with each other and with $x_0 \sim (\bar{x}_0, P_0)$. An estimate of $x_k \in R^{n_1}$ is desired, but the value of $\xi_k \in R^{n_2}$ is unimportant to us and an estimate for this portion of the state is not required. Suppose that H_2 has full column rank so that a left inverse H_2^+ exists,

$$H_2^+ H_2 = I. \tag{4.2-4}$$

[Note that $H_2^+ = (H_2^T H_2)^{-1} H_2^T$.]

We can reduce this from an $(n_1 + n_2)$-order problem to an n_1-order problem by substituting for ξ_k as follows (Simon and Stubberud 1969).

Define the *derived measurement* (cf., Section 2.7) as

$$z'_{k+1} = z_{k+1} - H_2 A_2 H_2^+ z_k. \tag{4.2-5}$$

Then we have

$$z'_{k+1} = H_1(A_1 x_k + w_k) + H_2(A_{21} x_k + A_2 \xi_k + \eta_k)$$
$$- H_2 A_2 H_2^+ H_1 x_k - H_2 A_2 \xi_k.$$

This becomes

$$z'_{k+1} = H' x_k + v_k, \tag{4.2-6}$$

where the new measurement matrix is

$$H' = H_1 A_1 + H_2 A_{21} - H_2 A_2 H_2^+ H_1 \qquad (4.2\text{-}7)$$

and the new measurement noise is

$$v_k = H_1 w_k + H_2 \eta_k. \qquad (4.2\text{-}8)$$

The new measurement noise is white noise with zero mean and a covariance of

$$R \triangleq \overline{v_k v_k^T} = H_1 Q_1 H_1^T + H_2 Q_2 H_2^T. \qquad (4.2\text{-}9)$$

Now the system model can be reduced to

$$x_{k+1} = A_1 x_k + w_k \qquad (4.2\text{-}10a)$$

$$z'_{k+1} = H' x_k + v_k, \qquad (4.2\text{-}10b)$$

with $w_k \sim (0, Q_1)$, $v_k \sim (0, R)$, $x_0 \sim (\bar{x}_0, P_0)$. The noises w_k and v_k are correlated with cross-correlation

$$\overline{v_j w_k^T} = \overline{(H_1 w_j + H_2 \eta_j) w_k^T} = H_1 Q \delta_{jk}, \qquad (4.2\text{-}11)$$

so that for optimality the modified Kalman gain of Section 2.7 should be used.

A Kalman filter of order n_1 designed using this reduced-order model will provide the *optimal* estimate $\hat{x}_k$ for the $(n_1 + n_2)$-order plant (4.2-3).

Exercise 4.2-2: Kinematic Modeling for Continuous Systems

As a variation on the above procedure, consider the plant

$$\dot{x} = A_1 x + A_{12} \xi + w \qquad (4.2\text{-}12a)$$

$$\dot{\xi} = f(x, \xi, u, \eta) \qquad (4.2\text{-}12b)$$

where $w(t) \sim (0, Q_1)$, $\eta(t) \sim (0, Q_2)$ are white noise processes uncorrelated with each other and with the initial states. The function $f(\cdot)$ is in general nonlinear and $u(t)$ is a deterministic input. See Kortüm (1976).

In the context of mechanical systems, (4.2-12a) is the set of linearized kinematic equations describing the behavior of the angles and angular velocities. Equation (4.2-12b) represents the nonlinear dynamic equations of motion, the Euler equations.

Suppose measurements are available of the form

$$z_1 = H_1 x + v_1 \qquad (4.2\text{-}13a)$$

$$z_2 = H_2 \xi + v_2, \qquad (4.2\text{-}13b)$$

where $v_1(t) \sim (0, R_1)$ and $v_2(t) \sim (0, R_2)$ are white noises uncorrelated with each other and with $w(t)$, $\eta(t)$, and the initial states. Assume that H_2 has a left inverse H_2^+. This corresponds to having good information about the dynamic state components ξ.

We desire an estimate of $x(t)$ but not of $\xi(t)$.

a.   Eliminate $\xi(t)$ to show that

$$\dot{x} = A_1 x + A_{12} H_2^+ z_2 + (w - A_{12} H_2^+ v_2),\qquad (4.2\text{-}14)$$

where $A_{12} H_2^+ z_2(t)$ can be treated as a known deterministic input.

b.   Find the spectral density matrix of the new process noise

$$w'(t) = w(t) - A_{12} H_2^+ v_2(t).\qquad (4.2\text{-}15)$$

c.   A Kalman filter can now be applied to the reduced model (4.2-14), (4.2-13a). Write down the Riccati equation, the Kalman gain, and the filter equation.

d.   Why is this filter suboptimal? (*Note:* This procedure is called "kinematic modeling." The dynamics (4.3-12b) can even be unknown.)   ∎

Let us now examine the effects of simply ignoring some of the states of a system. To do this we shall use the results of Exercises 4.1-2 and 4.1-4.

Let the actual plant be given by

$$x'_{k+1} = \bar{A} x'_k + \bar{G} w'_k \qquad (4.2\text{-}16a)$$

$$z'_k = \bar{H} x'_k + v'_k, \qquad (4.2\text{-}16b)$$

with $w'_k \sim (0, \bar{Q})$, $v'_k \sim (0, \bar{R})$ white and uncorrelated with each other and with $x'_0 \sim [E(x'_0), \bar{P}_0]$. Let the plant state x'_k be an n_p vector. (The following results can easily be modified to incorporate a deterministic input.)

To design a filter for this plant, suppose that we assume the model

$$x_{k+1} = A x_k + G w_k \qquad (4.2\text{-}17a)$$

$$z_k = H x_k + v_k, \qquad (4.2\text{-}17b)$$

with $w_k \sim (0, Q)$, $v_k \sim (0, R)$ white and uncorrelated with each other and with $x_0 \sim [E(x_0), P_0]$. Let the model state x_k be an n_m vector.

The object here is to obtain filters of reduced order by assuming a model of order n_m less than n_p, the order of the plant. Let us simplify things by supposing that to find a reduced-order model we just select certain components of the plant state x'_k to incorporate in (4.2-17), discarding the remaining components. Then we have

$$x_k = T x'_k, \qquad (4.2\text{-}18)$$

with the state transformation given without loss of generality by

$$T = [I_{n_m} \quad 0]. \qquad (4.2\text{-}19)$$

A right inverse of T is given by

$$T^+ = T^T = \begin{bmatrix} I_{n_m} \\ 0 \end{bmatrix}. \qquad (4.2\text{-}20)$$

In this case, the estimate $\hat{x}_k$ provided by a filter based on the model will be a (suboptimal) estimate for the portion of the plant state retained in the model.

According to Exercise 4.1-2 the correlation matrix S_k of the actual error

$$\tilde{x}_k = x'_k - T^T \hat{x}_k \qquad (4.2\text{-}21)$$

is found as follows. Let U_k be the correlation of plant state vector x'_k, and V_k the cross-correlation of $\tilde{x}_k$ with x'_k. Define the model mismatch matrices as

$$\Delta A = \bar{A} - T^T A T \qquad (4.2\text{-}21a)$$

$$\Delta H = \bar{H} - HT. \qquad (4.2\text{-}21b)$$

Then, to find S_k perform the time update

$$U_{k+1} = \bar{A} U_k \bar{A}^T + \bar{G} \bar{Q} \bar{G}^T \qquad (4.2\text{-}22a)$$

$$V^-_{k+1} = \Delta A U_k \bar{A}^T + T^T A T V_k \bar{A}^T + \bar{G} \bar{Q} \bar{G}^T \qquad (4.2\text{-}22b)$$

$$S^-_{k+1} = T^T A T S_k T^T A^T T + \bar{G} \bar{Q} \bar{G}^T + \Delta A U_k \Delta A^T$$
$$+ (T^T A T V_k \Delta A^T + \Delta A V_k^T T^T A^T T), \qquad (4.2\text{-}22c)$$

followed by the measurement update

$$U_k = U_k \qquad (4.2\text{-}23a)$$

$$V_k = (I - T^T K_k HT) V^-_k - T^T K_k \Delta H U_k \qquad (4.2\text{-}23b)$$

$$S_k = (I - T^T K_k HT) S^-_k (I - T^T K_k HT)^T + T^T K_k (\bar{R} + \Delta H U_k \Delta H^T) K_k^T T$$
$$- (I - T^T K_k HT) V^-_k \Delta H^T K_k^T T + T^T K_k \Delta H (V^-_k)^T (I - T^T K_k HT)^T \qquad (4.2\text{-}23c)$$

(Note that these equations are not as messy as they seem, since the multiplications by T and T^T simply correspond to "padding" a matrix with zero blocks!)

The dynamics of the a priori actual error ($\tilde{x}^-_k = x'_k - T^T \hat{x}^-_k$) are described by the error system

$$\tilde{x}^-_{k+1} = T^T A (I - K_k H) T \tilde{x}^-_k + (\Delta A - T^T A K_k \Delta H) x'_k + \bar{G} w'_k - T^T A K_k v'_k. \qquad (4.2\text{-}24)$$

The continuous counterparts to these equations are given in Exercise 4.1-4.

To carry out a *performance analysis* of the effects of disregarding the bottom portion of the plant state vector, it is only necessary to proceed as follows:

1. Compute the optimal error covariances using the Riccati equation for the full plant (4.2-16).
2. Design a Kalman filter for the reduced model (4.2-17).
3. Use the filter gain K_k found in step 2 in equations (4.2-22), (4.2-23) to compute the actual error covariance S_k resulting when the reduced filter in step 2 is applied to the full plant.

4. If S_k is unacceptably greater than the optimal covariance from step 1, select a different set of plant states to retain in the model and return to step 2.

Basically, we should retain in the model all the states we are directly interested in estimating and all states strongly coupled with them, and all unstable states.

An illustration of an analysis for a reduced-order filter is given in Example 4.1-5. In practice we would not find error covariances analytically, as we did in that example. Instead we would implement the equations for S_k or $S(t)$ in a subroutine and make some plots for comparison with the optimal error covariance.

A special case of reduced-order modeling occurs when the plant has the form

$$\begin{bmatrix} \dot{x}' \\ \dot{\xi}' \end{bmatrix} = \begin{bmatrix} A_1 & A_{12} \\ 0 & A_2 \end{bmatrix} \begin{bmatrix} x' \\ \xi' \end{bmatrix} + \begin{bmatrix} w' \\ \eta' \end{bmatrix} \qquad (4.2\text{-}25a)$$

$$z' = \bar{H}x' + v', \qquad (4.2\text{-}25b)$$

with $w'(t) \sim (0, Q_1)$, $\eta'(t) \sim (0, Q_2)$, and $v'(t) \sim (0, R)$ white and uncorrelated with each other and with the initial states. An estimate of $x'(t)$ is required, but we are not interested in $\xi'(t)$.

This system could arise, for example, when the subsystem of $\xi'(t)$ is a shaping filter for a portion of the process noise (Sections 2.7 and 3.6). See Huddle (1967), Huddle and Wismer (1968), and Darmon (1976).

The optimal Kalman gain is of the form

$$K(t) = \begin{bmatrix} K_1(t) \\ K_2(t) \end{bmatrix}. \qquad (4.2\text{-}26)$$

To design a reduced-order filter, assume the reduced-order model

$$\dot{x} = A_1 x + w \qquad (4.2\text{-}27a)$$

$$z = Hx + v, \qquad (4.2\text{-}27b)$$

with $w(t) \sim (0, Q_1)$, $v(t) \sim (0, R)$ white and uncorrelated. [In the case where $\xi'(t)$ is the state of a shaping filter, this corresponds to assuming the process noise is white when it is not.] Then $x(t)$ is related to the plant state by the transformation (4.2-19).

Discounting the lower portion of the plant state is equivalent to restricting the filter gain to the form

$$K(t) = \begin{bmatrix} K_1(t) \\ 0 \end{bmatrix}, \qquad (4.2\text{-}28)$$

which results in a suboptimal but simplified filter. Such a design has been called a *Kalman–Schmidt* filter (Schmidt 1966).

We could analyze the actual error by using the results of Exercise 4.1-4. It is quite enlightening, however, to use instead another approach which gives the same answers in a simplified form.

Let the error covariance for the plant $\bar{P}$ be partitioned conformably with $x'(t)$, $\xi'(t)$ as

$$\bar{P} = \begin{bmatrix} P_1 & P_{12} \\ P_{12}^T & P_2 \end{bmatrix}. \tag{4.2-29}$$

Applying a filter with gain of $K(t)$ to the full plant (4.2-25), the error covariance satisfies the full Riccati equation

$$\dot{\bar{P}} = (\bar{A} - K\bar{H})\bar{P} + \bar{P}(\bar{A} - K\bar{H})^T + \bar{Q} + K\bar{R}K^T. \tag{4.2-30}$$

Let us now restrict the gain $K(t)$ to have the suboptimal form in (4.2-28). Then by using the partitioning in (4.2-25) and (4.2-29), and the fact that

$$\bar{H} = [H \quad 0], \tag{4.2-31}$$

we can write (4.2-30) as

$$\dot{P}_1 = (A_1 - K_1 H)P_1 + P_1(A_1 - K_1 H)^T + A_{12}P_{12}^T + P_{12}A_{12}^T$$
$$+ K_1 R K_1^T + Q_1 \tag{4.2-32a}$$

$$\dot{P}_{12} = (A_1 - K_1 H)P_{12} + A_{12}P_2 + P_{12}A_2^T \tag{4.2-32b}$$

$$\dot{P}_2 = A_2 P_2 + P_2 A_2^T + Q_2. \tag{4.2-32c}$$

The interpretation of these *Kalman–Schmidt* equations is as follows. $P_1(t)$ is the actual error covariance for the estimate of $x'(t)$ using the suboptimal gain (4.2-28). $P_2(t)$ is the actual error covariance for the estimate of $\xi'(t)$. Since no correction is applied to this estimate based on the measurements, it is the same as the covariance of $\xi'(t)$ itself (see Section 1.1). $P_{12}(t)$ is the cross-covariance between the estimation error in $x'(t)$ and the state portion $\xi'(t)$.

The predicted error covariance based on the model (4.2-27) only is given by

$$\dot{P} = (A_1 - K_1 H)P + P(A_1 - K_1 H)^T + Q_1 + K_1 R K_1^T. \tag{4.2-33}$$

Define the degradation in performance incurred due to using the suboptimal gain as

$$E = P_1 - P. \tag{4.2-34}$$

Then we have, by subtracting (4.2-33) from (4.2-32a)

$$\dot{E} = (A_1 - K_1 H)E + E(A_1 - K_1 H)^T + A_{12}P_{12}^T + P_{12}A_{12}^T. \tag{4.2-35}$$

This is the error increase induced by the unmodeled states $\xi'(t)$. In some cases this may be small enough to warrant using the reduced-order filter implied by (4.2-28).

The gain $K_1(t)$ is usually selected to be the optimal (Kalman) gain based on the reduced-order model,

$$K_1 = PH^T R^{-1}. \tag{4.2-36}$$

Exercise 4.2-3:   Alternative Derivation of Kalman–Schmidt Error Covariance Equations

The equations (4.2-32) can also be derived using Exercise 4.1-4.

a.   Show that the model mismatch matrices are

$$\Delta A = \begin{bmatrix} 0 & A_{12} \\ 0 & A_2 \end{bmatrix}, \qquad \Delta H = 0. \tag{1}$$

b.   Write down the equations of Exercise 4.1-4 in terms of the sub blocks of the partitioned plant matrices, also partitioning the correlation matrices as

$$U = \begin{vmatrix} U_1 & U_{12} \\ U_{12}^T & U_2 \end{vmatrix}, \qquad V = \begin{vmatrix} V_1 & V_{12} \\ V_{21} & V_2 \end{vmatrix}, \qquad S = \begin{vmatrix} S_1 & S_{12} \\ S_{12}^T & S_2 \end{vmatrix}. \tag{2}$$

c.   Show that, for the correct choice of initial conditions,

$$S_{12} = V_{12} \tag{3}$$

$$S_2 = U_2 = V_2, \tag{4}$$

so that the simplified equations (4.2-32) are recovered.   ∎

4.3   USING SUBOPTIMAL GAINS

The Kalman filter is time varying even when the system is time invariant because the gain K depends on time. In the discrete filter this means that K_k must either be precomputed and stored as a sequence, or that the Riccati equation must be solved for each k during the measurement run. In the continuous filter implemented on an analog computer, it means that continuously variable gains must be used. In the continuous filter implemented on a digital computer, it means that a time-dependent gain must be used in the filter routine that is called by the integration algorithm.

In some applications it is desirable to approximate the optimal Kalman gain by a suboptimal gain that is easier to implement. One common choice is to use the steady-state gain K_∞; for this gain the Kalman filter is just the Wiener filter.

To examine the effects of using suboptimal gains, let us summarize some results from Sections 2.5 and 3.4.

The discrete Kalman filter is given by

$$\hat{x}_{k+1}^- = A(I - K_k H)\hat{x}_k^- + Bu_k + AK_k z_k. \tag{4.3-1}$$

The error system is

$$\tilde{x}_{k+1}^- = A(I - K_k H)\tilde{x}_k^- + Gw_k - AK_k v_k \tag{4.3-2}$$

and the associated error covariance is given by

$$P_{k+1}^- = A(I - K_kH)P_k^-(I - K_kH)^TA^T + AK_kRK_k^TA^T + GQG^T. \quad (4.3\text{-}3)$$

If the gain is selected as the optimal Kalman value

$$K_k = P_kH^T(HP_k^-H^T + R)^{-1} \quad (4.3\text{-}4)$$

then (4.3-3) reduces to the Riccati equation in Table 2.3-3. With a little bit of thought, however, it is clear that (4.3-3) is the error covariance of filter for *any* fixed gain sequence K_k. This is because, once K_k has been selected, (4.3-2) is a stochastic system with a fixed plant matrix. The error covariance, therefore, propagates according to the *Lyapunov* (i.e., not Riccati) equation (4.3-3). This equation can therefore be used to analyze the effects of selecting a suboptimal K_k.

If the steady-state Kalman gain K_∞ is used for all k, we obtain the (time-invariant) Wiener filter which is (4.3-1) with K_k replaced by K_∞. The Wiener filter has an associated error covariance of (4.3-3) with K_k replaced by K_∞.

The continuous estimate for any gain $K(t)$ is given by

$$\dot{\hat{x}} = [A - K(t)H]\hat{x} + Bu + K(t)z. \quad (4.3\text{-}5)$$

The error system is

$$\dot{\tilde{x}} = [A - K(t)H]\tilde{x} + Gw - K(t)v \quad (4.3\text{-}6)$$

with associated error covariance

$$\dot{P} = (A - KH)P + P(A - KH)^T + KRK^T + GQG^T. \quad (4.3\text{-}7)$$

If the optimal gain

$$K = PH^TR^{-1} \quad (4.3\text{-}8)$$

is used in (4.3-5), then (4.3-7) reduces to the Riccati equation in Table 3.1-1. It is important to realize, however, that (4.3-7) describes the error covariance associated with *any* gain $K(t)$ used in the filter (4.3-5). If $K(t)$ is a given gain function, then (4.3-7) is a Lyapunov equation, not a Riccati equation.

If the steady-state Kalman gain K_∞ is used, the result is the time-invariant Wiener filter, (4.3-5) with $K(t)$ replaced by K_∞. This filter has an associated error covariance of (4.3-7) with $K(t)$ replaced by K_∞.

The next example shows how 4.3-3 and 4.3-7 are used to analyze the effects of using simplified suboptimal filter gains.

Example 4.3-1: Suboptimal Filter Gains for Continuous Wiener Process

Let the system model be

$$\dot{x} = w \quad (1)$$

$$z = x + v \quad (2)$$

with $w(t) \sim (0, q)$ and $v(t) \sim (0, r)$ white and uncorrelated with $x(0) \sim (0, p_0)$.
A low-pass filter for estimating $x(t)$ from the data $z(t)$ is given by

$$\dot{\hat{x}} = K(z - \hat{x}) = -K\hat{x} + Kz. \qquad (3)$$

Let us discuss several ways to select the filter gain $K(t)$.

a. Kalman Filter

To find the optimal gain $K(t)$ we must solve the Riccati equation

$$\dot{p} = -\frac{p^2}{r} + q. \qquad (4)$$

According to Example 3.2-2d,

$$p(t) = \sqrt{qr} + \frac{\sqrt{qr}}{ce^{t/\tau} - \frac{1}{2}} \qquad (5)$$

where

$$c = \frac{1}{2}\frac{p_0 + \sqrt{qr}}{p_0 - \sqrt{qr}}, \qquad (6)$$

$$\tau = \frac{1}{2}\sqrt{\frac{r}{q}}. \qquad (7)$$

Hence $K = p/r$, or

$$K(t) = \sqrt{\frac{q}{r}}\left(1 + \frac{1}{ce^{t/\tau} - \frac{1}{2}}\right). \qquad (8)$$

This time-varying gain must be implemented to get the optimal filter. One way to do this is shown in Fig. 4.3-1.

b. Time-Invariant Filter

To simplify the implementation, let us use the Wiener filter. The steady-state covariance and gain are

$$p_\infty = \sqrt{qr} \qquad (9)$$

$$K_\infty = \sqrt{\frac{q}{r}}, \qquad (10)$$

so the Wiener filter is

$$\dot{\hat{x}} = -\sqrt{\frac{q}{r}}\hat{x} + \sqrt{\frac{q}{r}}z. \qquad (11)$$

An implementation is shown in Fig. 4.3-2.
To compare the performances of the Kalman and Wiener filters, we need to find the Wiener filter error covariance. This can be done by using (4.3-7) with $K = K_\infty$, which becomes

$$\dot{p} = -2K_\infty p + rK_\infty^2 + q \qquad (12)$$

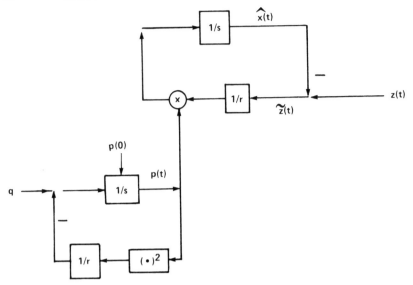

FIGURE 4.3-1   Optimal filter.

or

$$\dot{p} = -2 \sqrt{\frac{q}{r}} \, p + 2q.$$ (13)

The solution to (13) is

$$p(t) = e^{-t/\tau} p_0 + \int_0^t e^{-(t-\lambda)/\tau} 2q \, d\lambda$$

or

$$p(t) = \sqrt{qr} + (p_0 - \sqrt{qr}) e^{-t/\tau}.$$ (14)

The optimal and suboptimal error covariances (5) and (14) are sketched in Fig. 4.3-3. Note that their steady-state values are identical. By comparing the two error covariances, we can now decide if the simplified structure of the suboptimal filter justifies its degraded performance in our particular application.

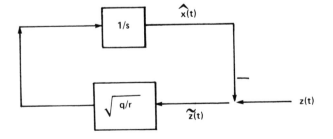

FIGURE 4.3-2   Suboptimal time-invariant filter.

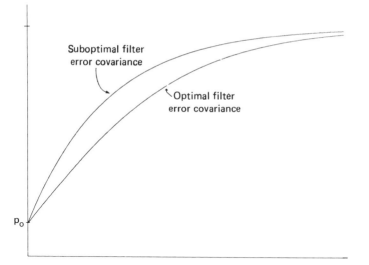

FIGURE 4.3-3 Error covariances for Kalman and Wiener filters.

The curves in Fig. 4.3-3 were obtained by writing subroutines $F(t, p, \dot{p})$ to implement (4) and (13), which were then integrated by the driver routine TRESP in Appendix B. ∎

4.4 SCALAR MEASUREMENT UPDATING

If the measurement z_k is a scalar, then the discrete Kalman filter measurement update is considerably simplified. In this case we have, from Table 2.3-2,

$$K_k = \frac{P_k^- H^T}{HP_k^- H^T + R} \tag{4.4-1a}$$

$$P_k = (I - K_k H)P_k^- \tag{4.4-1b}$$

$$\hat{x}_k = \hat{x}_k^- + K_k(z_k - H\hat{x}_k^-), \tag{4.4-1c}$$

where no matrix inversions are needed since $(HP_k^- H^T + R)$ is a scalar.

In general, $z_k \in R^p$ is not a scalar, but we can always perform the single p-vector measurement update with p sequential scalar updates by *pre-whitening* the data. This results in a dramatic reduction in computing time if p is large. (In a vector update, the number of operations depends on p^3. In sequential scalar updates, it depends on p.)

To do this, we proceed as in Section 1.4. The measurement equation is

$$z_k = Hx_k + v_k, \tag{4.4-2}$$

$v_k \sim (0, R)$. Factor R into its square roots U:

$$R = UU^T. \tag{4.4-3}$$

To save operations, it is convenient to require that U be upper triangular, which we can always do. We can find U by the *upper triangular Cholesky decomposition* in Bierman (1977). Now invert the triangular matrix U. Then a decomposition of R^{-1} is

$$R^{-1} = U^{-T}U^{-1}. \tag{4.4-4}$$

As in Section 1.4 we now define new data, measurement matrix, and noise by

$$z_k^w = U^{-1}z_k \tag{4.4-5}$$

$$H^w = U^{-1}H \tag{4.4-6}$$

$$v_k^w = U^{-1}v_k. \tag{4.4-7}$$

Then (4.4-2) becomes

$$z_k^w = H^w x_k + v_k^w. \tag{4.4-8}$$

Since $v_k^w \sim (0, I)$, the components of z_k^w are statistically uncorrelated so that z_k^w is a *white* data vector. The operation (4.4-5) corresponds to *prefiltering* the data by U^{-1} prior to using it in the Kalman filter.

Suppose the data have been prewhitened. Then according to the results in Section 1.4, at each time R we can incorporate *each of the p components of z_k separately* by a scalar update like (4.4-1). To do this, in (4.4-1) we replace R by 1, z_k by the ith component of the whitened data z_k^w, and H by the ith row of H^w. Then we cycle through (4.4-1) for $i = 1, \ldots, p$. This procedure is repeated at each time k.

If the noise covariance is time varying (R_k), then we need to do the Cholesky decomposition $R_k = U_k U_k^T$ at each time k to get $H_k^w = U_k^{-1}H$ at each time. In the time-invariant case we only need to factor R and find H^w at the beginning of the estimation run.

A stabilized [e.g., (2.3-15)] version of scalar measurement update (4.4-1) is given in Bierman (1977). The algorithm uses symmetry to minimize the number of operations.

Scalar sequential processing of the data when $R = I$ should be compared to our discussion in Section 4.2 on breaking the optimal filter into smaller "subfilters" when the plant matrices A, Q, R, and P_0 are block diagonal.

PROBLEMS

Section 4.1

4.1-1: Divergence in the Continuous Kalman Filter

a. Let the model be the random constant

$$\dot{x} = 0 \tag{1}$$

$$z = x + v, \tag{2}$$

with $v \sim (0, 1)$ white and uncorrelated with $x(0) \sim (0, 1)$. Show that the optimal estimate is given by

$$\dot{\hat{x}} = \frac{1}{t+1}(z - \hat{x}) \tag{3}$$

with predicted error variance

$$p(t) = \frac{1}{t+1}. \tag{4}$$

b. Show that the state transition matrix of (3) is

$$\phi(t, t_0) = \frac{1}{t - t_0 + 1} \tag{3}$$

(Hint: Solve $(d/dt)\phi(t, 0) = [-1/(t+1)]\phi(t, 0)$, with $\phi(0, 0) = 1$.) Hence find the estimate $\hat{x}(t)$ if the data are $z(t) = tu_{-1}(t)$.

c. Now suppose that the actual plant is the Wiener process

$$\dot{x} = w \tag{6}$$

$$z = x + v \tag{7}$$

with $w \sim (0, \varepsilon)$ and $v \sim (0, 1)$ white and uncorrelated with $x(0) \sim (0, 1)$. Show that if the filter (3) is used then the actual error variance is

$$s(t) = \frac{1}{t+1} + \frac{\varepsilon}{3}(t+1) - \frac{\varepsilon}{3(t+1)^2}, \tag{8}$$

which is unbounded for any $\varepsilon \neq 0$ (Anderson 1973).

4.1-2: Inaccurate Modeling of Newton's System as Constant Unknown. The model used to design a filter for an unknown x is

$$\dot{x} = 0$$

$$z = x + v,$$

$v(t) \sim (0, r)$. In fact, the unknown has dynamics described by

$$\dot{x} = \varepsilon x_2$$

$$\dot{x}_2 = 0$$

for a small ε.

a. Find Kalman gain and predicted error covariance.
b. Write down the actual error dynamics.
c. Use Exercise 4.1-3 to determine the actual error covariance.
d. Select another first-order model that can be used to design a filter in which divergence does not occur. For this model, repeat parts a through c.

4.1-3: Exponential Data Weighting. Use exponential data weighting to correct the filter divergence experienced in:

a. Example 4.1-5.
b. Problem 4.1-1.
c. Problem 4.1-2.

4.1-4: Filter with a Margin of Stability. Redo Example 2.5-1 if the steady-state filter must be guaranteed to have poles inside $|z| = 0.98$.

4.1-5: Filter with a Margin of Stability. Redo Problem 3.2-2 if the steady-state filter must be guaranteed to have poles to the left of $s = -0.1$.

Section 4.2

4.2-1: Decoupled Filter

a. Let the model be decoupled so that $A = \text{diag}[A_1, A_2]$, $Q = \text{diag}[Q_1, Q_2]$, $R = \text{diag}[R_1, R_2]$, $P(0) = \text{diag}[P_1(0), P_2(0)]$. Show that $P(t)$ is block diagonal for all t, and that the optimal filter decomposes into two independent subfilters.

b. Repeat a for the discrete-time filter.

4.2-2: Block Triangular Filter Structure. Suppose the system is block triangular of the form

$$\begin{bmatrix} \dot{x}_1 \\ \dot{x}_2 \end{bmatrix} = \begin{bmatrix} A_1 & A_{12} \\ 0 & A_2 \end{bmatrix} \begin{bmatrix} x_1 \\ x_2 \end{bmatrix} + \begin{bmatrix} w_1 \\ w_2 \end{bmatrix} \tag{1a}$$

$$\begin{bmatrix} z_1 \\ z_2 \end{bmatrix} = \begin{bmatrix} H_1 & H_{12} \\ 0 & H_2 \end{bmatrix} \begin{bmatrix} x_1 \\ x_2 \end{bmatrix} + \begin{bmatrix} v_1 \\ v_2 \end{bmatrix}, \tag{1b}$$

$w_1(t) \sim (0, Q_1)$, $w_2(t) \sim (0, Q_2)$, $v_1(t) \sim (0, R_1)$, $v_2(t) \sim (0, R_2)$, $x_1(0) \sim [x_1(0), P_1(0)]$, $x_2(0) \sim [x_2(0), P_2(0)]$. Then subsystem 2 is not influenced by subsystem 1, but only drives it. In this case we might try the simplified filter structure

$$\dot{\hat{x}}_2 = A_2\hat{x}_2 + K_2(z_2 - H_2\hat{x}_2) \tag{2a}$$

$$\dot{\hat{x}}_1 = A_1\hat{x}_1 + A_{12}\hat{x}_2 + K_1(z_1 - H_1\hat{x}_1 - H_{12}\hat{x}_2), \tag{2b}$$

where (2a) does not depend on (2b) but only drives it.

a. Write the Riccati equation in terms of the submatrices.
b. Is the filter (2) ever optimal?

4.2-3: Effect of Mismodeling of Process Noise. Let

$$\dot{x} = \varepsilon w \tag{1}$$

$$z = x + v, \tag{2}$$

where $x(0) \sim (0, 1)$ and $v(t) \sim (0, 1)$ is white and uncorrelated with the nonwhite process noise which has spectral density of

$$\Phi_w(\omega) = \frac{1}{\omega^2 + 1}. \tag{3}$$

a. Find a shaping filter for $w(t)$ and augment the state equation.
b. Design a filter for (1) and (2) assuming incorrectly that $w(t)$ is white $(0, 1)$. Find error covariance and optimal gain.
c. Making the same erroneous assumption as in part b, find K_∞ and use this suboptimal gain in a filter for $x(t)$. Find suboptimal error variance. Assume $\varepsilon = 0.1$ if simplification is desired.
d. Now we use this K_∞ on the actual overall plant found in part a. Solve (4.2-32) for the actual error variance $p_1(t)$.
e. Write the optimal Kalman filter equations for the full plant in part a. Find optimal steady-state error covariance, and compare to the steady-state value of $p_1(t)$ from part d.

Section 4.3

4.3-1: Suboptimal Filtering. In Problem 2.3-6:
a. Find the steady-state Kalman gain K_∞ if $r_k = r$ is constant.
b. Redo Problem 2.3-6 part a, determining for $k = 0$ through 5 the estimate and (suboptimal) error covariance if K_∞ is used for the data processing instead of the optimal K_k.

4.3-2: Suboptimal Filtering. Let

$$\dot{x} = 0.1x + w,$$

$$z = x + v,$$

with $w(t) \sim (0, 2)$, $v(t) \sim (0, 1)$ uncorrelated white noise processes and $x(0) \sim (0, 1)$.
a. Find optimal error covariance and Kalman gain.
b. Find steady-state Kalman gain K_∞. Now use a suboptimal filter

$$\dot{\hat{x}} = (A - KH)\hat{x} + Kz \tag{1}$$

with $K = K_\infty$. Find the resulting error covariance.
c. Now a constant feedback gain K is selected in (1) to make the filter stable with a pole at $s = -1$. Find this gain, and compute the error covariance resulting when this suboptimal filter is used on the data.
d. Plot your three error covariances on the same graph.

4.3-3: Suboptimal Filtering. For Newton's system in Example 3.3-2, two suboptimal filters are proposed.
a. A filter is built using K_∞. Find the resulting suboptimal error covariance.

b. Use Ackermann's formula to find a filter gain that places both the filter poles at $s = -\sqrt{\gamma}$. Compute the error covariance of this suboptimal filter.

Section 4.4

4.4-1: Scalar Measurement Updating

$$x_{k+1} = \begin{bmatrix} 1 & 0.5 \\ 0 & 1 \end{bmatrix} x_k + \begin{bmatrix} 0 \\ 1 \end{bmatrix} u_k + w_k$$

$$z_k = \begin{bmatrix} 1 & 1 \\ 1 & -1 \end{bmatrix} x_k + v_k$$

where

$$Q = \begin{bmatrix} 1 & -0.5 \\ -0.5 & 1 \end{bmatrix}, \qquad R = \begin{bmatrix} 2 & 1 \\ 1 & 1 \end{bmatrix}, \qquad x(0) \sim (0, I).$$

The input is $u_k = u_{-1}(k)$ and the data are

$$z_1 = \begin{bmatrix} 1 \\ -1 \end{bmatrix}, \qquad z_2 = \begin{bmatrix} 2.5 \\ 1.5 \end{bmatrix}, \qquad z_3 = \begin{bmatrix} 4.5 \\ -1.5 \end{bmatrix}.$$

a. Determine $\hat{x}_k$ and P_k for $k = 0, 1, 2, 3$.
b. Repeat using prewhitening of the data and scalar measurement updating.

5

ESTIMATION FOR NONLINEAR SYSTEMS

We have derived the optimal estimator for linear stochastic systems with Gaussian statistics. It is known as the Kalman filter, and is itself a linear system with a gain matrix that depends on the solution to a matrix quadratic equation.

The optimal estimation problem for nonlinear systems is in general very complicated, and only in a few special cases do algorithms exist which are easy to implement or understand.

In this chapter we find time and measurement updates for the entire hyperstate (i.e., the conditional probability density function of the unknown given the data). We then use these to find update equations for the estimate and error covariance which are in general not computationally realizable. It is demonstrated that these reduce to the Kalman filter in the linear Gaussian case.

To obtain a computationally viable algorithm for general nonlinear systems, we make some approximations that result in the *extended Kalman filter*.

5.1 UPDATE OF THE HYPERSTATE

In this section are derived general equations for time and measurement updates of the hyperstate $f_{x/Z}$, where x is the unknown and Z is all of the available data. Since most practical applications today use sampled time signals, we deal with the case of discrete measurements.

First discrete systems, and then continuous systems are considered.

Discrete Systems

Consider a general dynamical discrete-time system with state x_k. The system is assumed to be Markov so that the conditional PDF of x_{k+1} given the entire previous history of the system is simply f_{x_{k+1}/x_k}, the conditional PDF given the previous state. There are prescribed measurements z_k of the unknown state x_k. The measurement process has no memory and z_k depends only on x_k, so that the measurement is completely described probabilistically by the conditional PDF f_{z_k/x_k}. This situation is shown in Fig. 5.1-1, where $f_{x_k/x_{k-1}}$ and f_{z_k/x_k} are assumed known.

As an illustration of such a characterization of systems using PDFs, consider the special case of linear Gaussian measurements $z_k \in R^p$,

$$z_k = Hx_k + v_k, \qquad (5.1\text{-}1)$$

where $v_k \sim N(0, R)$, or

$$f_{v_k}(v_k) = \frac{1}{\sqrt{(2\pi)^p |R|}} e^{-(1/2)v_k^T R^{-1} v_k}. \qquad (5.1\text{-}2)$$

This situation can alternatively be represented by specifying the conditional PDF

$$f_{z_k/x_k}(z_k/x_k) = f_{v_k}(z_k - Hx_k) = \frac{1}{\sqrt{(2\pi)^p |R|}} e^{-(1/2)(z_k - Hx_k)^T R^{-1}(z_k - Hx_k)}. \qquad (5.1\text{-}3)$$

The characterizations (5.1-1)–(5.1-2) and (5.1-3) are equivalent.

The optimal mean-square estimate of x_k is the conditional mean given all previous data (Section 1.1). Let us represent the data up through time k by

$$Z_k = \{z_1, z_2, \ldots, z_k\}. \qquad (5.1\text{-}4)$$

It is desired to find a recursive optimal estimation procedure by finding an expression for the optimal estimate $\overline{x_{k+1}/Z_{k+1}}$ in terms of $\overline{x_k/Z_k}$, where overbar denotes expected value. To determine an error bound we would also like an expression for the error covariance $P_{x_{k+1}/Z_{k+1}}$ in terms of P_{x_k/Z_k}. Both of these expressions can in principle be determined from a recursive expression for the conditional PDF $f_{x_{k+1}/Z_{k+1}}$ in terms of f_{x_k/Z_k}. The quantity f_{x_k/Z_k} is the *hyperstate* of the system.

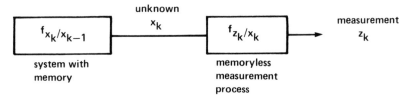

FIGURE 5.1-1 Characterization of general nonlinear stochastic system using conditional PDFs.

We can use the PDF chain rule and its integral the Chapman–Kolmogorov equation (Papoulis 1965) to find quite easily a recursion for the hyperstate. (See Sorenson 1970 and Bryson and Ho 1975.)

According to the chain rule we have

$$f_{x_{k+1}, z_{k+1}/Z_k} = f_{x_{k+1}/z_{k+1}, Z_k} f_{z_{k+1}/Z_k}$$
$$= f_{z_{k+1}/x_{k+1}, Z_k} f_{x_{k+1}/Z_k}. \tag{5.1-5}$$

Rearranging and taking into account the memoryless nature of the measurement process there results the difference equation

$$f_{x_{k+1}/Z_{k+1}} = \frac{f_{z_{k+1}/x_{k+1}}}{f_{z_{k+1}/Z_k}} f_{x_{k+1}/Z_k}. \tag{5.1-6}$$

This represents the *measurement update* of the hyperstate.

To find the time update from f_{x_k/Z_k} to f_{x_{k+1}/Z_k}, use the Chapman–Kolmogorov equation to write

$$f_{x_{k+1}/Z_k} = \int f_{x_{k+1}/x_k, Z_k} f_{x_k/Z_k} \, dx_k.$$

By the Markov system assumption we obtain the time update

$$f_{x_{k+1}/Z_k} = \int f_{x_{k+1}/x_k} f_{x_k/Z_k} \, dx_k, \tag{5.1-7}$$

which is a prediction of f_{x_{k+1}/Z_k} in terms of f_{x_k/Z_k}.

Combining the time and measurement updates yields the desired recursion for the hyperstate

$$f_{x_{k+1}/Z_{k+1}} = \frac{\int f_{z_{k+1}/x_{k+1}} f_{x_{k+1}/x_k} f_{x_k/Z_k} \, dx_k}{f_{z_{k+1}/Z_k}}. \tag{5.1-8}$$

According to the Chapman–Kolmogorov equation, the normalizing constant can be expressed as

$$f_{z_{k+1}/Z_k} = \int f_{z_{k+1}/x_{k+1}, x_k, Z_k} f_{x_{k+1}/x_k, Z_k} f_{x_k/Z_k} \, dx_k \, dx_{k+1}.$$

Taking into account *both* the memorylessness of the measurements *and* the Markov property results in

$$f_{z_{k+1}/Z_k} = \int f_{z_{k+1}/x_{k+1}} f_{x_{k+1}/x_k} f_{x_k/Z_k} \, dx_k \, dx_{k+1}. \tag{5.1-9}$$

The hyperstate recursion then becomes

$$f_{x_{k+1}/Z_{k+1}} = \frac{\int f_{z_{k+1}/x_{k+1}} f_{x_{k+1}/x_k} f_{x_k/Z_k} \, dx_k}{\int f_{z_{k+1}/x_{k+1}} f_{x_{k+1}/x_k} f_{x_k/Z_k} \, dx_k \, dx_{k+1}} \tag{5.1-10}$$

The next example shows the power of this recursion by demonstrating that it yields the Kalman filter as a special case!

Example 5.1-1:   *Derivation of Discrete Kalman Filter from Hyperstate Recursion*

Let the system be given by

$$x_{k+1} = Ax_k + Gw_k \tag{1}$$

$$z_k = Hx_k + v_k, \tag{2}$$

with $w_k \sim N(0, Q)$ and $v_k \sim N(0, R)$ white and uncorrelated with each other and with $x_0 \sim N(\bar{x}_0, P_0)$. Suppose $x_k \in R^n$, $z_k \in R^p$.

     a.   *Time Update*

Assume that $x_k/Z_k \sim N(\hat{x}_k, P_k)$, or

$$f_{x_k/Z_k} = \frac{1}{\sqrt{(2\pi)^n |P_k|}} \, e^{-(1/2)(x_k - \hat{x}_k)^T P_k^{-1}(x_k - \hat{x}_k)}. \tag{3}$$

We could use $f_{x_{k+1}/x_k} = f_{w_k}[G^{-1}(x_{k+1} - Ax_k)]$ and perform the integration in (5.1-7). [If G is singular we must use characteristic functions. See Sorenson (1970a, b).] However since x_{k+1} is a sum of Gaussian RVs (see the problems) we know that

$$x_{k+1}/Z_k \sim N(A\hat{x}_k, AP_kA^T + GQG^T), \tag{4}$$

or

$$f_{x_{k+1}/Z_k} = \frac{1}{\sqrt{(2\pi)^n |P_{k+1}^-|}} \, e^{-(1/2)(x_{k+1} - \hat{x}_k^-)^T (P_{k+1}^-)^{-1}(x_{k+1} - \hat{x}_{k+1}^-)} \tag{5}$$

where the predicted estimate and error covariance are

$$\hat{x}_{k+1}^- = A\hat{x}_k \tag{6}$$

$$P_{k+1}^- = AP_kA^T + GQG^T. \tag{7}$$

     b.   *Measurement Update*

We have a priori

$$x_{k+1}/Z_k \sim N(\hat{x}_{k+1}^-, P_{k+1}^-), \tag{8}$$

and we must use (5.1-6) to determine the PDF of the a posteriori RV x_{k+1}/Z_{k+1}.

     First let us find the normalizing constant. We could evaluate the expression (5.1-9). However, since z_{k+1} is a sum of Gaussian RVs we know that

$$z_{k+1}/Z_k \sim N(H\hat{x}_{k+1}^-, HP_{k+1}^-H^T + R). \tag{9}$$

We can also say [see (5.1-3)]

$$z_{k+1}/x_{k+1} \sim N(Hx_{k+1}, R). \tag{10}$$

Using (8)–(10) in (5.1-6) yields

$$f_{x_{k+1}/Z_{k+1}} =$$

$$\frac{\sqrt{(2\pi)^p |HP_{k+1}^- H^T + R|}}{\sqrt{(2\pi)^p |R|}\sqrt{(2\pi)^n |P_{k+1}^-|}} \exp\left\{-\frac{1}{2}[(z_{k+1} - Hx_{k+1})^T R^{-1}(z_{k+1} - Hx_{k+1})\right.$$

$$+ (x_{k+1} - \hat{x}_{k+1}^-)^T (P_{k+1}^-)^{-1}(x_{k+1} - \hat{x}_{k+1}^-)$$

$$\left. - (z_{k+1} - H\hat{x}_{k+1}^-)^T (HP_{k+1}^- H^T + R)^{-1}(z_{k+1} - H\hat{x}_{k+1}^-)]\right\}. \tag{11}$$

After completing the square in the exponent and some slightly tedious work we realize that this can be written as

$$f_{x_{k+1}/Z_{k+1}} = \frac{1}{\sqrt{(2\pi)^n |P_{k+1}|}} e^{-(1/2)(x_{k+1} - \hat{x}_{k+1})^T P_{k+1}^{-1}(x_{k+1} - \hat{x}_{k+1})}, \tag{12}$$

where the updated estimate and error covariance are

$$\hat{x}_{k+1} = \hat{x}_{k+1}^- + P_{k+1} H^T R^{-1}(z_{k+1} - H\hat{x}_{k+1}^-) \tag{13}$$

$$P_{k+1} = P_{k+1}^- - P_{k+1}^- H^T (HP_{k+1}^- H^T + R)^{-1} HP_{k+1}^-. \tag{14}$$

(See Example 1.1-2.)

Equations (6), (7), (13), and (14) are exactly the Kalman filter equations.

A simple alternative derivation of (13), (14) is worth mentioning. Equation (5.1-6) can be written as

$$f_{x_{k+1}/z_{k+1}} = \frac{f_{z_{k+1}/x_{k+1}} f_{x_{k+1}}}{f_{z_{k+1}}} \tag{15}$$

where all PDFs are interpreted as being conditioned on the previous data Z_k. This is Bayes' rule.

Now, consider the augmented RV

$$\begin{bmatrix} x_{k+1} \\ z_{k+1} \end{bmatrix} \sim N\left(\begin{bmatrix} \hat{x}_{k+1}^- \\ H\hat{x}_{k+1}^- \end{bmatrix}, \begin{bmatrix} P_{k+1}^- & P_{k+1}^- H^T \\ HP_{k+1}^- & HP_{k+1}^- H^T + R \end{bmatrix}\right), \tag{16}$$

with means and covariances conditioned on Z_k. We want to find an alternate to (15) for determining the conditional mean and covariance of x_{k+1} given the new data z_{k+1} [and all previous data; this is implicit in (8)]. But this is exactly the problem we solved in Example 1.1-3, which yields (13) and (14) immediately. ∎

Continuous Systems

Now let us discuss a nonlinear continuous-time system with discrete measurements. A full analysis requires the use of *Itô calculus*. We shall avoid this by ignoring the mathematical niceties and taking some liberties with notation in order to make clear the essential ideas. For a rigorous treatment see Jazwinski (1970).

Let there be prescribed the nonlinear time-varying system

$$\dot{x} = a(x, t) + g(x, t)w \tag{5.1-11}$$

with discrete measurements at times t_k given by

$$z_k = h[x(t_k), k] + v_k, \tag{5.1-12}$$

with $w(t) \sim (0, Q)$ and $v_k \sim (0, R)$ uncorrelated white noises. Let $x(t) \in R^n$.

The hyperstate measurement update can easily be found by using the same techniques as in the discrete case. It is

$$f_{x(t_k)/Z_k} = \frac{f_{z_k/x(t_k)}}{f_{z_k/Z_{k-1}}} f_{x(t_k)/Z_{k-1}} \tag{5.1-13}$$

with normalization constant

$$f_{z_k/Z_{k-1}} = \int f_{z_k/x} f_{x/Z_{k-1}} \, dx. \tag{5.1-14}$$

The time update is not so easy to discover, and it is here that the Itô calculus is needed (Jazwinski 1970). We merely state the result that between measurements the hyperstate satisfies the *Fokker–Planck* (or Kolmogorov) diffusion equation

$$\frac{\partial}{\partial t} f_{x(t)/Z_k} = -\text{trace} \frac{\partial f_{x(t)/Z_k} a}{\partial x} + \frac{1}{2} \sum_{i,j=1}^{n} \frac{\partial^2 f_{x(t)/Z_k}[gQg^T]_{ij}}{\partial x_i \, \partial x_j}; \qquad t_k \le t < t_{k+1} \tag{5.1-15}$$

where $[M]_{ij}$ is the (i, j)th element of a matrix M. This is very similar to the heat equation.

The Fokker–Planck equation has only been solved in a few simple cases, one of which is the following.

Example 5.1-2: Fokker–Planck Equation for Wiener Process

Let

$$\dot{x}(t) = w(t) \tag{1}$$

with $w(t) \sim N(0, 1)$ white. Suppose $x(0)$ is initially zero with probability 1, that is,

$$f_{x(t)}(x, 0) = \delta(x). \tag{2}$$

a. Mean and Variance

To find the mean and variance of $x(t)$ we can use Example 3.1-1 to see that

$$\dot{\bar{x}}(t) = 0, \qquad \bar{x}(0) = 0, \tag{3}$$

so

$$\bar{x}(t) = 0. \tag{4}$$

Furthermore the Lyapunov equation is

$$\dot{p}(t) = 1, \qquad p(0) = 0, \tag{5}$$

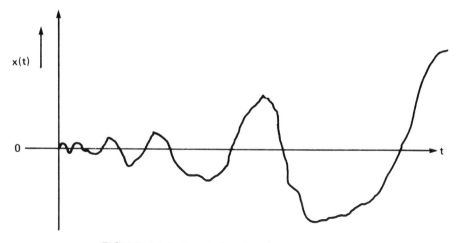

FIGURE 5.1-2 Sample function of a Wiener process.

so

$$p(t) = t. \tag{6}$$

Thus, the variance of a Wiener process increases linearly with t, which expresses our increasing uncertainty regarding $x(t)$.

A sample function of the Wiener process $\{x(t)\}$ is shown in Fig. 5.1-2.

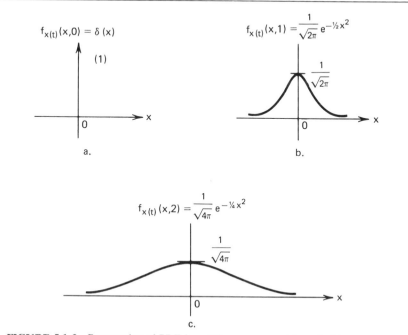

FIGURE 5.1-3 Propagation of PDF for a Wiener process. (a) $t = 0$. (b) $t = 1$. (c) $t = 2$.

b. Fokker–Planck Equation

In this simple case the Fokker–Planck equation for the unconditional PDF $f_{x(t)}(x, t)$ is

$$\frac{\partial f_{x(t)}}{\partial t} = \frac{1}{2} \frac{\partial^2 f_{x(t)}(x, t)}{\partial x^2}; \qquad t \geq 0, \tag{7}$$

with $f_{x(t)}(x, 0) = \delta(x)$.

Since the white noise $w(t)$ is normal, so is $x(t)$. Thus, according to (4) and (6)

$$f_{x(t)}(x, t) = \frac{1}{\sqrt{2\pi t}} e^{-x^2/2t}. \tag{8}$$

It is straightforward to verify that (8) satisfies (7).

A sketch of $f_{x(t)}(x, t)$ for several values of t is shown in Fig. 5.1-3.   ■

5.2 GENERAL UPDATE OF MEAN AND COVARIANCE

Here we discuss the exact time and measurement updates for the mean and covariance of the nonlinear continuous system with discrete measurements (5.1-11), (5.1-12). First we derive a useful preliminary result.

To simplify the appearance of our equations as far as possible, let us define the following shorthand notation for the a posteriori and a priori conditional expected values:

$$\hat{x}(t) = \varepsilon[x(t)] \triangleq E[x(t)/(z_k; t_k \leq t)] \tag{5.2-1a}$$

$$\hat{x}^-(t) = \varepsilon^-[x(t)] \triangleq E[x(t)/(z_k; t_k < t)]. \tag{5.2-1b}$$

Thus $\hat{x}(t_k)$ incorporates the measurements $z_k, z_{k-1}, \ldots$, while $\hat{x}^-(t_k)$ incorporates only $z_{k-1}, z_{k-2}, \ldots$. [The only reason we introduce $\varepsilon(\cdot)$ is to avoid using very large carats over entire expressions!] Note that the carat has the same meaning here as it has in the past; it simply denotes the optimal estimate.

Time Update

Let $\phi[x(t)]$ be an arbitrary twice differentiable scalar function of $x \in R^n$. To determine a time update between measurements for $\phi[x(t)]$ due to (5.1-11) we proceed as follows (Jazwinski 1970). By definition

$$\hat{\phi}[x(t)] = \int \phi(x) f_{x/Z_k} \, dx; \qquad t_k \leq t. \tag{5.2-2}$$

Therefore

$$\frac{d\hat{\phi}[x(t)]}{dt} = \int \phi(x) \frac{\partial f_{x/Z_k}}{\partial t} \, dx. \tag{5.2-3}$$

Taking into account the Fokker–Planck equation this becomes

$$\frac{d\hat{\phi}[x(t)]}{dt} = -\int \text{trace}\left[\phi(x)\frac{\partial fa}{\partial x}\right] dx + \frac{1}{2}\int \phi(x) \sum_{i,j=1}^{n} \frac{\partial^2 f[gQg^T]_{ij}}{\partial x_i \partial x_j} dx,$$

where $f \triangleq f_{x(t)/Z_k}$. Integrating by parts, assuming f is well behaved at infinity, and using some matrix identities this yields (see the problems)

$$\frac{d\hat{\phi}[x(t)]}{dt} = \int \phi_x^T af_{x/Z_k} dx + \frac{1}{2}\int \text{trace}(gQg^T\phi_{xx})f_{x/Z_k} dx. \quad (5.2\text{-}4)$$

By definition this is

$$\frac{d\hat{\phi}[x(t)]}{dt} = \varepsilon(\phi_x^T a) + \frac{1}{2}\text{trace }\varepsilon(gQg^T\phi_{xx}). \quad (5.2\text{-}5)$$

This is an important preliminary result which we can use to determine exact time updates for the mean and covariance for the nonlinear system (5.1-11).

To wit, first let $\phi(x_i) = x_i$, the ith component of $x \in R^n$. Then gradient $\phi_{x_i} = 1$ and Hessian $\phi_{x_i x_i} = 0$. Therefore

$$\frac{d\hat{x}_i}{dt} = \varepsilon(a_i) = \hat{a}_i, \quad (5.2\text{-}6)$$

with $a_i(x, t)$ the ith component of $a(x, t)$. Therefore, a time update between measurements for the conditional mean is provided by

$$\dot{\hat{x}} = \hat{a}(x, t). \quad (5.2\text{-}7)$$

To find a time update for the conditional covariance

$$P(t) \triangleq P_{x(t)/Z_k} = \varepsilon[(x - \hat{x})(x - \hat{x})^T] = \varepsilon(xx^T) - \hat{x}\hat{x}^T \quad (5.2\text{-}8)$$

we first let $\phi(y) = x_i x_j$, with $y = [x_i \quad x_j]^T$ a temporary auxiliary random vector. Then gradient $\phi_y = [x_j \quad x_i]^T$ and Hessian $\phi_{yy} = \begin{bmatrix} 0 & 1 \\ 1 & 0 \end{bmatrix}$. Therefore, by (5.2-5),

$$\frac{d}{dt}\varepsilon(x_i x_j) = \varepsilon(a_i x_j + x_i a_j) + \frac{1}{2}([gQg^T]_{ij} + [gQg^T]_{ji})$$

$$= \varepsilon(a_i x_j + x_i a_j) + \varepsilon([gQg^T]_{ij}), \quad (5.2\text{-}9)$$

where we have used the symmetry of gQg^T. (To show this, first find an expression for $\dot{y}$.) Now,

$$\frac{d}{dt}(\hat{x}_i \hat{x}_j) = \hat{x}_i \frac{d}{dt}\hat{x}_j + \hat{x}_j \frac{d}{dt}\hat{x}_i,$$

so by (5.2-6)

$$\frac{d}{dt}(\hat{x}_i \hat{x}_j) = \hat{x}_i \hat{a}_j + \hat{a}_i \hat{x}_j. \quad (5.2\text{-}10)$$

By combining (5.2-8)–(5.2-10) and employing some bookkeeping to keep track of indices we finally achieve the desired error covariance time update

$$\dot{P} = [\varepsilon(xa^T) - \hat{x}\hat{a}^T] + [\varepsilon(ax^T) - \hat{a}\hat{x}^T] + \varepsilon(gQg^T). \qquad (5.2\text{-}11)$$

This componentwise derivation is tedious, but it is required because differentiating a matrix with respect to a vector yields a three-dimensional object (a third-order tensor).

Equations (5.2-7) and (5.2-11) are the exact estimate and error covariance time updates between measurements for the nonlinear system (5.1-11). They are in general unsolvable, since they are *not* ordinary differential equations. To see this, note that (5.2-7) is really

$$\frac{d\hat{x}}{dt} = \int a(x, t) f_{x(t)/Z_k} \, dx; \qquad t_k \leq t, \qquad (5.2\text{-}12)$$

which depends on *all* the conditional moments of $x(t)$. Note the important point that in general

$$\hat{a}(x, t) \neq a(\hat{x}, t). \qquad (5.2\text{-}13)$$

In the linear case

$$\dot{x} = a(x, t) + g(x, t)w = Ax + Gw \qquad (5.2\text{-}14)$$

we do have $\hat{a}(x, t) = \varepsilon(Ax) = A\hat{x} = a(\hat{x}, t)$, so the mean update is the familiar

$$\dot{\hat{x}} = A\hat{x}. \qquad (5.2\text{-}15)$$

Furthermore, in the linear case (5.2-11) reduces to the familiar Lyapunov equation since

$$\dot{P} = [\varepsilon(xx^TA^T) - \hat{x}\hat{x}^TA^T] + [\varepsilon(Axx^T) - A\hat{x}\hat{x}^T] + GQG^T$$
$$= PA^T + AP + GQG^T. \qquad (5.2\text{-}16)$$

Measurement Update

It is quite easy to determine how an arbitrary *matrix* function $\Phi[x(t)]$ is updated at a measurement due to (5.1-12). Multiplying the PDF update (5.1-13) by $\Phi[x(t)]$ and integrating with respect to $x(t_k)$ there results

$$\varepsilon\{\Phi[x(t_k)]\} = \frac{\varepsilon^-\{\Phi[x(t_k)]f_{z_k/x(t_k)}\}}{\varepsilon^-[f_{z_k/x(t_k)}]}. \qquad (5.2\text{-}17)$$

From this it follows immediately that the exact measurement updates for the estimate and error covariance are given by

$$\hat{x}(t_k) = \frac{\varepsilon^-[x(t_k)f_{z_k/x(t_k)}]}{\varepsilon^-[f_{z_k/x(t_k)}]} \qquad (5.2\text{-}18)$$

and

$$P(t_k) = \frac{\varepsilon^-[x(t_k)x^T(t_k)f_{z_k/x(t_k)}]}{\varepsilon^-[f_{z_k/x(t_k)}]} - \hat{x}(t_k)\hat{x}^T(t_k). \qquad (5.2\text{-}19)$$

Unfortunately, these expressions are of no use algorithmically. Furthermore, judging by Example 5.1-1b which was for the linear Gaussian case, any attempt to use them to derive more convenient measurement updates for $\hat{x}$ and P will quickly become embroiled in some lengthy manipulations. Consequently we select an alternate approach. For further discussion of (5.2-18) and (5.2-19) see Jazwinski (1970).

Linear Measurement Update

To find more convenient expressions for measurement updating of the estimate and error covariance in the nonlinear case, we shall restrict the estimate to be a linear function of the data (Gelb 1974).

Define the a priori and a posteriori estimation errors as

$$\tilde{x}_k^- \triangleq x(t_k) - \hat{x}^-(t_k) \qquad (5.2\text{-}20)$$

$$\tilde{x}_k \triangleq x(t_k) - \hat{x}(t_k). \qquad (5.2\text{-}21)$$

Then the a priori and a posteriori error covariance are

$$P^-(t_k) \triangleq E[\tilde{x}_k^-(\tilde{x}_k^-)^T] \qquad (5.2\text{-}22)$$

$$P(t_k) \triangleq E[\tilde{x}_k \tilde{x}_k^T]. \qquad (5.2\text{-}23)$$

For notational ease define

$$x_k \triangleq x(t_k) \qquad (5.2\text{-}24)$$

and suppress the k argument of $h(\cdot, \cdot)$. The nonlinear measurement is then written as

$$z_k = h(x_k) + v_k, \qquad (5.2\text{-}25)$$

with white noise $v_k \sim (0, R)$ uncorrelated with $x(t)$.

To make this problem tractable, let us now restrict the a posteriori estimate to be a linear function of the data, so that

$$\hat{x}_k = b_k + K_k z_k \qquad (5.2\text{-}26)$$

for some deterministic b_k and K_k which are yet to be found. Then the estimation error update is

$$\tilde{x}_k = x_k - b_k - K_k[h(x_k) + v_k] = -b_k - K_k h(x_k) - K_k v_k + \tilde{x}_k^- + \hat{x}_k^-. \qquad (5.2\text{-}27)$$

It is desired that $\hat{x}_k$ be unbiased if $\hat{x}_k^-$ is, so by taking expected values

$$0 = -b_k - K_k \hat{h}^-(x_k) + \hat{x}_k^-. \qquad (5.2\text{-}28)$$

Solving for b_k and substituting into (5.2-26) yields the familiar form

$$\hat{x}_k = \hat{x}_k^- + K_k[z_k - \hat{h}^-(x_k)]. \tag{5.2-29}$$

Taking (5.2-28) into account, (5.2-27) becomes

$$\tilde{x}_k = \tilde{x}_k^- - K_k[h(x_k) - \hat{h}^-(x_k)] - K_k v_k. \tag{5.2-30}$$

Using this, an expression for the error covariance $P(t_k)$ in terms of the as yet undetermined K_k can now be written as

$$\begin{aligned}
P(t_k) = P^-(t_k) &+ K_k E\{[h(x_k) - \hat{h}^-(x_k)][h(x_k) - \hat{h}^-(x_k)]^T\} K_k^T \\
&- K_k E\{[h(x_k) - \hat{h}^-(x_k)](\tilde{x}_k^-)^T\} \\
&- E\{\tilde{x}_k^-[h(x_k) - \hat{h}^-(x_k)]^T\} K_k^T + K_k R K_k^T.
\end{aligned} \tag{5.2-31}$$

The gain K_k is now selected to minimize $P(t_k)$. Differentiating $P(t_k)$ with respect to K_k and solving for K_k results in the "Kalman-like" gain

$$\begin{aligned}
K_k = E\{(x_k - \hat{x}_k^-)[h(x_k) - \hat{h}^-(x_k)]^T\} \\
\times (E\{[h(x_k) - \hat{h}^-(x_k)][h(x_k) - \hat{h}^-(x_k)]^T\} + R)^{-1}.
\end{aligned} \tag{5.2-32}$$

Substituting this into (5.2-31) and simplifying yields

$$P(t_k) = P^-(t_k) - K_k E\{[h(x_k) - \hat{h}^-(x_k)](x_k - \hat{x}_k^-)^T\}. \tag{5.2-33}$$

The complete linear estimate update due to nonlinear measurements is given by (5.2-32), (5.2-33), and (5.2-29). These equations are computationally intractable because they are not ordinary difference equations. They depend on all conditional moments of $x(t)$ to compute $\hat{h}^-[x(t_k)]$. In the next subsection we make first-order approximations to derive a computationally viable algorithm known as the *extended Kalman filter*.

It can quickly be demonstrated that in the case of linear measurements

$$z_k = H x_k + v_k \tag{5.2-34}$$

our equations reduce to the Kalman filter measurement update. In general $\hat{h}^-[x(t_k)] \neq h[\hat{x}^-(t_k)]$, however, in the linear case

$$\hat{h}^-[x(t_k)] = \varepsilon^-(H x_k) = H \hat{x}_k^- = h[\hat{x}^-(t_k)]. \tag{5.2-35}$$

Substituting this into (5.2-32), (5.2-33), (5.2-29) yields the measurement update in Table 3.3-2.

5.3 EXTENDED KALMAN FILTER

Let us discuss continuous nonlinear systems of the form

$$\dot{x} = a(x, t) + G(t) w \tag{5.3-1a}$$

with measurements at discrete times t_k given by

$$z_k = h[x(t_k), k] + v_k, \tag{5.3-1b}$$

where $w(t) \sim (0, Q)$ and $v_k \sim (0, R)$ are white noise processes uncorrelated with each other and with $x(0) \sim (\bar{x}_0, P_0)$. For simplicity we assume the process noise matrix $G(t)$ is independent of $x(t)$. The case $g[x(t), t]$ is covered in the problems.

Exact time and measurement updates for the hyperstate $f_{x(t)/Z_k}$, where $Z_k = \{z_k, z_{k-1}, \ldots,\}$ is all the data available through time t_k, are given by (5.1-15) and (5.1-13), respectively. Using these equations we derived the exact time and measurement updates for the estimate (5.2-7) and (5.2-18), and error covariance (5.2-11) and (5.2-19). To make the measurement updates assume a more convenient form, we next restricted the estimate to a linear dependence on the data z_k. This resulted in the measurement updates (5.2-29) for the estimate and (5.2-32), (5.2-33) for the error covariance.

All these update equations are in general intractable, since they depend on the entire hyperstate, or equivalently on all the moments of $x(t)$. We demonstrated at several points that in the linear Gaussian case the Kalman filter was recovered.

What we would like to do now is to make some approximations to $a(x, t)$ and $h(x, k)$ to obtain a computationally viable algorithm for filtering in nonlinear systems. We shall see that the result is an "extended" version of the Kalman filter. See Gelb (1974).

Approximate Time Update

The exact time update for nonlinear $a(x, t)$ is given by (5.2-7), (5.2-11). In order to find a time update that can be conveniently programmed on a computer, expand $a(x, t)$ in a Taylor series about $\hat{x}(t)$, the current state estimate:

$$a(x, t) = a(\hat{x}, t) + \frac{\partial a}{\partial x}\bigg|_{x=\hat{x}} (x - \hat{x}) + \cdots. \tag{5.3-2}$$

Taking conditional expectations yields

$$\hat{a}(x, t) = a(\hat{x}, t) + O(2), \tag{5.3-3}$$

where $O(2)$ denotes terms of order 2. As a first-order approximation to $\hat{a}(x, t)$ we can therefore take $a(\hat{x}, t)$. Update (5.2-7) is thus replaced by

$$\dot{\hat{x}} = a(\hat{x}, t). \tag{5.3-4}$$

To determine a first-order approximation to (5.2-11), substitute for $a(x, t)$ using the first two terms of (5.3-2) and then take the indicated expected values. Representing the Jacobian as

$$A(x, t) \triangleq \frac{\partial a(x, t)}{\partial x} \tag{5.3-5}$$

this results in

$$\dot{P} = A(\hat{x}, t)P + PA^T(\hat{x}, t) + GQG^T. \tag{5.3-6}$$

Equations (5.3-4) and (5.3-6) represent an approximate, computationally feasible time update for the estimate and error covariance. The estimate simply propagates according to the nonlinear dynamics, and the error covariance propagates like that of a linear system with plant matrix $A(\hat{x}, t)$.

Note that Jacobian $A(\cdot, t)$ is evaluated for each t at the *current estimate*, which is provided by (5.3-4), so that there is coupling between (5.3-4) and (5.3-6). To eliminate this coupling it is possible to introduce a further approximation and solve not (5.3-6) but instead

$$\dot{P}(t) = A[\hat{x}(t_k), t]P(t) + P(t)A^T[\hat{x}(t_k), t] + G(t)QG^T(t); \qquad t_k \le t < t_{k+1}. \tag{5.3-7}$$

In this equation, the Jacobian is evaluated once using the estimate $\hat{x}(t_k)$ after updating at t_k to include z_k. This is used as the plant matrix for the time propagation over the entire interval until the next measurement time t_{k+1}.

We shall soon present an example demonstrating how to accomplish the time update (5.3-4), (5.3-6) very easily on a digital computer using a Runge–Kutta integrator.

Approximate Measurement Update

The optimal linear measurement update for nonlinear measurement matrix $h[x(t_k), k]$ was found to be given by (5.2-29), (5.2-32), and (5.2-33). In order to find a measurement update that can be conveniently programmed, expand $h(x_k)$ in a Taylor series about $\hat{x}_k^-$, the a priori estimate at time t_k:

$$h(x_k) = h(\hat{x}_k^-) + \frac{\partial h}{\partial x}\bigg|_{x=\hat{x}_k^-} (x_k - \hat{x}_k^-) + \cdots. \tag{5.3-8}$$

Represent the Jacobian as

$$H(x, k) = \frac{\partial h(x, k)}{\partial x}. \tag{5.3-9}$$

To find a first-order approximate update, substitute for $h(x_k)$ in (5.2-29), (5.2-32), (5.2-33) using the first two terms of (5.3-8). The result is

$$\hat{x}_k = \hat{x}_k^- + K_k[z_k - h(\hat{x}_k^-)] \tag{5.3-10}$$

$$K_k = P^-(t_k)H^T(\hat{x}_k^-)[H(\hat{x}_k^-)P^-(t_k)H^T(\hat{x}_k^-) + R]^{-1} \tag{5.3-11}$$

$$P(t_k) = [I - K_k H(\hat{x}_k^-)]P^-(t_k). \tag{5.3-12}$$

These equations represent an approximate, computationally feasible linear measurement update for the estimate and the error covariance. The residual is computed using the nonlinear measurement function $h(\cdot, k)$ evaluated at the a priori estimate $\hat{x}_k^-$. The error covariance is found using the Jacobian matrix $H(\hat{x}_k^-)$.

The EKF

The approximate time and measurement updates we have derived are collected for reference in Table 5.3-1. They comprise what is known as the *extended Kalman filter* (*EKF*) for a continuous system with discrete measurements. If Q and R are time varying, it is only necessary to use their

TABLE 5.3-1   Continuous-Discrete Extended Kalman Filter

System model and measurement model:

$$\dot{x} = a(x, u, t) + G(t)w$$

$$z_k = h[x(t_k), k] + v_k$$

$$x(0) \sim (\bar{x}_0, P_0), \ w(t) \sim (0, Q), \ v_k \sim (0, R)$$

Assumptions:

$\{w(t)\}$ and $\{v_k\}$ are white noise processes uncorrelated with $x(0)$ and with each other.

Initialization:

$$P(0) = P_0, \ \hat{x}(0) = \bar{x}_0.$$

Time update:

    estimate: $\dot{\hat{x}} = a(\hat{x}, u, t)$
    error covariance: $\dot{P} = A(\hat{x}, t)P + PA^T(\hat{x}, t) + GQG^T$

Measurement update:

    Kalman gain:

$$K_k = P^-(t_k)H^T(\hat{x}_k^-)[H(\hat{x}_k^-)P^-(t_k)H^T(\hat{x}_k^-) + R]^{-1}$$

    error covariance: $P(t_k) = [I - K_kH(\hat{x}_k^-)]P^-(t_k)$
    estimate: $\hat{x}_k = \hat{x}_k^- + K_k[z_k - h(\hat{x}_k^-, k)]$

Jacobians:

$$A(x, t) = \frac{\partial a(x, u, t)}{\partial x}$$

$$H(x) = \frac{\partial h(x, k)}{\partial x}$$

---

values $Q(t)$ and R_k in the filter. A deterministic input $u(t)$ is included in the equations for completeness.

An important advantage of the continuous-discrete EKF is that the optimal estimate is available continuously at all times, including times *between* the measurement times t_k.

Note that if $a(x, u, t)$ and $h(x, k)$ are linear, then the EKF reverts to the continuous-discrete Kalman filter in Section 3.7.

Given the continuous nature of dynamical systems and the requirement of microprocessors for discrete data, the continuous-discrete EKF is usually the most useful formulation for modern control purposes. The EKF approximation can also be applied to continuous systems with *continuous* measurements. For the sake of completeness, the continuous-continuous EKF is given in Table 5.3-2. A discrete-discrete Kalman filter could also be written down very easily.

Since the time and measurement updates depend on Jacobians evaluated *at the current estimate*, the error covariance and Kalman gain K_k cannot be computed off-line for the EKF. They must be computed in real time as the data become available.

In some cases the nominal trajectory $x_N(t)$ of the state $x(t)$ is known a priori. An example is a robot arm for welding a car door which always moves along the same path. In this case the EKF equations can be used with $\hat{x}(t)$ replaced by $x_N(t)$. Since $x_N(t)$ is known beforehand, the error covariance and K_k can now be computed off-line before the measurements are taken. This procedure of linearizing about a known nominal trajectory results in what is called the *linearized Kalman filter*. If a controller is employed to keep the state $x(t)$ on the nominal trajectory $x_N(t)$, the linearized Kalman filter performs quite well.

Higher-order approximations to the optimal nonlinear updates can also be derived by retaining higher-order terms in the Taylor series expansions. For details, see Jazwinski (1970) and Gelb (1974).

It should be realized that the measurement times t_k need not be equally spaced. The time update is performed over any interval during which no data are available. When data become available, a measurement update is performed. This means that the cases of intermittent or missing measurements, and pure prediction in the absence of data can easily be dealt with using the EKF.

A block diagram of a software implementation of the continuous-discrete EKF is shown in Fig. 3.7-1. The continuous-time update is written into subroutine F(TIME, X, XP) which is called by a Runge–Kutta integrator. The discrete measurement update is written into subroutine MEASUP(TIME, X, Z) which is called whenever data are available. A driver program which implements the EKF along the lines of Fig. 3.7-1 is given in Appendix B (CDKAL).

The implementation of the continuous-continuous Kalman filter is even easier. It is only necessary to write a subroutine F(TIME, X, XP) for use

TABLE 5.3-2   Continuous-Continuous Extended Kalman Filter

System model and measurement model:

$$\dot{x} = a(x, u, t) + G(t)w$$

$$z = h(x, t) + v$$

$$x(0) \sim (\bar{x}_0, P_0), \; w(t) \sim (0, Q), \; v(t) \sim (0, R)$$

Assumptions:

$\{w(t)\}$ and $\{v(t)\}$ are white noise processes uncorrelated with $x(0)$ and with each other.

Initialization:

$$P(0) = P_0, \; \hat{x}(0) = \bar{x}_0$$

Estimate update:

$$\dot{\hat{x}} = a(\hat{x}, u, t) + K[z - h(\hat{x})]$$

Error covariance update:

$$\dot{P} = A(\hat{x}, t)P + PA^T(\hat{x}, t) + GQG^T - PH^T(\hat{x}, t)R^{-1}H(\hat{x}, t)P$$

Kalman gain:

$$K = PH^T(\hat{x}, t)R^{-1}$$

Jacobians:

$$A(x, t) = \frac{\partial a(x, u, t)}{\partial x}$$

$$H(x, t) = \frac{\partial h(x, t)}{\partial x}$$

---

with program TRESP in Appendix B to integrate the estimate and error covariance dynamics simultaneously. The vector X in the subroutine argument will contain the components of both $\hat{x}(t)$ and $P(t)$.

The next examples illustrate how to write subroutines in order to use the EKF. They also make the point that the filter implementation can be considerably simplified by doing some *preliminary analysis*!

Example 5.3-1:   FM Demodulation

a.   System and Measurement

Let message signal $s(t)$ be normal with zero mean and variance σ^2. Suppose it has a first-order Butterworth spectrum with a bandwidth of α (cf. Anderson and Moore 1979),

$$\Phi_s(\omega) = \frac{2\alpha\sigma^2}{\omega^2 + \alpha^2}. \tag{1}$$

The FM modulated signal that is transmitted is

$$y(t) = \sqrt{2}\sigma \sin[\omega_c t + \theta(t)] \tag{2}$$

where ω_c is the carrier frequency and the phase is related to the message by

$$\theta(t) = \int_0^t s(\tau)\, d\tau. \tag{3}$$

It is desired to use the EKF to estimate $s(t)$ from samples of the received signal, which is $z(t) = y(t) + v(t)$ with $v(t) \sim (0, r)$ a white noise added by the communication channel.

Defining state $x = [s \quad \theta]^T$, the process (1), (3) can be modeled as (see Fig. 2.7-2).

$$\dot{x} = \begin{bmatrix} -\alpha & 0 \\ 1 & 0 \end{bmatrix} x + \begin{bmatrix} 1 \\ 0 \end{bmatrix} w \triangleq Ax + Gw \tag{4}$$

with white noise $w(t) \sim (0, 2\alpha\sigma^2)$. The discrete nonlinear time-varying measurements are

$$z_k = \sqrt{2}\sigma \sin(\omega_c Tk + Cx_k) + v_k \triangleq h(x_k, k) + v_k, \tag{5}$$

with $C \triangleq [0 \quad 1]$ and T the sampling period.

b. Time Update

The time update is straightforward since the dynamics (4) are linear. Therefore, the error covariance propagates according to

$$\dot{P} = AP + PA^T + GQG^T. \tag{6}$$

Letting

$$P \triangleq \begin{bmatrix} p_1 & p_2 \\ p_2 & p_4 \end{bmatrix} \tag{7}$$

we can write (6) as the set of linear scalar equations

$$\dot{p}_1 = -2\alpha p_1 + 2\alpha\sigma^2 \tag{8a}$$

$$\dot{p}_2 = -\alpha p_2 + p_1 \tag{8b}$$

$$\dot{p}_4 = 2p_2. \tag{8c}$$

It is quite easy to solve these analytically; however, we shall deliberately refrain from doing so to make the point that an analytical solution is not required to implement the Kalman filter.

The estimate propagates by

$$\dot{\hat{x}} = A\hat{x}, \tag{9}$$

or

$$\dot{\hat{s}} = -\alpha\hat{s} \tag{10a}$$

$$\dot{\hat{\theta}} = \hat{s}. \tag{10b}$$

c. Measurement Update

The measurement is nonlinear, so the EKF update in Table 5.3-1 is required here. The measurement Jacobian is

$$H(x) = \frac{\partial h(x,k)}{\partial x} = \sqrt{2}\sigma C \cos(\omega_c Tk + Cx). \tag{11}$$

Letting $H(\hat{x}_k^-) \triangleq [h_1 \quad h_2]$ we have

$$h_1 = 0 \tag{12a}$$

$$h_2 = \sqrt{2}\sigma \cos(\omega_c Tk + \hat{\theta}_k^-). \tag{12b}$$

The updated Kalman gain is found as follows. Let

$$\delta \triangleq H(\hat{x}_k^-)P^-(t_k)H^T(\hat{x}_k^-) + r. \tag{13}$$

Defining

$$P^-(t_k) \triangleq \begin{bmatrix} p_1^- & p_2^- \\ p_2^- & p_4^- \end{bmatrix} \tag{14}$$

(time subscripts on the scalar quantities are not needed since they will be updated by FORTRAN replacement in the software), we have after simplifying

$$\delta = 2\sigma^2 p_4^- \cos^2(\omega_c Tk + \hat{\theta}_k^-) + r = h_2^2 p_4^- + r. \tag{15}$$

Hence

$$K_k \triangleq \begin{bmatrix} k_1 \\ k_2 \end{bmatrix} = \begin{bmatrix} p_2^- h_2/\delta \\ p_4^- h_2/\delta \end{bmatrix}. \tag{16}$$

The updated covariance is given by

$$P(t_k) = \begin{bmatrix} 1 & -k_1 h_2 \\ 0 & 1 - k_2 h_2 \end{bmatrix} P^-(t_k), \tag{17}$$

or

$$p_1 = p_1^- - k_1 h_2 p_2^- = p_1^- - \frac{h_2^2(p_2^-)^2}{\delta} \tag{18a}$$

$$p_2 = (1 - k_2 h_2)p_2^- = \frac{rp_2^-}{\delta} \tag{18b}$$

$$p_4 = (1 - k_2 h_2)p_4^- = \frac{rp_4^-}{\delta}. \tag{18c}$$

Finally, the estimate update is (omitting the time subscript k)

$$\hat{s} = \hat{s}^- + k_1[z - \sqrt{2}\sigma \sin(\omega_c Tk + \hat{\theta}^-)] \tag{19a}$$

$$\hat{\theta} = \hat{\theta}^- + k_2[z - \sqrt{2}\sigma \sin(\omega_c Tk + \hat{\theta}^-)]. \tag{19b}$$

```
                SUBROUTINE F(TIME,X,XP)

C    FM DEMODULATION CONTINUOUS DYNAMICS FOR TIME UPDATE
C    CALLED BY SUBROUTINE RUNKUT

C    X(1)-X(2)  STATE ESTIMATE
C    X(3)-X(5)  ERROR COVARIANCE ENTRIES P(1),P(2),P(4)
C    XP         TIME DERIVATIVES OF X

                REAL X(*), XP(*)
                DATA AL,SIGSQ/1.,1./

C    ERROR COVARIANCE UPDATE EQUATION

                XP(3)= 2.*AL*( -X(3) + SIGSQ )
                XP(4)= -AL*X(4) + X(3)
                XP(5)= 2.*X(4)

C    ESTIMATE EQUATION

                XP(1)= -AL*X(1)
                XP(2)= X(1)

                RETURN
                END


                SUBROUTINE MEASUP(TIME,X,Z)

C    FM DEMODULATION DISCRETE MEASUREMENT UPDATE

C    X(1)-X(2)  STATE ESTIMATE
C    X(3)-X(5)  ERROR COVARIANCE ENTRIES P(1),P(2),P(4)
C    Z          DATA SAMPLE

                REAL X(*)
                DATA SIG,WC,R/1.,1.,1./

                RT2SIG= 1.414214*SIG

C    MEASUREMENT JACOBIAN

                ARG= WC*TIME + X(2)
                H2= RT2SIG*COS(ARG)

C    KALMAN GAIN

                DEL= R + X(5)*H2**2
                AK1= X(4)*H2/DEL
                AK2= X(5)*H2/DEL

C    ERROR COVARIANCE

                X(3)= X(3) - AK1*H2*X(4)
                X(4)= X(4)*( 1. - AK2*H2 )
                X(5)= X(5)*( 1. - AK2*H2 )

C    ESTIMATE

                RES= Z - RT2SIG*SIN(ARG)


                X(1)= X(1) + AK1*RES
                X(2)= X(2) + AK2*RES

                RETURN
                END
```

FIGURE 5.3-1 Time update and measurement update subroutines for FM demodulation.

d. Filter Implementation

The time update (8), (10) is implemented between measurements with an integration routine such as Runge–Kutta. The measurement update (12), (15), (16), (18), (19) is performed whenever data become available. Note that it is *not* necessary for measurements to be taken at regular intervals. For irregular or intermittent measurements at times t_k, it is only necessary to change the time dependence Tk in (12b) and (19) to t_k. Note also that the EKF yields the optimal estimate at all times, even between the measurement times t_k.

The preliminary analysis we performed resulted in scalar updates, which are easier to program and faster to run than matrix updates. Fortran implementations of the time update and measurement update for use with CDKAL in Appendix B are given in Fig. 5.3-1. The vector X in that figure contains both the estimates and the error covariance entries. We have avoided using two sets of error covariances, one for $P(t_k)$ and one for $P^-(t_k)$, by paying attention to the order of operations. Thus, in the program (18a) must be performed before (18b) since it needs the a priori value p_2^-, which is updated in (18b).

We would expect the error covariance to behave much like the error covariance in Example 3.7-1.   ■

Example 5.3-2: Satellite Orbit Estimation

a. Satellite Equations of Motion

The equations of motion of a satellite in a planar orbit about a point mass are

$$\ddot{r} = r\dot{\theta}^2 - \frac{\mu}{r^2} + w_r \tag{1}$$

$$\ddot{\theta} = \frac{-2\dot{r}\dot{\theta}}{r} + \frac{1}{r} w_\theta, \tag{2}$$

where r is the radial distance of the satellite from the mass and θ is its angle from a reference point on the orbit (usually perigree, the closest point of the satellite's orbit to the attracting mass M). White process noises $w_r(t) \sim (0, q_r)$ and $w_\theta(t) \sim (0, q_\theta)$ represent disturbance accelerations. The gravitational constant of the point mass M is $\mu \triangleq GM$, with G the universal gravitation constant. See Maybeck (1979), Bryson and Ho (1975), and Jazwinski (1970).

b. Orbit Estimation

An elliptic orbit in a plane is completely specified if we know

$$a, \text{ the semimajor axis} \tag{3}$$

$$e, \text{ the eccentricity} \tag{4}$$

$$t_p, \text{ the time of perigree passage} \tag{5}$$

$$\theta_p, \text{ the angle between perigree and a reference axis.} \tag{6}$$

Within a specified orbit, the satellite's position is completely specified if we know $\theta(t)$. The variables a, e, θ_p, and $\theta(t)$ are the *Keplerian orbital elements*.

The orbital elements ot a satellite can be determined from its state $x \triangleq [r \;\; \dot{r} \;\; \theta \;\; \dot{\theta}]^T$. For example,

$$a^2 = \frac{2}{r} - \frac{\dot{r}^2}{\mu} \tag{7}$$

$$e^2 = \left(1 - \frac{r}{2}\right)^2 + \frac{(r\dot{r})^2}{a\mu}. \tag{8}$$

Another useful parameter is the period

$$T = 2\pi \sqrt{\frac{a^3}{\mu}}. \tag{9}$$

The problem of orbit estimation is thus equivalent to the problem of determining the instantaneous state $x(t)$ of the satellite, which we can do using the continuous-discrete EKF.

c. *Time Update*

The estimate for $x = [r \;\; \dot{r} \;\; \theta \;\; \dot{\theta}]^T$ propagates between measurements according to the deterministic (noise-free) version of (1), (2)

$$\frac{d\hat{r}}{dt} = \hat{\dot{r}} \tag{10a}$$

$$\frac{d\hat{\dot{r}}}{dt} = \hat{r}\hat{\dot{\theta}}^2 - \frac{\mu}{\hat{r}^2} \tag{10b}$$

$$\frac{d\hat{\theta}}{dt} = \hat{\dot{\theta}} \tag{10c}$$

$$\frac{d\hat{\dot{\theta}}}{dt} = \frac{-2\hat{\dot{\theta}}\hat{\dot{r}}}{\hat{r}}. \tag{10d}$$

To determine the error covariance time update we require the Jacobian

$$A(x, t) = \begin{bmatrix} 0 & 1 & 0 & 0 \\ \dot{\theta}^2 + 2\mu/r^3 & 0 & 0 & 2r\dot{\theta} \\ 0 & 0 & 0 & 1 \\ 2\dot{r}\dot{\theta}/r^2 & -2\dot{\theta}/r & 0 & -2\dot{r}/r \end{bmatrix}. \tag{11}$$

In addition, the process noise matrix is

$$g(x, t) = \begin{bmatrix} 0 & 0 \\ 1 & 0 \\ 0 & 0 \\ 0 & 1/r \end{bmatrix}, \tag{12}$$

which is a function of the state, so we must use (see the problems)

$$\dot{P} = A(\hat{x}, t)P + PA^T(\hat{x}, t) + g(\hat{x}, t)Qg^T(\hat{x}, t)$$

$$+ \frac{1}{2} \sum_{i,j=1}^{n} \frac{\partial g}{\partial x_i}\bigg|_{x=\hat{x}} Q \frac{\partial g^T}{\partial x_j}\bigg|_{x=\hat{x}} P_{ij}(t), \tag{13}$$

where $x \in R^n$ and p_{ij} is the (i, j)th element of P. We could approximate by omitting the last term, but we shall include it for completeness since it presents no real complication. The only nonzero term in the double sum is

$$\frac{1}{2}\frac{\partial g}{\partial r}\bigg|_{x=\hat{x}} Q \frac{\partial g^T}{\partial r}\bigg|_{x=\hat{x}} p_{11}(t) = \begin{bmatrix} 0 & 0 & 0 & 0 \\ 0 & 0 & 0 & 0 \\ 0 & 0 & 0 & 0 \\ 0 & 0 & 0 & p_{11}q_\theta/\hat{r}^4 \end{bmatrix}. \tag{14}$$

By substituting (11) and (14) in (13) and simplifying, 10 scalar differential equations are obtained for the $n(n+1)/2$ distinct components of P. Defining auxiliary variables

$$a_1 \triangleq \hat{\theta}^2 + \frac{2\mu}{\hat{r}^3} \tag{15a}$$

$$a_2 \triangleq \frac{2\hat{r}\hat{\theta}}{\hat{r}^2} \tag{15b}$$

$$a_3 \triangleq \frac{-2\hat{\theta}}{\hat{r}} \tag{15c}$$

$$a_4 \triangleq 2\hat{r}\hat{\theta} \tag{15d}$$

$$a_5 \triangleq \frac{-2\hat{r}}{\hat{r}}, \tag{15e}$$

we have

$$\dot{p}_{11} = 2p_{12} \tag{16a}$$

$$\dot{p}_{12} = a_1 p_{11} + p_{22} + a_4 p_{14} \tag{16b}$$

$$\dot{p}_{13} = p_{14} + p_{23} \tag{16c}$$

$$\dot{p}_{14} = a_2 p_{11} + a_3 p_{12} + a_5 p_{14} + p_{24} \tag{16d}$$

$$\dot{p}_{22} = 2a_1 p_{12} + 2a_4 p_{24} + q_r \tag{16e}$$

$$\dot{p}_{23} = a_1 p_{13} + a_4 p_{34} + p_{24} \tag{16f}$$

$$\dot{p}_{24} = a_2 p_{12} + a_3 p_{22} + a_5 p_{24} + a_1 p_{14} + a_4 p_{44} \tag{16g}$$

$$\dot{p}_{33} = 2p_{34} \tag{16h}$$

$$\dot{p}_{34} = a_2 p_{13} + a_3 p_{23} + a_5 p_{34} + p_{44} \tag{16i}$$

$$\dot{p}_{44} = 2a_2 p_{14} + 2a_3 p_{24} + 2a_5 p_{44} + \frac{q_\theta}{\hat{r}^2} + \frac{p_{11}q_\theta}{\hat{r}^4}. \tag{16j}$$

The time update equations (10), (15), (16) can easily be programmed in a subroutine F(TIME, X, XP) for use with CDKAL in Appendix B (cf. Fig. 5.3-1). It is hoped that we are making the point that, even for fairly complicated nonlinear systems, some preliminary analysis makes the EKF very straightforward to implement!

d. Measurement Update

We have deliberately made no mention yet of measurements. The time update is dependent only on the satellite dynamics, and whatever measurements are used, the subroutine F(TIME, X, XP) does not change.

Several measurement schemes can be used, some of which combine measurements by both ground tracking stations and instruments on board the satellite. We shall discuss a simple scheme in which range r and range-rate $\dot{r}$ are measured by a station on the earth's surface. Thus the data are

$$z_k = \begin{bmatrix} 1 & 0 & 0 & 0 \\ 0 & 1 & 0 & 0 \end{bmatrix} x_k + \begin{bmatrix} v_1(k) \\ v_2(k) \end{bmatrix} \tag{17}$$

with $v_1(k) \sim (0, \sigma_r^2)$, $v_2(k) \sim (0, \sigma_{\dot{r}}^2)$. Suppose data are taken at times t_k, so that $x_k \triangleq x(t_k)$. Suppose also that r and $\dot{r}$ are determined by independent means (e.g., radar ranges and doppler range rates) so that $v_1(k)$ and $v_2(k)$ are uncorrelated.

The measurements are linear, so the EKF measurement update in Table 5.3-1 reduces to the regular discrete Kalman filter update. This means that the stabilized measurement update software in Bierman (1977), for example, can be used. Since

$$R = \begin{bmatrix} \sigma_r^2 & 0 \\ 0 & \sigma_{\dot{r}}^2 \end{bmatrix} \tag{18}$$

is diagonal, scalar updates can be performed with no prewhitening of the data. Thus, at each time t_k we would perform a two-step update to include separately the two independent measurements

$$z_1(k) = [1 \quad 0 \quad 0 \quad 0] x_k + v_1(k) \tag{19}$$

and

$$z_2(k) = [0 \quad 1 \quad 0 \quad 0] x_k + v_2(k). \tag{20}$$

e. Discussion

The sample times t_k need not be uniform. We propagate through time by integrating the function F(TIME, X, XP) until data are available, at which point we call a measurement update routine to incorporate the data. If, at a given t_k, only one of the measurements (19), (20) is available, we can include it individually with a call only to the appropriate scalar measurement update routine. ∎

These examples clearly demonstrate the power and convenience of the EKF. It is a scheme that can be tailored to each application, and its modular structure can be exploited by using modular programming techniques that make it easy to modify the measurement scheme with a minimum of software redevelopment. It provides an optimal estimate at all times, and does not depend on a uniform data arrival rate.

PROBLEMS

Section 5.1

5.1-1: Sum of Random Variables. Let Z be a linear combination of RVs X and Y so that

$$Z = AX + BY$$

with A and B deterministic matrices. Show that the mean and covariance of Z are

$$\bar{Z} = A\bar{X} + B\bar{Y}$$

$$P_Z = AP_XA^T + AP_{XY}B^T + BP_{YX}A^T + BP_YB^T$$

with P_X, P_Y the covariances of X and Y, and P_{XY} the cross-covariance.

5.1-2: Fokker–Planck Equation for Wiener Process. Verify that $f_{x(t)}$ in Example 5.1-2 satisfies the Fokker–Planck equation.

5.1-3: Fokker–Planck Equation for Markov Process. Let

$$\dot{x} = ax + w$$

with $w(t) \sim N(0, q)$ white and independent of $x(0) \sim N(\bar{x}_0, p_0)$.

a. Determine the mean and covariance, and hence the unconditional PDF $f(x, t)$ of $x(t)$.

b. Write the Fokker–Planck equation and verify that the PDF satisfies it.

c. Now measurements are taken every T sec so that

$$z = x + v,$$

with $v(t) \sim N(0, r)$ independent of $w(t)$ and $x(0)$. Determine the measurement update of $f(x, t^-)$ to $f(x, t)$ at $t = kT$.

d. Between measurements the PDF of $x(t)$ satisfies the Fokker–Planck equation. Sketch the covariance of $x(t)$ over several measurement periods.

5.1-4: Let

$$x_{k+1} = x_k + w_k$$

$$z_k = x_k + v_k,$$

with $w_k \sim N(0, 1)$ and $v_k \sim N(0, 1)$ white and independent of each other and of $x_0 \sim N(\bar{x}_0, 2)$. Use (5.1-7) and (5.1-6) to find time and measurement updates for the hyperstate.

5.1-5: Measurement Update for Uniform Noise. Repeat Problem 5.1-4 if v_k is zero mean white noise uniformly distributed on $[-\frac{1}{2}, \frac{1}{2}]$.

5.1-6:   Update with Nonlinear Measurement.   Repeat Problem 5.1-4 if

$$z_k = \ln\left(\frac{1}{x_k}\right) + v_k$$

with v_k white noise with PDF

$$f_{v_k}(v) = \begin{cases} e^{-v}, & v \geq 0 \\ 0, & v < 0 \end{cases},$$

and w_k uniform on $[0, 1]$.

Section 5.2

5.2-1:   Prove equation (5.2-4).

5.2-2:   Time Update of Moments.   Let

$$\dot{x} = ax + w$$

with $w(t) \sim (0, q)$ white and $x(0) \sim (\bar{x}_0, p_0)$. Use (5.2-5) to compute analytically the following.

a. $\hat{x}(t)$.
b. $\varepsilon[x^2(t)]$ and the covariance $p(t)$.
c. $\varepsilon[x^3(t)]$. Suppose $\varepsilon[x^3(0)] = d$ is given.

5.2-3:   Nonlinear System Dynamics.   Let

$$\dot{x} = ax^2 + w$$

with $w(t) \sim (0, q)$ and $x(0) \sim (\bar{x}_0, p_0)$. Find differential equations for $\hat{x}(t)$ and the covariance $p(t)$. How can these be solved?

5.2-4:   Nonlinear Measurements.   Let

$$\dot{x} = w$$

$$z_k = \ln\left(\frac{1}{x_k}\right) + v$$

with $w(t)$ uniform on $[0, 1]$ and $v(t)$ independent white noise and

$$f_v(v) = \begin{cases} e^{-v}, & v \geq 0 \\ 0, & v < 0 \end{cases}.$$

The measurements are taken every T sec. Go as far as you can in determining $\hat{x}(t)$ and $p(t)$ using the linearized measurement updates (5.2-29) and (5.2-33).

Section 5.3

5.3-1: Nonlinear Process Noise-Weighting Matrix. Suppose that a nonlinear system is described by

$$\dot{x}(t) = a(x, t) + g(x, t)w, \tag{1}$$

with $w(t) \sim (0, Q)$. Show that to a first-order approximation

$$\varepsilon(gQg^T) = g(\hat{x}, t)Qg^T(\hat{x}, t) + \frac{1}{2}\sum_{i,j=1}^{n}\frac{\partial g}{\partial x_i}\bigg|_{x=\hat{x}} Q \frac{\partial g^T}{\partial x_j}\bigg|_{x=\hat{x}} p_{ij}(t), \tag{2}$$

where $p_{ij}(t)$ is element (i, j) of $P(t)$.

5.3-2: EKF. Let

$$\dot{x} = x^2 + w$$

$$z = \frac{1}{x} + v$$

with $w(t) \sim (0, 1)$ and $v(t) \sim (0, 1)$ independent white noises, and $x(0) \sim (0, 1)$.

a. Write the continuous-discrete EKF equations if measurements are taken every second.
b. Solve for and sketch the EKF error covariance.
c. Write subroutines for use with CDKAL.

5.3-3: Repeat Problem 5.3-2 using the continuous-continuous EKF, except that you now need to write subroutine F(TIME, X, XP) for use with TRESP in Appendix B.

5.3-4: EKF for Pendulum. The equation of motion of a simple pendulum of length l is

$$\ddot{\theta} = -\frac{g}{l}\sin\theta + w.$$

The horizontal displacement of the bob is measured so that

$$z = l(1 - \cos\theta) + v.$$

Suppose that $w(t) \sim N(0, 0.02)$, $v(t) \sim N(0, 1)$, and $\theta(0) \sim (0, 0.02)$.
a. Write the EKF equations if data are taken every $T = (\pi/5)\sqrt{l/g}$ sec. You can let $g/l = 1$ if you desire.
b. Solve for and sketch $p(t)$ over several measurement periods.

5.3-5: EKF for Rocket Dynamics. The equations of motion of a rocket are

$$\dot{r} = w$$

$$\frac{d}{dt}\dot{m} = 0$$

$$\dot{v} = -\frac{vw}{r} + \frac{\dot{m}V_e \cos\phi}{m_0 + \dot{m}t} + n_1$$

$$\dot{w} = \frac{v^2}{r} - \frac{\mu}{r^2} + \frac{\dot{m}V_e \sin\phi}{m_o + \dot{m}t} + n_2$$

where $\phi(t)$ = thrust direction angle, $\gamma(t)$ = flight path angle, total velocity $V(t)$ has radial component $w(t)$ and tangential component $v(t)$, $\mu = GM$ is the gravitational constant of the attracting center, V_e = constant exhaust velocity, m_0 = initial mass, $m(t)$ = rocket mass, and $r(t)$ = radius from the attracting mass M. Independent white process noises $n_i(t) \sim (0, q_i)$ have been added to compensate for unmodeled dynamics. Take the state as $x = [v \quad w \quad r \quad \dot{m}]^T$.

Suppose that $\dot{m}$ and r are measured every second so that

$$z_k^1 = \dot{m}_k + v_k^1$$

$$z_k^2 = r_k + v_k^2$$

where $v_k^1 \sim (0, r_1)$, $v_k^2 \sim (0, r_2)$.

a. Find the Jacobian and write the EKF error covariance propagation equation.

b. Write subroutines to implement the continuous-discrete EKF using CDKAL in Appendix B.

c. Write subroutines to implement the continuous-continuous EKF using TRESP in Appendix B.

5.3-6: Spacecraft Reentry Trajectory Estimation. The state of a spacecraft in the three-dimensional reentry problem is

$$x = [v \quad \gamma \quad \psi \quad h \quad \phi \quad \theta]^T, \tag{1}$$

where v is velocity magnitude, γ is flight-path angle measured from the local horizontal, ψ is heading angle measured from east toward north in the horizon plane, h is the altitude above the earth's surface, ϕ is the latitude, and θ is the longitude (Jazwinski 1970). If m is the (constant) vehicle mass and r is its distance from the earth's center, the state equations are

$$\dot{v} = \frac{F_{xv}}{mv} - \frac{\mu}{r^2}\sin\gamma + \Delta\dot{v}_{\text{rot}} \tag{2}$$

$$\dot{\gamma} = \frac{F_{zv}}{mv} + \frac{v\cos\gamma}{r} - \frac{\mu}{vr^2}\cos\gamma + \Delta\dot{\gamma}_{\text{rot}} \tag{3}$$

$$\dot{\psi} = \frac{F_{yv}}{mv} - \frac{v}{r}\cos\gamma\cos\psi\tan\phi + \Delta\dot{\psi}_{rot} \tag{4}$$

$$\dot{h} = v\sin\gamma \tag{5}$$

$$\dot{\phi} = \frac{v}{r}\cos\gamma\sin\psi \tag{6}$$

$$\dot{\theta} = \frac{v}{r\cos\phi}\cos\gamma\cos\psi. \tag{7}$$

The earth's gravitational constant is $\mu = GM$, with G the universal gravitation constant and M the mass of the earth. F_{xv}, F_{zv}, F_{yv} are aerodynamic forces, and the last terms in (2)–(4) are due to the earth's rotation. Assume a nonrotating earth so that these three terms are zero.

The aerodynamic forces are not well known since they depend on winds, air density, temperature, and so on. To make the problem tractable, assume that they are uncorrelated white noise processes with covariances of σ_x^2, σ_z^2, σ_y^2.

We intend to apply the EKF to estimate the state x and find the error covariance.

a. Find the time update equations. (Note that the noise matrix is $g(x, t)$, a function of x!) Write a software implementation F(TIME, X, XP).

b. Suppose we make discrete measurements of h, ϕ, and θ from a ground-tracking station, and of v using instruments aboard the vehicle. Assume all measurements are independent, and add white measurement noise to each one. Write four separate sets of measurement update equations, one to incorporate each of these pieces of data. Write separate measurement update subroutines for each of these measurements. Then whenever one of these measurements becomes available, we can simply call the appropriate subroutine to incorporate it into the state estimate.

c. Write subroutine F(TIME, X, XP) to implement the continuous-continuous EKF using program TRESP.

5.3-7: Radar Tracking in x–y Coordinates. An aircraft is flying straight and level with a constant velocity V in an easterly direction as shown in Fig. P5.3-1. It is tracked by a radar that provides measurements of range r and azimuth θ every 5 sec.

a. Let the state x be $[x_1 \ \dot{x}_1 \ x_2 \ \dot{x}_2]$. Write the state equations, adding acceleration disturbances $w_1(t) \sim (0, q_1)$ and $w_2(t) \sim (0, q_2)$ to the northerly and easterly motion, respectively.

b. Write the (nonlinear) measurement equation for data points $z_k = [r(kT) \ \theta(kT)]^T$, adding independent measurement noises $v_1(k) \sim (0, \sigma_r^2)$, $v_2(k) \sim (0, \sigma_\theta^2)$.

Compare to Example 2.4-3. In r–θ coordinates with straight line motion by the target we get nonlinear state equations with linear

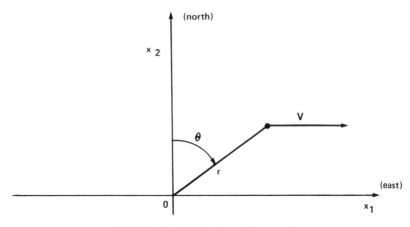

FIGURE P5.3-1   Radar tracking problem.

measurements. In $x-y$ coordinates with straight line target motion we get linear equations with nonlinear measurements!

c. Find the EKF time update for the estimate and error covariance. Simplify by preliminary analysis and write a subroutine F(TIME, X, XP) to implement the time update.

d. Find the EKF measurement update, simplify, and implement in a subroutine MEASUP(TIME, X, Z). Since the measurements of r and θ are assumed independent, two scalar updates can be performed at each sample time t_k to incorporate separately the effects of the two measurements. Write update subroutines for each measurement.

e. (Optional). Use the EKF driver program CDKAL in Appendix B to simulate a tracking run. (You will find that one of the problems with simulations is generating the data z_k.)

5.3-8:   Converting Nonlinear Systems to Linear Systems by Using Two Sets of Coordinates.   Consider the radar tracking situation in Fig. P5.3-1. To obtain linear state equations we can use $x-y$ coordinates. To obtain linear measurements we must use $r-\theta$ coordinates.

Explore the possibility of formulating the continuous-time Kalman filter with the time update in $x-y$ coordinates and the measurement update in $r-\theta$ coordinates. You will have to find a state-space transformation to apply to the state estimate before and after each measurement. It will also be necessary to determine how a state-space transformation affects the error covariance.

5.3-9:   Tracking a Body in Free Fall.   The state of a body falling vertically through the atmosphere is $x = [h \quad \dot{h} \quad \beta]$, where h is its height above the earth's surface and β is its *ballistic coefficient*. The ballistic coefficient is included as a state because it is not well known and so it must be estimated.

The state equations are

$$\frac{d\dot{h}}{dt} = D - g \tag{1a}$$

$$\frac{d\beta}{dt} = w(t), \tag{1b}$$

where $g = 9.8 \text{ m/sec}^2$ is acceleration due to gravity, $w(t) \sim (0, q)$ is white noise, and drag is given by

$$D = \frac{\rho \dot{h}^2}{2\beta}. \tag{2}$$

Atmospheric density is

$$\rho = \rho_0 e^{-h/c} \tag{3}$$

with $\rho_0 = 1220 \text{ g/m}^3$ the density at sea level and $c = 10263 \text{ m}$ a decay constant. Equation (3) is good to about 6000 m in the troposphere.

Suppose range measurements are taken every second as in Fig. P5.3-2. Thus

$$z_k = \sqrt{r_1^2 + (h_k - r_2)^2} + v_k \tag{4}$$

with $h_k \triangleq h(kT)$ and $v_k \sim (0, \sigma_r^2)$. (See Gelb 1974.)

It is desired to use the EKF to estimate the state.

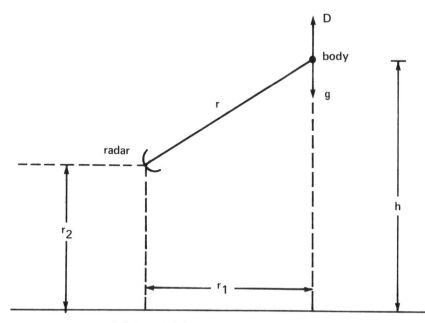

FIGURE P5.3-2 Tracking a free-falling body.

a. Write down the time update equations and simplify them by preliminary analysis. Write a subroutine F(TIME, X, XP) to update the estimate and covariance.

b. Write the measurement update and simplify by preliminary analysis. Write a subroutine MEASUP(TIME, X, Z) to implement this.

c. (Optional). Run a simulation using $q = 1000 \text{ g}^2/\text{m}^2\text{-sec}^6$, $h(0) \sim N(10{,}000 \text{ m}, \quad 500 \text{ m}^2)$, $\dot{h}(0) \sim N(-500 \text{ m/sec}, \quad 100 \text{ m}^2/\text{sec}^2)$, $\beta(0) \sim N(6 \times 10^7 \text{ g/m-sec}^2, \ 2 \times 10^{13} \text{ g}^2/\text{m}^2\text{-sec}^4)$. Use initial error variances of $p_{11}(0) = 50 \text{ m}^2$, $p_{22}(0) = 200 \text{ m}^2/\text{sec}^2$, $p_{33}(0) = 2 \times 10^{12} \text{ g}^2/\text{m}^2\text{-sec}^4$. Let $\sigma_r^2 = 5 \text{ m}^2/\text{Hz}$, and the radar tracker distances be $r_1 = 1000 \text{ m}$, $r_2 = 500 \text{ m}$. The measured data are taken every second beginning at $t_1 = 1$ sec, and are given in meters by $z_k = 9055, 8560, 7963, 7467, 7000, 6378, 5885, 5400, 4928, 4503$.

d. The acceleration due to gravity is not a constant but varies as

$$g = g_0 \left(\frac{R}{R+h} \right)^2 \tag{5}$$

where R is the radius of the earth and $g_0 = 9.8 \text{ m/sec}^2$. Is it worth accounting for this variation in this situation? How would we account for this if we wanted to?

5.3-10: Biological Treatment of Waste Sewage. An industrial plant's effluent waste is fed into a pool of bacteria that transforms the waste into nonpolluting forms. If the bacteria concentration is x_1 and the pollutant concentration is x_2, a state model is given by (Luenberger 1979)

$$\dot{x}_1 = \frac{a x_1 x_2}{x_2 + K} + w_1 \tag{1a}$$

$$\dot{x}_2 = -\frac{b x_1 x_2}{x_2 + K} + w_2 \tag{1b}$$

with $w_1(t) \sim N(0, q_1)$, $w_2(t) \sim N(0, q_2)$ uncorrelated white noises. Let measurements of x_2 be taken every hour, and suppose they are affected by the bacteria concentration so that

$$z_k = x_2(k) + c x_1^2(k) + v_k, \tag{2}$$

with noise $v_k \sim N(0, r)$.

a. Determine the EKF time update and simplify.

b. Determine the EKF measurement update and simplify.

c. Write subroutines F(TIME, X, XP) and MEASUP(TIME, X, Z) to implement the EKF.

PART 2
OPTIMAL STOCHASTIC CONTROL

6

STOCHASTIC
CONTROL FOR STATE
VARIABLE SYSTEMS

In Part 1 we showed how to estimate the state vector of a stochastic system given the noise-corrupted measurements $z(t)$ of the state. We shall now show how to use this state estimate for the optimal control of stochastic systems. Since nature is inherently noisy and seldom gives us information in the form in which we need it, the techniques we shall discuss here are usually required for the optimal control of actual physical systems. The deterministic optimal control theory is generally useful only in some ideal noise-free situations where the entire state can be exactly measured.

Although our presentation stands on its own, a background in the optimal control of deterministic systems given full state information will make the next two chapters more meaningful.

6.1 DYNAMIC PROGRAMMING APPROACH

Dynamic programming is based on Bellman's *principle of optimality* (Bellman 1957, Bellman and Dreyfus, 1962, Bellman and Kalaba, 1965):

> An optimal policy has the property that no matter what the previous decisions (i.e. controls) have been, the remaining decisions must constitute an optimal policy with regard to the state resulting from those previous decisions. (6.1-1)

283

We shall see that the principle of optimality serves to limit the number of potentially optimal control strategies that must be investigated. It also implies that optimal control strategies must be determined by working *backward* from the final stage; the optimal control problem is inherently a *backward-in-time problem*, in direct contrast to the optimal filtering problem which is a forward-in-time problem.

Dynamic programming can easily be used to find optimal controls for nonlinear stochastic systems. In this section we shall assume full state information; that is, that the entire state x_k is exactly measurable at time k. Thus, we only consider the effects of process noise, not measurement noise. First we cover discrete, and then continuous, systems.

Discrete-Time Systems

Consider the nonlinear stochastic plant

$$x_{k+1} = f^k(x_k, u_k) + Gw_k, \qquad k \geq i \qquad (6.1\text{-}2)$$

where w_k is a white noise sequence. Superscript k indicates time dependence. The initial state x_i is a random variable. To make the plant exhibit a desired behavior over a time interval $[i, N]$, we could try to select the control input u_k to minimize the *performance index*

$$J_i(x_i) = \phi(N, x_N) + \sum_{k=i}^{N-1} L^k(x_k, u_k), \qquad (6.1\text{-}3)$$

where ϕ and L^k are deterministic functions. However, since x_k is a random variable due to the random process noise w_k and initial state x_i, so also is the cost $J_i(x_i)$. To define a control problem that does not depend on the particular x_i or sequence w_k, let us select u_k to minimize not $J_i(x_i)$, but its expected value, the *expected cost*

$$j_i = E[J_i(x_i)]. \qquad (6.1\text{-}4)$$

We shall assume that x_k is exactly measurable at time k.

Example 6.1-1: Some Useful Performance Indices

To clarify the problem formulation it is worthwhile to discuss some common performance indices (6.1-3) that we can select for the given system (6.1-2).

a. Minimum-Time Problems

Suppose we want to find the control u_k to drive the system from the given initial state x_i to a desired final state $x \in R^n$ in minimum time. Then we could select the performance index

$$J = N - i = \sum_{k=i}^{N-1} 1 \qquad (1)$$

and specify the *boundary condition*

$$x_N = x. \qquad (2)$$

In this case $\phi = N - i$ and $L = 0$, or equivalently $\phi = 0$ and $L = 1$.

b. Minimum Fuel Problems

To find the scalar control u_k to drive the system from x_i to a desired final state x at a fixed time N using minimum fuel we could use

$$J = \sum_{k=i}^{N-1} |u_k|,\tag{3}$$

since the fuel burned is proportional to the magnitude of the control vector. Then $\phi = 0$ and $L^k = |u_k|$. The boundary condition $x_N = x$ would again apply.

c. Minimum Energy Problems

Suppose we want to find u_k to minimize the expected energy of the final state and all intermediate states, and also of the control used to achieve this. Let the final time N again be fixed. Then we could use

$$J = \frac{1}{2} s x_N^T x_N + \frac{1}{2} \sum_{k=i}^{N-1} (q x_k^T x_k + r u_k^T u_k),\tag{4}$$

where q, r, and s are scalar weighting factors. Then $\phi = \frac{1}{2} s x_N^T x_N$ and $L = \frac{1}{2}(q x_k^T x_k + r u_k^T u_k)$ are quadratic functions.

Minimizing the expected energy corresponds in some sense to keeping the state and the control close to zero. If it is more important to us that the intermediate state be small, then we should choose q *large* to weight it heavily in J, which we are trying to minimize. If it is more important that the control energy be small, then we should select a large value of r. If we are more interested in a small final state then s should be large.

For more generality, we could select weighting *matrices* Q, R, S instead of scalars. ∎

At this point, several things should be clearly understood. First, the system dynamics (6.1-2) are *given* by the physics of the problem, while the performance index (6.1-3) is what we *choose* to achieve the desired system response. Second, to achieve different control objectives, different types of performance indices J may be selected. Finally, the optimal control problem is characterized by *compromises and tradeoffs*, with different weighting factors in J resulting in different balances between conformability with performance objectives and magnitude of the required optimal controls.

Returning to the problem at hand, using Bayes' rule we may write (6.1-4) as

$$j_i = E_{x_i}\{E[J_i(x_i)/x_i]\},\tag{6.1-5}$$

where E_{x_i} is the expected value with respect to x_i. We shall find it very useful at times to consider the initial state x_i fixed. Let us therefore define

$$\bar{J}_i(x_i) = E[J_i(x_i)/x_i],\tag{6.1-6}$$

the expected cost with fixed initial state, or *conditional expected cost*.

Then

$$j_i = E_{x_i}[\bar{J}_i(x_i)].\tag{6.1-7}$$

It can be shown (Åström 1970) that, if asterisks denote optimal values,

$$j_i^* = \min_{u_k}(j_i) = E_{x_i}\{\min_{u_k}[\bar{J}_i(x_i)]\} = E_{x_i}[\bar{J}_i^*(x_i)]; \qquad (6.1\text{-}8)$$

that is, minimization with respect to the control sequence u_k, $i \le k \le N-1$, and expectation with respect to x_i commute. Thus, to minimize j_i we can equivalently minimize $\bar{J}_i(x_i)$ for each x_i (cf. (1.1-8)). This is the approach we shall take.

To see what the principle of optimality (6.1-1) looks like for this stochastic system, we proceed as follows (see Fel'dbaum 1965 and Åström 1970). Suppose we have computed the optimal conditional expected cost $\bar{J}_{k+1}^*(x_{k+1})$ and the associated optimal control sequence u_{k+1}^*, $u_{k+2}^*, \ldots, u_{N-1}^*$ for each admissible value of x_{k+1}. That is, the optimal control problem on the interval $[k+1, N]$ has been solved.

To solve the optimal control problem on $[k, N]$, first decompose $J_k(x_k)$ into two parts as

$$J_k(x_k) = L^k(x_k, u_k) + J_{k+1}(x_{k+1}). \qquad (6.1\text{-}9)$$

We assume throughout that x_k is fixed (i.e. known by some measurement process) at time k. Suppose first that the transition

$$\Delta x_k = x_{k+1} - x_k \qquad (6.1\text{-}10)$$

is deterministic (i.e., fixed). Then so is x_{k+1}, since x_k is exactly known. Taking expected values of (6.1-9) conditioned on Δx_k fixed, or equivalently on x_{k+1} fixed, yields

$$E[J_k(x_k)/x_{k+1}] = E[L^k(x_k, u_k)/x_{k+1}] + E[J_{k+1}(x_{k+1})/x_{k+1}]. \qquad (6.1\text{-}11)$$

The quantity $L^k(x_k, u_k)$ is a deterministic function of the control u_k which we are trying to determine, since x_k is given. Therefore,

$$E[J_k(x_k)/x_{k+1}] = L^k(x_k, u_k) + \bar{J}_{k+1}(x_{k+1}). \qquad (6.1\text{-}12)$$

According to the optimality principle (6.1-1), the only costs that need be considered at time k in determining u_k^* are those *the last portions of which are optimal*. This means that we may restrict our attention to the *admissible costs*

$$E[J_k(x_k)/x_{k+1}] = L^k(x_k, u_k) + \bar{J}_{k+1}^*(x_{k+1}). \qquad (6.1\text{-}13)$$

We now realize that in fact x_{k+1} is not deterministic. The transition Δx_k depends on the process noise w_k, so it is a random variable. According to (6.1-10), so is x_{k+1}. Using Bayes' rule and denoting expectation with respect to x_{k+1} as $E_{x_{k+1}}$, we obtain

$$\bar{J}_k(x_k) = L^k(x_k, u_k) + E_{x_{k+1}}[\bar{J}_{k+1}^*(x_{k+1})], \qquad (6.1\text{-}14)$$

since $L^k(x_k, u_k)$ is deterministic. Referring to the principle of optimality again we see that at time k the optimal conditional cost is

$$\bar{J}_k^*(x_k) = \min_{u_k}\{L^k(x_k, u_k) + E_{x_{k+1}}[\bar{J}_{k+1}^*(x_{k+1})]\}, \qquad (6.1\text{-}15a)$$

and the optimal control u_k^* is the value of u_k achieving the minimum.

This equation is the principle of optimality for the discrete nonlinear stochastic control problem with complete state information. Optimal controls found using this equation have the form of a *state-variable feedback*.

Because of the one-to-one relation (6.1-10) between Δx_k and x_{k+1}, we can replace the expectation in (6.1-15a) by $E_{\Delta x_k}$, which is sometimes more convenient.

Note also that since $L^k(x_k, u_k)$ is deterministic, (6.1-15a) can be written as

$$\bar{J}_k^*(x_k) = \min_{u_k} E_{x_{k+1}}[L^k(x_k, u_k) + \bar{J}_{k+1}^*(x_{k+1})]. \qquad (6.1\text{-}15b)$$

Examining (6.1-3) and (6.1-6), it is evident that the boundary condition for (6.1-15) is

$$\bar{J}_N^*(x_N) = E[\phi(N, x_N)/x_N] = \phi(N, x_N), \qquad (6.1\text{-}16)$$

if we assume that x_N is known at time $k = N$.

In terms of $\bar{J}_i^*(x_i)$ the minimal expected cost is

$$j_i^* = E_{x_i}[\bar{J}_i^*(x_i)]. \qquad (6.1\text{-}17)$$

Example 6.1-2: Dynamic Programming for a Scalar Discrete Stochastic System

To illustrate the use of (6.1-15), consider the stochastic plant

$$x_{k+1} = x_k + u_k + w_k, \qquad (1)$$

where process noise w_k is white with the PDF shown in Fig. 6.1-1. The performance

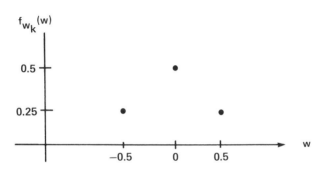

FIGURE 6.1-1 Process noise probability density function.

index selected is

$$J_0(x_0) = (x_N - 1)^2 + \frac{1}{2} \sum_{k=0}^{N-1} u_k^2. \tag{2}$$

For simplicity, let $N = 2$.

The *admissible controls* are

$$u_k = -0.5, 0, 0.5, \tag{3}$$

and the *admissible state values* are half-integers:

$$x_k = 0, \pm0.5, \pm1, \ldots. \tag{4}$$

It is desired to find the control input u_k to minimize $j_0 = E[J_0(x_0)]$, the expected cost. This will tend to keep the final state x_N near 1, but we would also like to ensure that the *final state constraint*

$$0 \le x_N \le 1.5 \tag{5}$$

holds. That is, the final state must belong to the *target set*

$$x_N \in (0, 0.5, 1, 1.5). \tag{6}$$

To solve this optimal control problem using (6.1-15), we can construct a decision grid in tabular form, showing the decision process at each stage separately.

First, set $k = N = 2$ to determine the final costs $\bar{J}_N(x_N)$. Since at $k = N$ no decision needs to be made, these costs are also equal to $\bar{J}_N^*(x_N)$. This step is simple, since we must only compute the deterministic quantity

$$\bar{J}_2^*(x_2) = (x_2 - 1)^2 \tag{7}$$

for each admissible value of x_2 in (6). The result is shown in Table 6.1-1.

Now decrement to $k = N - 1 = 1$ to find u_1^*. For each admissible value of state x_1, we must apply each admissible value of the control as u_1, and compute the costs

$$J_1(x_1) = \frac{u_1^2}{2} + \bar{J}_2^*(x_2) \tag{8}$$

for each resulting x_2. [Note: This is actually $J_1(x_1)/x_2$.] We also need to compute the probability $P(x_2)$ of each value of x_2 arising when u_1 is applied to x_1. Then (6.1-14) can be used to compute the admissible expected costs

$$\bar{J}_1(x_1) = E_{x_2}[J_1(x_1)] = \sum_{x_2} J_1(x_1)P(x_2). \tag{9}$$

Finally, we must use (6.1-15), comparing the $\bar{J}_1(x_1)$ for each value of u_1 to determine

TABLE 6.1-1   Final Costs

x_2	$\bar{J}_2(x_2) = (x_2 - 1)^2 = \bar{J}_2^*(x_2)$
1.5	0.25
1.0	0.0
0.5	0.25
0.0	1.0

the optimal expected cost

$$\bar{J}_1^*(x_1) = \min_{u_1}[\bar{J}_1(x_1)], \tag{10}$$

and to select u_1^*.

See Table 6.1-2. To accomplish all of this, suppose first that $x_1 = 1.5$. Then $u_1 = 0.5$ is clearly a bad choice since it will result in

$$x_2 = x_1 + u_1 + w_1 = 2 + w_1, \tag{11}$$

which does not satisfy (5) if $w_1 = 0$ or 0.5. Therefore, control $u_1 = 0.5$ is *inadmissible* if $x_1 = 1.5$. Let us next examine the control $u_1 = 0$.

If $x_1 = 1.5$ and $u_1 = 0$, then

$$x_2 = x_1 + u_1 + w_1 = 1.5 + w_1, \tag{12}$$

TABLE 6.1-2   First Stage Optimization

x_1	u_1	x_2	$P(x_2)$	$J_1(x_1) =$ $u_1^2/2 + \bar{J}_2^*(x_2)$	$\bar{J}_1(x_1)$	u_1^*	$\bar{J}_1^*(x_1)$
2.0	−0.5	2.0	0.25	—			
		1.5	0.5				
		1.0	0.25				
1.5	0.0	2.0	0.25	—			
		1.5	0.5				
		1.0	0.25				
	−0.5	1.5	0.25	0.375	0.25	−0.5	0.25
		1.0	0.5	0.125			
		0.5	0.25	0.375			
1.0	0.0	1.5	0.25	0.25	0.125	0.0	0.125
		1.0	0.5	0.0			
		0.5	0.25	0.25			
	−0.5	1.0	0.25	0.125	0.5		
		0.5	0.5	0.375			
		0.0	0.25	1.125			
0.5	0.5	1.5	0.25	0.375	0.25	0.5	0.25
		1.0	0.5	0.125			
		0.5	0.25	0.375			
	0.0	1.0	0.25	0.0	0.375		
		0.5	0.5	0.25			
		0.0	0.25	1.0			
0.0	0.5	1.0	0.25	0.125	0.5	0.5	0.5
		0.5	0.5	0.375			
		0.0	0.25	1.125			
−0.5	0.5	0.5	0.25	—			
		0.0	0.5				
		−0.5	0.25				

so that, using Fig. 6.1-1,

$$x_2 = \begin{cases} 2.0 & \text{with probability } 0.25 \\ 1.5 & \text{with probability } 0.5 \\ 1.0 & \text{with probability } 0.25 \end{cases} \tag{13}$$

This information is recorded in Table 6.1-2. Since $x_2 = 2$ is not admissible, and since it occurs with probability 0.25 if we select $u_1 = 0$, this control is also inadmissible if $x_1 = 0$. We symbolize this by a line in the $J_1(x_1)$ column, which simply means that further computations using $u_1 = 0$ are not required, since we cannot apply that control if $x_1 = 1.5$.

Now apply $u_1 = -0.5$ when $x_1 = 1.5$. The result is

$$x_2 = x_1 + u_1 + w_1 = 1.0 + w_1, \tag{14}$$

so that, using the PDF of w_1, the possible values of x_2 and their probabilities $P(x_2)$ are

$$x_2 = \begin{cases} 1.5 & \text{with probability } 0.25 \\ 1.0 & \text{with probability } 0.5 \\ 0.5 & \text{with probability } 0.25 \end{cases} \tag{15}$$

This is recorded in the table. Since all values of x_2 are in the target set, $u_1 = -0.5$ is an admissible control, so we can proceed to complete the information required in the table.

The cost (8) is given by

$$J_1(x_1 = 1.5) = \frac{u_1^2}{2} + \bar{J}_2^*(x_2) = 0.125 + \bar{J}_2^*(x_2) \tag{16}$$

$$= \begin{cases} 0.375 & \text{if } x_2 = 1.5 \\ 0.125 & \text{if } x_2 = 1.0, \\ 0.375 & \text{if } x_2 = 0.5 \end{cases}$$

where $\bar{J}_2^*(x_2)$ is found from Table 6.1-1. Now, the expected cost (9) is found in order to remove the dependence on the particular value of w_1:

$$\bar{J}_1(x_1 = 1.5) = E_{x_2}[J_1(x_1)]$$
$$= 0.375 \times 0.25 + 0.125 \times 0.5 + 0.375 \times 0.25 = 0.25. \tag{17}$$

At this point, $\bar{J}_1(x_1)$ has been determined for $x_1 = 1.5$ for all admissible values of u_1 (i.e., only one value: $u_1 = -0.5$). According to (10), then, $u_1^* = -0.5$, and $\bar{J}_1^*(x_1 = 1.5) = 0.25$.

We now assume $x_1 = 1.0$, and determine x_2, $P(x_2)$, $J_1(x_1 = 1.0)$ and $\bar{J}_1(x_1 = 1.0)$ for values of u_1 equal to 0 and -0.5 (clearly $u_1 = 0.5$ is inadmissible, since if the noise $w_1 = 0.5$, it results in $x_2 = x_1 + u_1 + w_1 = 2$). We obtain (see Table 6.1-2)

$$\bar{J}_1(x_1 = 1) = \begin{cases} 0.125 & \text{if } u_1 = 0 \\ 0.5 & \text{if } u_1 = -0.5 \end{cases}. \tag{18}$$

Performing the minimization (10), we select

$$u_1^* = 0, \qquad \bar{J}_1^*(x_1 = 1) = 0.125. \tag{19}$$

Similarly, we fill in the table for $x_1 = 0.5$ and $x_1 = 0$, using only control values which ensure that (5) is satisfied.

We have also filled in the table for $x_1 = 2$ and $x_1 = -0.5$. Note, however, that $x_1 = 2$ and $x_1 = -0.5$ are not admissible states, since no control drives them to a final state satisfying (5) with certainty.

Now decrement k to zero in order to find the optimal costs $\bar{J}_0^*(x_0)$ and u_0^* for each possible initial state x_0. In exactly the manner just described, Table 6.1-3 is completed. The required values of $\bar{J}_1^*(x_1)$ are taken from Table 6.1-2. Any combination of the state x_0 and control u_0 which results in $x_1 > 1.5$ or $x_1 < 0$ is not allowed since according to Table 6.1-2 these are inadmissible values of x_1.

The tables we have constructed contain the optimal feedback control law. To display it more clearly, a graph like Fig. 6.1-2 can be constructed, which shows for each state value the optimal control and *cost-to-go*. The arrows indicate $x_{k+1} = x_k + u_k$ for the given state and control; they ignore the effect of the noise. Such a graph could be called a "stochastic field of extremals" (Bryson and Ho 1975) since it has the form of a vector field. Changing the weightings in the performance index (2) will change the form of the graph (see the problems).

The figure indicates the optimal conditional cost $\bar{J}_0^*(x_0)$ for each x_0. To compute the optimal expected cost j_0^*, we require some statistics for initial state x_0. Let us suppose it is known that

$$x_0 = \begin{cases} 0 & \text{with probability } 0.5 \\ 0.5 & \text{with probability } 0.25 \\ 1.0 & \text{with probability } 0.125 \\ 1.5 & \text{with probability } 0.125. \end{cases} \qquad (20)$$

TABLE 6.1-3   Zeroth Stage Optimization

x_0	u_0	x_1	$P(x_1)$	$J_0(x_0) = u_0^2/2 + \bar{J}_1^*(x_1)$	$\bar{J}_0(x_0)$	u_0^*	$\bar{J}_0^*(x_0)$
1.5	−0.5	1.5	0.25	0.375	0.3125	−0.5	0.3125
		1.0	0.5	0.25			
		0.5	0.25	0.375			
1.0	0.0	1.5	0.25	0.25	0.1875	0.0	0.1875
		1.0	0.5	0.125			
		0.5	0.25	0.25			
	−0.5	1.0	0.25	0.25	0.40625		
		0.5	0.5	0.375			
		0.0	0.25	0.625			
0.5	0.5	1.5	0.25	0.375	0.3125		
		1.0	0.5	0.25			
		0.5	0.25	0.375			
	0.0	1.0	0.25	0.125	0.28125	0.0	0.28125
		0.5	0.5	0.25			
		0.0	0.25	0.5			
0.0	0.5	1.0	0.25	0.25	0.40625	0.5	0.40625
		0.5	0.5	0.375			
		0.0	0.25	0.625			

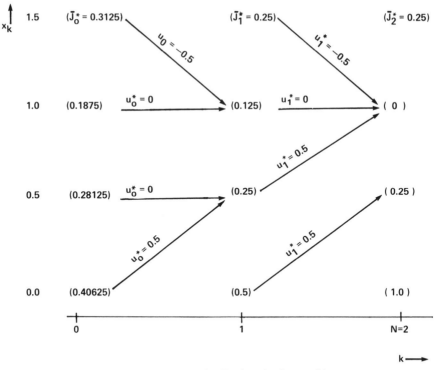

FIGURE 6.1-2   State feedback optimal control law.

Then, performing the expectation with respect to x_0 yields

$$j_0^* = E_{x_0}[\bar{J}_0^*(x_0)] = \sum_{x_0} \bar{J}^*(x_0)P(x_0) = 0.3359. \qquad (21)$$

∎

We have not yet discussed a very important aspect of optimal control for stochastic systems. We know how to find the optimal control minimizing an expected cost using (6.1-15), and the example demonstrated how easy it is to ensure that the final state x_N belongs to a specified target set. However, once the optimal control has been constructed, if we are given an initial state x_i the actual value of the final state is still not known! Given an initial state x_i, it is impossible to say with certainty what value x_N will take on if the optimal control u_k^* is applied, since w_k is unknown. Unlike the deterministic control problem, the optimal control sequence and state trajectory cannot be exactly predicted given x_i. This is because the extremal field is now stochastic, depending as it does on the values of the process noise w_k.

Given the *transition probabilities* that were computed to determine u_k^* [i.e., the $P(x_2)$ in Table 6.1-2 and the $P(x_1)$ in Table 6.1-3], we can determine the *transition PDF* $f(x_N/x_i)$ using the Chapman–Kolmogorov equation. Let $i = 0$ for notational simplicity. Since the process (6.1-2) is

Markov, we have $f(x_2/x_1, x_0) = f(x_2/x_1)$ and so on, so that

$$f(x_N/x_0) = \sum_{x_1, x_2, \ldots, x_{N-1}} f(x_N/x_{N-1}) \cdots f(x_2/x_1)f(x_1/x_0). \quad (6.1\text{-}18)$$

Let us demonstrate using Example 6.1-1.

Example 6.1-1 (cont.)

Suppose that it is required to determine the possible values of final state x_2 and their probabilities if $x_0 = 0.5$ and the optimal control scheme of Fig. 6.1-2 is used.
   Equation (6.1-18) becomes

$$f(x_2/x_0 = 0.5) = \sum_{x_1} f(x_2/x_1)f(x_1/x_0 = 0.5). \quad (22)$$

The PDFs $f(x_2/x_1)$ and $f(x_1/x_0)$ are found using Tables 6.1-2 and 6.1-3, respectively. Note that what we called $P(x_2)$ and $P(x_1)$ in those tables are actually the *conditional* probabilities $P(x_2/x_1)$ and $P(x_1/x_0)$. These are called *transition probabilities*.
   If $x_0 = 0.5$ and $u_0^* = 0$ is applied, then according to Table 6.1-3, x_1 takes on values of

$$x_1 = \begin{cases} 1.0 & \text{with probability } 0.25 \\ 0.5 & \text{with probability } 0.5 \\ 0 & \text{with probability } 0.25. \end{cases} \quad (23)$$

The conditional PDF $f(x_1/x_0 = 0.5)$ is therefore as shown in Fig. 6.1-3. Note that due to (1) ($u_0^* = 0$)

$$f(x_1/x_0 = 0.5) = f_{w_k}(w - 0.5). \quad (24)$$

Equation (22) can now be written as

$$f(x_2/x_0 = 0.5) = \frac{1}{4} f(x_2/x_1 = 1) + \frac{1}{2} f(x_2/x_1 = 0.5) + \frac{1}{4} f(x_2/x_1 = 0). \quad (25)$$

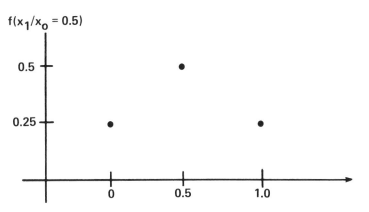

FIGURE 6.1-3 Transition PDF for $x_0 = 0.5$.

Next we determine the transition PDFs $f(x_2/x_1)$, assuming u_1^* is applied, from Table 6.1-2. The results are

$$f(x_2/x_1 = 1) = \begin{cases} \frac{1}{4}, & x_2 = 1.5 \\ \frac{1}{2}, & x_2 = 1.0 \\ \frac{1}{4}, & x_2 = 0.5 \end{cases} \tag{26}$$

and so on. Using these values in (25) yields

$$f(x_2/x_0 = 0.5) = \begin{cases} \frac{3}{16}, & x_2 = 1.5 \\ \frac{7}{16}, & x_2 = 1.0. \\ \frac{5}{16}, & x_2 = 0.5 \\ \frac{1}{16}, & x_2 = 0 \end{cases} \tag{27}$$

Thus, for example, the probability that $x_2 = 1$ given that $x_0 = 0.5$ and the optimal control is used is

$$P(x_2 = 1/x_0 = 0.5) = \tfrac{7}{16}. \tag{28}$$

∎

We shall now discuss an alternative to the performance index (6.1-3). Suppose it is desired to *maximize* the probability that x_N takes on a value in a given target set S (which might contain a single value r_N, the desired or "reference" final state). Then we could define the performance index

$$J_i(x_i) = P(x_N \in S/x_i). \tag{6.1-19}$$

Let us determine the form of the optimality principle in this case.

Using the Chapman–Kolmogorov equation, decompose $J_k(x_k)$ into two parts as

$$J_k(x_k) = \sum_{x_{k+1}} P(x_N \in S/x_{k+1})P(x_{k+1}/x_k, u_k)$$

$$= \sum_{x_{k+1}} J_{k+1}(x_{k+1})P(x_{k+1}/x_k, u_k), \tag{6.1-20}$$

where $P(x_{k+1}/x_k, u_k)$ is the transition probability from x_k to x_{k+1} if the control u_k is applied.

According to (6.1-1), we only need to consider costs the last portions of which are optimal:

$$J_k(x_k) = \sum_{x_{k+1}} P(x_{k+1}/x_k, u_k)J_{k+1}^*(x_{k+1}). \tag{6.1-21}$$

Maximizing over these admissible costs yields

$$J_k^*(x_k) = \max_{u_k} \sum_{x_{k+1}} P(x_{k+1}/x_k, u_k)J_{k+1}^*(x_{k+1}). \tag{6.1-22}$$

This is the principle of optimality for the plant (6.1-2) with performance index (6.1-19). The optimal control u_k^* is the value of u_k for which the maximum is attained.

We shall not present an example illustrating (6.1-22), but its use is very similar to that of (6.1-15). See Elbert (1984).

Continuous-Time Systems

The discrete stochastic principle of optimality (6.1-15) can be applied to continuous systems if they are first discretized and quantized. Note that the process noise values must also be quantized. This approach to the optimal control of stochastic systems is very suitable for programming on a digital computer, although the *curse of dimensionality* (Bellman and Kalaba 1965) is a limitation on the size of the systems that can be dealt with and the accuracy that can be achieved.

We shall now derive a partial differential equation that is satisfied by the optimal conditional expected cost in a continuous stochastic control problem. The steps are similar to those leading up to (6.1-15). See Fel'dbaum (1965).

Consider the nonlinear stochastic plant

$$\dot{x} = a(x, u, t) + w, \tag{6.1-23}$$

where $w(t) \sim N(0, \sigma^2)$ is white noise and $x(t_0)$ is random. For simplicity we consider the scalar case $x \in R$. Let us associate with (6.1-23) the performance index

$$J[x(t_0), t_0] = \phi[x(T), T] + \int_{t_0}^{T} L(x, u, t)\, dt, \tag{6.1-24}$$

where ϕ and L are deterministic functions.

We want to find the optimal control that minimizes the expected cost

$$j(t_0) = E\{J[x(t_0), t_0]\} \tag{6.1-25}$$

and ensures that the final state satisfies

$$\psi[x(T), T] = 0 \tag{6.1-26}$$

for a given deterministic function ψ. Final time T is fixed. The entire state $x(t)$ is assumed known by measurement at time t.

Using Bayes' rule we have

$$j(t_0) = E_{x(t_0)}\{\bar{J}[x(t_0), t_0]\}, \tag{6.1-27}$$

where the *conditional expected cost* assuming $x(t_0)$ is fixed is defined as

$$\bar{J}[x(t_0), t_0] = E\{J[x(t_0), t_0]/x(t_0)\}. \tag{6.1-28}$$

It can be shown [see (1.1-8)] that

$$j^*(t_0) = E_{x(t_0)}\{\bar{J}^*[x(t_0), t_0]\}, \tag{6.1-29}$$

so that the optimal control problem can be solved by minimizing the conditional cost. To solve this modified problem, we shall find a partial differential equation that can be solved for $\bar{J}^*(x, t)$.

Before we do this, the conditional PDF of the increment Δx given x and t must be determined, since it will soon be needed.

From the state equation, for small Δt we may write

$$\Delta x = a\Delta t + w\Delta t, \tag{6.1-30}$$

with the process $w\Delta t \sim N(0, \sigma^2\Delta t)$. Since $w\Delta t = \Delta x - a\Delta t$ we have (function of a random variable)

$$f(\Delta x/x, t) = \frac{1}{\sqrt{2\pi\sigma^2\Delta t}} e^{-(\Delta x - a\Delta t)^2/2\sigma^2\Delta t}, \tag{6.1-31}$$

so that Δx with x, t fixed is $N(a\Delta t, \sigma^2\Delta t)$. The PDF in (6.1-31) is conditional since $a(x, u, t)$ depends on x and t [and, to be complete, on $u(t)$]. Since Δx depends only on the current state and not previous states, the process $x(t)$ is Markov.

Now, let us focus our attention on a fixed current state and time (x, t). Let Δx be a deterministic (i.e., fixed) increment in x occurring over a small time Δt. Then (6.1-24) can be decomposed into two parts as

$$J(x, t) = \int_t^{t+\Delta t} L(x, u, \tau) \, d\tau + J(x + \Delta x, t + \Delta t). \tag{6.1-32}$$

Taking expected values conditioned on Δx being fixed yields

$$E[J(x, t)/\Delta x] = E\left[\int_t^{t+\Delta t} L(x, u, \tau) \, d\tau/\Delta x\right] + E\left[J(x + \Delta x, t + \Delta x)/\Delta x\right], \tag{6.1-33}$$

but the integral is a deterministic function of the control $u(\tau)$, $t \le \tau \le \Delta\tau$, since $x(t)$ is fixed. Therefore,

$$E[J(x, t)/\Delta x] = \int_t^{t+\Delta t} L(x, u, \tau) \, d\tau + \bar{J}(x + \Delta x, t + \Delta t). \tag{6.1-34}$$

(Note: x and Δx fixed implies $x + \Delta x$ fixed.)

According to Bellman's principle (6.1-1), the only costs we must consider at time t in determining the optimal policy are those of the form

$$E[J(x, t)/\Delta x] = \int_t^{t+\Delta t} L(x, u, \tau) \, d\tau + \bar{J}^*(x + \Delta x, t + \Delta t). \tag{6.1-35}$$

Now we consider the fact that Δx is not fixed; it depends on the process noise, and so it is a random variable. Its conditional PDF is given by

(6.1-31). Therefore, using Bayes' rule and denoting expectation with respect to Δx as $E_{\Delta x}$, we obtain

$$\bar{J}(x, t) = E_{\Delta x}\left[\int_t^{t+\Delta t} L(x, u, \tau)\, d\tau + \bar{J}^*(x + \Delta x, t + \Delta t)\right]$$

$$= \int_t^{t+\Delta t} L(x, u, \tau)\, d\tau + E_{\Delta x}[\bar{J}^*(x + \Delta x, t + \Delta t)]. \qquad (6.1\text{-}36)$$

Principle (6.1-1) therefore yields

$$\bar{J}^*(x, t) = \min_{\substack{u(\tau) \\ t \leq \tau \leq t+\Delta t}} \left[\int_t^{t+\Delta t} L(x, u, \tau)\, d\tau \right.$$

$$\left. + \int \bar{J}^*(x + \Delta x, t + \Delta t) f(\Delta x/x, t)\, d(\Delta x)\right], \qquad (6.1\text{-}37)$$

with $f(\Delta x/x, t)$ given by (6.1-31). This is the continuous stochastic principle of optimality assuming complete state information.

To find a method for determining analytic expressions for the optimal conditional cost $\bar{J}^*(x, t)$ and control $u^*(t)$, let us derive a partial differential equation satisfied by $\bar{J}^*(x, t)$.

Take a first-order approximation to the first integral in (6.1-37), and perform a Taylor series expansion of $\bar{J}^*(x + \Delta x, t + \Delta t)$ about (x, t). Then (recall $x \in R$)

$$\bar{J}^*(x, t) = \min_{\substack{u(\tau) \\ t \leq \tau \leq t+\Delta t}} \left[L(x, u, t)\Delta t \right.$$

$$+ \int \bar{J}^*(x, t) f(\Delta x/x, t)\, d(\Delta x)$$

$$+ \int \bar{J}_x^* \Delta x f(\Delta x/x, t)\, d(\Delta x) + \int \bar{J}_t^* \Delta t f(\Delta x/x, t)\, d(\Delta x)$$

$$+ \frac{1}{2} \int \bar{J}_{xx}^* (\Delta x)^2 f(\Delta x/x, t)\, d(\Delta x)$$

$$+ \frac{1}{2} \int \bar{J}_{tt}^* (\Delta t)^2 f(\Delta x/x, t)\, d(\Delta x)$$

$$\left. + \int \bar{J}_{xt}^* \Delta x \Delta t f(\Delta x/x, t)\, d(\Delta x) + \cdots \right]. \qquad (6.1\text{-}38)$$

Now observe that Δt, $\bar{J}^*$, and the partials $\bar{J}_x^*$, $\bar{J}_t^*$, $\bar{J}_{xx}^*$, $\bar{J}_{tt}^*$, $\bar{J}_{xt}^*$ can be removed from the integrals, and that the remaining mathematical expectations are either the area under the PDF (i.e., equal to 1) or the moments of $\Delta x \sim N(a\Delta t, \sigma^2 \Delta t)$. Therefore,

$$\bar{J}^*(x, t) = \min_{\substack{u(\tau) \\ t \le \tau \le t+\Delta t}} \left[L(x, u, t)\Delta t + \bar{J}^*(x, t) + \bar{J}^*_x a\Delta t \right.$$

$$\left. + \bar{J}^*_t \Delta t + \frac{1}{2} \bar{J}^*_{xx} \sigma^2 \Delta t + \frac{1}{2} \bar{J}^*_{tt} (\Delta t)^2 + \bar{J}^*_{xt} \Delta t(a\Delta t) + \cdots \right]. \tag{6.1-39}$$

[To first order, $E(\Delta x^2) = \sigma^2 \Delta t$.] Removing from under the minimization the quantities $\bar{J}^*(x, t)$ and $\bar{J}^*_t \Delta t$, which are independent of $u(\tau)$ for $t \le \tau \le t + \Delta t$, and neglecting second-order terms in Δt yields

$$0 = \bar{J}^*_t \Delta t + \min_{\substack{u(\tau) \\ t \le \tau \le t+\Delta t}} \left[L(x, u, t)\Delta t + \bar{J}^*_x a\Delta t + \frac{1}{2} \bar{J}^*_{xx} \sigma^2 \Delta t \right]. \tag{6.1-40}$$

Finally, we divide by Δt and let it become small to get

$$-\frac{\partial \bar{J}^*}{\partial t} = \min_{u(t)} \left[L + \frac{\partial \bar{J}^*}{\partial x} a + \frac{1}{2} \frac{\partial^2 \bar{J}^*}{\partial x^2} \sigma^2 \right]. \tag{6.1-41}$$

This is the *stochastic Hamilton–Jacobi–Bellman* (*HJB*) *equation* for the case of completely known state $x(t)$. In the vector case

$$\dot{x} = a(x, u, t) + Gw, \tag{6.1-42}$$

with $x \in R^n$ and white process noise $w(t) \sim (0, Q')$, it generalizes to

$$-\frac{\partial \bar{J}^*}{\partial t} = \min_{u(t)} \left[L + \left(\frac{\partial \bar{J}^*}{\partial x} \right)^T a + \frac{1}{2} \text{trace} \left(\frac{\partial^2 \bar{J}^*}{\partial x^2} GQ'G^T \right) \right]. \tag{6.1-43}$$

The stochastic HJB equation is solved backward for the optimal conditional cost with a boundary condition of

$$\bar{J}^*[x(T), T] = \phi[x(T), T] \text{ on the hypersurface } \psi[x(T), T] = 0. \tag{6.1-44}$$

Then $j^*(t)$ is given using (6.1-29) for general t.

Once $u(t)$ has been eliminated, the stochastic HJB equation becomes a *Kolmogorov equation*. In principle, it can be solved for analytic expressions for the optimal conditional cost and control, yielding a state feedback control law. In practice, the HJB equation has been solved for only a few cases. Fortunately, these cases are quite important, as the next example shows.

Example 6.1-3: HJB Derivation of Stochastic Linear Quadratic Regulator

Let the linear stochastic plant

$$\dot{x} = Ax + Bu + Gw \tag{1}$$

with white process noise $w(t) \sim N(0, Q')$ and $x(t_0)$ random have the associated *quadratic performance index*

$$J[x(t_0), t_0] = \frac{1}{2} x^T(T)S(T)x(T) + \frac{1}{2} \int_{t_0}^{T} (x^T Q x + u^T R u) \, dt, \tag{2}$$

with the final state $x(T)$ free. It is desired to minimize the expected cost

$$j(t_0) = E\{J[x(t_0), t_0]\} \tag{3}$$

if the final time T is fixed. The control $u(t)$ can depend on $x(t)$, which is assumed exactly known at time t. Assume $Q \geq 0$, $R > 0$.

a. Solution of HJB Equation

The stochastic HJB equation for the optimal conditional cost is

$$-\bar{J}_t^* = \min_{u(t)} \left[\tfrac{1}{2} x^T Q x + \tfrac{1}{2} u^T R u + (\bar{J}_x^*)^T (Ax + Bu) + \tfrac{1}{2} \text{trace}(\bar{J}_{xx}^* GQ'G^T) \right]. \tag{4}$$

Since the input is not constrained, the minimization can be performed by differentiating, so that

$$0 = Ru + B^T \bar{J}_x^* \tag{5}$$

and the optimal control is given in terms of $\bar{J}_x^*$ by

$$u = -R^{-1} B^T \frac{\partial \bar{J}^*}{\partial x}. \tag{6}$$

Using (6) to eliminate $u(t)$ in (4) yields the Kolmogorov equation

$$-\bar{J}_t^* = \tfrac{1}{2} x^T Q x - \tfrac{1}{2} (\bar{J}_x^*)^T BR^{-1} B^T \bar{J}_x^* + (\bar{J}_x^*)^T Ax + \tfrac{1}{2} \text{trace} \, (\bar{J}_{xx}^* GQ'G^T) \tag{7}$$

Let us initially make the naive assumption that $\bar{J}^*(x, t)$ has the same form as it does for the deterministic case (Bryson and Ho 1975, Lewis 1986). That is, assume

$$\bar{J}^*(x, t) = \tfrac{1}{2} x^T S x \tag{8}$$

for some as yet unknown deterministic symmetric matrix function $S(t)$. Then

$$\bar{J}_x^* = Sx, \qquad \bar{J}_t^* = \tfrac{1}{2} x^T \dot{S} x, \qquad \bar{J}_{xx}^* = S. \tag{9}$$

Using these in (7) results in

$$-\tfrac{1}{2} x^T \dot{S} x = \tfrac{1}{2} x^T Q x + x^T SAx - \tfrac{1}{2} x^T SBR^{-1} B^T Sx + \tfrac{1}{2} \text{trace} \, (SGQ'G^T), \tag{10}$$

$$0 = \tfrac{1}{2} x^T (\dot{S} + Q + A^T S + SA - SBR^{-1} B^T S)x + \tfrac{1}{2} \text{trace} \, (SGQ'G^T), \tag{11}$$

where we have replaced $x^T S A x$ by the symmetric form $\frac{1}{2} x^T (A^T S + SA) x$, which has the same value (Appendix A).

Unfortunately, (11) has no solution for $S(t)$ which is independent of $x(t)$. Evidently the assumption (8) is not valid. $\bar{J}^*(x, t)$ must contain a term whose partial derivative yields a term like $\frac{1}{2}$ trace$(SGQ'G^T)$. Let us therefore assume that

$$\bar{J}^*(x, t) = \frac{1}{2} \left[x^T S x + \text{trace} \int_t^T SGQ'G^T \, d\tau \right]. \tag{12}$$

Then $\bar{J}_x^*$ and $\bar{J}_{xx}^*$ are as in (9), but now

$$\bar{J}_t^* = \frac{1}{2}[x^T \dot{S} x - \text{trace}(SGQ'G^T)]. \tag{13}$$

Using these new partials in the HJB equation (7) results in

$$0 = \frac{1}{2} x^T (\dot{S} + A^T S + SA - SBR^{-1}B^T S + Q) x, \tag{14}$$

since the trace terms cancel. Now we can select $S(t)$ as the solution of

$$- \dot{S} = A^T S + SA - SBR^{-1}B^T S + Q, \tag{15}$$

with boundary condition (see (6.1-44)) of

$$S(T) \text{ given.} \tag{16}$$

This is the same Riccati equation appearing in the deterministic LQ regulator!

The optimal control can be computed as a state variable feedback, since (6) and (9) yield

$$u = -R^{-1}B^T S x. \tag{17}$$

The optimal conditional cost (12) on $[t, T]$ is given in terms of the state $x(t)$ at time t. To eliminate the need for knowing $x(t)$, we can take expectations with respect to $x(t)$ to get the optimal expected cost

$$j^*(t) = \frac{1}{2} E_x \left[x^T S x + \text{trace} \int_t^T SGQ'G^T \, d\tau \right] \tag{18}$$

$$= \frac{1}{2} \left[S(t)X(t) + \text{trace} \int_t^T SGQ'G^T \, d\tau \right].$$

We have used trace $(AB) = \text{trace}(BA)$ and the fact that $j^*(t)$ is a scalar. We have also defined

$$X(t) = E[x(t)x^T(t)] \tag{19}$$

the *mean-square value of* $x(t)$ [i.e., correlation of $x(t)$]. The optimal expected cost over $[t_0, T]$ is thus

$$j^*(t_0) = \frac{1}{2} \left[S(t_0)X(t_0) + \text{trace} \int_{t_0}^T SGQ'G^T \, d\tau \right]. \tag{20}$$

The effect of the process noise $w(t)$ is now clear: the solution to the stochastic optimal control problem, (15) and (17), is exactly the same as in the deterministic case. The only difference is that the term

$$\frac{1}{2} \text{trace} \int_{t_0}^T SGQ'G^T \, d\tau \tag{21}$$

must be added to the optimal cost. This represents the increase in cost due to the disturbing process noise with covariance Q'.

b. *Average Closed-Loop Behavior* (Bryson and Ho 1975)

In the deterministic case, as $(Q - R)$ tends to infinity, the LQ regulator will drive the state to zero as $(T - t_0)$ becomes large. In the stochastic case, this is not true. Due to the influence of the process noise $w(t)$, the state is continually disturbed from zero, and at steady state the behavior of the state $x(t)$ is determined by a balance between the disturbing effect of $w(t)$ and the regulating effect of the optimal control. The net result is that the mean-square value $X(t)$ of the state does not go to zero at steady state.

To find the mean-square state $X(t)$, write the closed-loop system under the influence of the optimal control (17) as

$$\dot{x} = (A - BK)x + Gw, \tag{22}$$

where $w(t) \sim (0, Q')$ and

$$K(t) = R^{-1}B^T S(t) \tag{23}$$

is the Kalman (control) gain. This is a stochastic system, and so from Example 3.1-1 the mean-square state is given by the Lyapunov equation

$$\dot{X} = (A - BK)X + X(A - BK)^T + GQ'G^T \tag{24}$$

with initial condition

$$X(t_0) = E[x(t_0)x^T(t_0)] = P_0 + x_0\bar{x}_0^T, \tag{25}$$

with $x_0 \sim (\bar{x}_0, P_0)$.

Note that the Riccati equation (15) is solved backward in time. Since it is independent of $x(t)$ it can be solved off-time. Then $K(t)$ is found using (23). Next, $X(t)$ can be determined using (24) by integrating *forward* in time given the initial value $X(t_0)$. Finally, the optimal expected cost is given by (20).

We can therefore determine both the expected cost and the mean-square value of the state *before* we actually apply the optimal control to the plant. Thus the performance of the control can be evaluated before we build the controller.

We can also determine the mean-square value of the control before it is applied, for by (17) and (23)

$$E(uu^T) = E(Kxx^T K^T) = KX(t)K^T. \tag{26}$$

At steady state, the mean-square value of $x(t)$ is found by solving the algebraic Lyapunov equation

$$0 = (A - BK)X + X(A - BK)^T + GQ'G^T. \tag{27}$$

c. *Non-Gaussian Process Noise*

The regulator (15), (17) is the optimal solution of the control problem (1)–(3) if $w(t)$ is normal. If $w(t)$ is not normal, then it provides the best control $u(t)$ which is a *linear* function of the state $x(t)$. See Kwakernaak and Sivan (1972) and Åstrom (1970). ∎

Example 6.1-4:   *Optimal Feedback Control and Simulation of a Scalar Stochastic System*

To show how to apply the results of the previous example and how to simulate stochastic control systems on a computer, let us consider the scalar plant

$$\dot{x} = ax + bu + gw \tag{1}$$

with white Gaussian process noise $w(t) \sim N(0, q')$ and $x(0) \sim N(\bar{x}_0, p_0)$. The performance index is

$$J = \frac{1}{2} s(T)x^2(T) + \frac{1}{2}\int_0^T (qx^2 + ru^2)\, dt. \tag{2}$$

It is desired to find a control to minimize the expected cost

$$j(0) = E(J), \tag{3}$$

and to find the resulting mean-square value of the state. The state $x(t)$ is assumed exactly measurable at each time t.

The LQ regulator is

$$u(t) = -K(t)x(t) \tag{4}$$

with Kalman gain

$$K(t) = s(t)b/r \tag{5}$$

where

$$-\dot{s} = 2as - b^2 s^2/r + q, \qquad t \le T \tag{6}$$

with $s(T)$ given. This could easily be solved by separation of variables as in Example 3.2-2.

The mean-square state, $X(t) = E[x^2(t)]$, resulting when this control is applied is given by the Lyapunov equation

$$\dot{X} = 2(a - bK)X + g^2 q', \qquad t \ge 0 \tag{7}$$

with $X(0) = p_0 + \bar{x}_0 \bar{x}_0^T$. The optimal expected cost on $[0, T]$ is

$$j^*(0) = \frac{1}{2}\left[s(0)X(0) + g^2 q' \int_0^T s(\tau)\, d\tau \right]. \tag{8}$$

Implementation of the stochastic LQ regulator is very easy. We can use program CTLQR in Appendix B. First the Riccati equation (6) is integrated backward to find $s(t)$, and the gain $K(t)$ is found and stored. Since the Runge–Kutta integrator in CTLQR works forward in time, we actually implement

$$\dot{s} = 2as - b^2 s^2/r + q, \qquad t \ge 0 \tag{9}$$

with $s(0)$ given, and then reverse the resulting function $s(t)$ in time. (Note if $\tau = T - t$ then $d/d\tau = -d/dt$.) Subroutines FRIC(TIME, S, $\dot{S}$) and FBGAIN(K, S, AK) to implement (9) and (5) are shown in Fig. 6.1-4.

The Lyapunov equation (7) must be integrated forward in time. Since the simulation using control (4) on the plant also runs forward in time, we can integrate (1) and (7) simultaneously. To avoid changing the driver program CTLQR we could proceed as follows.

```
C    SUBROUTINES FOR USE WITH CTLQR
C    CONTROL OF STOCHASTIC SCALAR SYSTEM

C    SUBROUTINE TO COMPUTE RICCATI EQUATION SOLUTION

           SUBROUTINE FRIC(TIME,S,SP)
           PARAMETER (NX= 30)
           REAL S(*), SP(*)
           COMMON/PLANT/A,B
           COMMON/WEIGHTS/Q(NX,NX),R

           BSQ= B**2
           SP(1)= 2.*A*S(1) - BSQ*S(1)**2/R + Q(1,1)

           RETURN
           END

C    SUBROUTINE TO COMPUTE OPTIMAL FEEDBACK GAINS

           SUBROUTINE FBGAIN(K,S,AK)
           PARAMETER (NX= 30)
           REAL S(*), AK(NX,1)
           COMMON/PLANT/A,B
           COMMON/WEIGHTS/Q(NX,NX),R

           AK(1,K)= B*S(1)/R

           RETURN
           END

C    SUBROUTINE TO COMPUTE OPTIMAL CONTROL INPUT
C    (U(2) IS USED TO PASS THE KALMAN GAIN TO SUBROUTINE F(TIME,X,XP)

           SUBROUTINE CONUP(K,X,AK)
           PARAMETER (NX= 30)
           REAL X(*), AK(NX,1)
           COMMON/COMMAND/U(2)

           U(1)= -AK(1,K)*X(1)
           U(2)= AK(1,K)

           RETURN
           END

C    SCALAR SYSTEM STATE EQUATIONS
C    AND LYAPUNOV EQUATION FOR MEAN SQUARE VALUE OF STATE

           SUBROUTINE F(TIME,X,XP)
           PARAMETER (NX= 30)
           REAL X(*), XP(*)
           COMMON/COMMAND/U(2)
           COMMON/PLANT/A,B
           COMMON/WEIGHTS/Q(NX,NX), R
           DATA Q(1,1),R,QP/1.,1.,5./
           DATA A,B,G/.05,1.,1./
           DATA TWOPI/6.2831853/

C    MANUFACTURE GAUSSIAN PROCESS NOISE

           W= QP * SQRT( -2.*ALOG(RANF()) ) * COS( TWOPI*RANF() )

C    PLANT DYNAMICS AND LYAPUNOV EQUATION

           XP(1)= A*X(1) + B*U(1) + G*W
           XP(2)= 2*(A-B*U(2))*X(2) + G**2*QP

           RETURN
           END
```

FIGURE 6.1-4 Subroutines for use with CTLQR for stochastic LQ regulator.

303

Implement the plant dynamics (1) and also (7) in subroutine F(TIME, X, Ẋ), where the vector in the argument X is integrated forward by the Runge–Kutta routine. To do this, assign the components of argument X as $X(1) = x(t)$ the state, and $X(2) = X(t)$ the mean-square value of the state. This subroutine is also shown in Fig. 6.1-4.

The Kalman gain $K(t)$ is needed for (7). To avoid adding it as an argument to F(TIME, X, Ẋ), which would entail changing the driver program CTLQR, we can simply pass it as U(2), the second component of the "control input vector," which is already in COMMON. U(1) is the plant input $u(t)$. Therefore, subroutine CONUP(K, X, AK) must compute $u(t)$ by (4) from the current state and also store $U(2) = K(t)$ at each Runge–Kutta step. This control update subroutine is shown in Fig. 6.1-4.

A word on generating the normal process $w(t) \sim N(0, q')$ is in order. FORTRAN routine RANF( ) generates a random variable z that is uniformly distributed on $[0, 1]$. To convert this to a normally distributed RV, we use the line of code in Fig. 6.1-4 that defines a function of z that is normal with mean of zero and variance of q'.

The results of a computer simulation using this software are shown in Fig. 6.1-5. The final state weighting was $s(T) = 0$, and the control weighting was $r = 1$. Note that as state weighting q increases, the steady-state values of the mean and mean-square of the state $x(t)$ both decrease. The Runge–Kutta step size was 25 msec. ∎

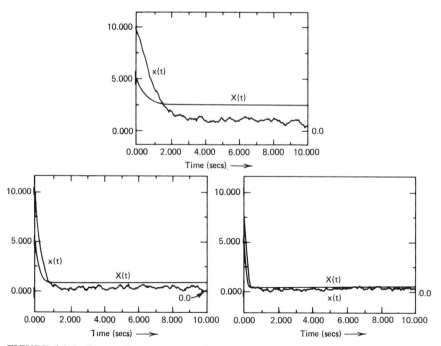

FIGURE 6.1-5 State and mean-square value of state for stochastic LQ regulator. (a) $q = 1$. (b) $q = 10$. (c) $q = 100$.

6.2 CONTINUOUS-TIME LINEAR QUADRATIC GAUSSIAN PROBLEM

In this section we discuss the linear stochastic plant.

$$\dot{x} = Ax + Bu + Gw \qquad (6.2\text{-}1)$$

with white process noise $w(t) \sim N(0, Q')$ and $x(t_0) \sim N(\bar{x}_0, P_0)$ a random variable. The associated performance index is the quadratic form

$$J[x(t_0), t_0] = \frac{1}{2} x^T(T)S(T)x(T) + \frac{1}{2} \int_{t_0}^{T} (x^T Q x + u^T R u) \, dt \qquad (6.2\text{-}2)$$

with symmetric weighting matrices $S(T) \geq 0$, $Q \geq 0$, $R > 0$. The plant and weighting matrices can be functions of time. It is desired to determine the control $u^*(t)$ on $[t_0, T]$ which minimizes the expected cost

$$j(t_0) = E\{J[x(t_0), t_0]\}, \qquad (6.2\text{-}3)$$

with $x(T)$ free and T fixed.

This is called the *linear quadratic Gaussian (LQG) control problem*.

Two cases must be considered. In the case of *complete state information*, the state vector $x(t)$ is exactly known at time t, so $u(t)$ can be expressed as a function of $x(t)$; that is, as a state variable feedback. In the case of *incomplete state information*, the state is measured in the presence of Gaussian noise to produce the observation, or measurement, vector

$$z = Hx + v, \qquad (6.2\text{-}4)$$

with measurement noise $v(t) \sim N(0, R')$. Now the control $u(t)$ can only depend on the *estimate* $\hat{x}(t)$ of the state given the measurements.

The results we are about to derive represent the optimal control if the noise(s) are Gaussian. In the situation where the noise statistics are not Gaussian, they yield the best *linear* control law (Åström 1970, Kwakernaak and Sivan 1972).

Complete State Information

Here we rederive the results of Example 6.1-3 by a different approach.

Suppose that the state $x(t)$ of (6.2-1) is exactly measurable at time t. Then admissible feedback control laws are of the form

$$u(t) = K[x(t), t] \qquad (6.2\text{-}5)$$

where K is a general nonlinear function. Admissible *linear* feedback control laws have the form

$$u(t) = K(t)x(t). \qquad (6.2\text{-}6)$$

We shall see that *the* optimal control is the linear state variable feedback (6.2-6) if all noises are normal.

As a preliminary step we would like to find an expression for $(d/dt)E(x^T S x)$, where $S(t)$ is a symmetric function of time and $x(t)$ described by (6.2-1) is stochastic with a PDF of $f_x(x)$. Equation (5.2-5) provides an expression for $(d/dt)E\{\phi[x(t)]\}$, where $\phi[x(t)]$ is an arbitrary twice-differentiable scalar function of $x(t)$. The quadratic form $x^T(t)S(t)x(t)$ is of the form $\phi[x(t), t]$. Carrying out a derivation like the one leading to (5.2-5) for the case where ϕ is an explicit function of x *and* of t, we obtain instead of (5.2-5):

$$\frac{d}{dt} E[\phi(x, t)] = \frac{d}{dt} \int \phi(x, t) f_x(x) \, dx$$

$$= \int \left[\frac{\partial \phi(x, t)}{\partial t} f_x(x) + \phi(x, t) \frac{\partial f_x(x)}{\partial t} \right] dx$$

$$= E\left[\frac{\partial \phi}{\partial t} \right] + \int \phi(x, t) \frac{\partial f_x(x)}{\partial t} \, dx. \tag{6.2-7}$$

Continuing the derivation, the final result is clearly (note that $a(x, t) = Ax + Bu$)

$$\frac{d}{dt} E[\phi(x, t)] = E(\phi_t) + E[\phi_x^T(Ax + Bu)] + \frac{1}{2} \text{trace } E(GQ'G^T \phi_{xx}). \tag{6.2-8}$$

Substituting $\phi(x, t) = x^T S x$ we obtain

$$\frac{d}{dt} E(x^T S x) = E(x^T \dot{S} x) + 2E[x^T S(Ax + Bu)] + \text{trace } E(GQ'G^T S)$$

$$= E(x^T \dot{S} x) + E[x^T S(Ax + Bu) + (Ax + Bu)^T S x]$$

$$+ \text{trace}(SGQ'G^T), \tag{6.2-9}$$

where we have used a symmetric form for the scalar $x^T S(Ax + Bu)$ (Appendix A). This holds for any symmetric matrix $S(t)$.

Now note the identity

$$\frac{1}{2} \int_{t_0}^{T} \frac{d}{dt} E(x^T S x) \, dt = \frac{1}{2} E[x^T(T)S(T)x(T)] - \frac{1}{2} E[x^T(t_0)S(t_0)x(t_0)]. \tag{6.2-10}$$

Using these two preliminary results, we can solve the LQG control problem with complete state information. Add zero in the form of the left side of (6.2-10) minus its right side to (6.2-2), and then use (6.2-9), to obtain for the expected cost

$$j(t_0) = \frac{1}{2} E[x^T(t_0)S(t_0)x(t_0)] + \frac{1}{2} E \int_{t_0}^{T} [x^T(\dot{S} + A^T S + SA + Q)x$$

$$+ x^T SBu + u^T B^T S x + u^T R u] \, dt + \frac{1}{2} \text{trace} \int_{t_0}^{T} SGQ'G^T \, dt. \tag{6.2-11}$$

Suppose that $S(t)$ is selected to satisfy the Riccati equation

$$-\dot{S} = A^T S + SA - SBR^{-1}B^T S + Q \qquad (6.2\text{-}12)$$

for $t \leq T$. Then (6.2-11) becomes

$$j(t_0) = \frac{1}{2} E[x^T(t_0)S(t_0)x(t_0)] + \frac{1}{2} E \int_{t_0}^{T} [x^T SBR^{-1}B^T Sx$$

$$+ x^T SBu + u^T B^T Sx + u^T Ru]\, dt + \frac{1}{2}\, \text{trace} \int_{t_0}^{T} SGQ'G^T\, dt, \qquad (6.2\text{-}13)$$

and the first integrand is now the perfect square of a weighted norm (McReynolds 1966), so that

$$j(t_0) = \frac{1}{2} E[x^T(t_0)S(t_0)x(t_0)] + \frac{1}{2} E \int_{t_0}^{T} \|R^{-1}B^T Sx + u\|_R^2\, dt$$

$$+ \frac{1}{2}\, \text{trace} \int_{t_0}^{t} SGQ'G^T\, dt. \qquad (6.2\text{-}14)$$

Note that the Riccati equation arises quite naturally in completing the square in the first integrand.

Since only the second term depends on $u(t)$, the minimum value of $j(t_0)$ is

$$j^*(t_0) = \frac{1}{2} E[x^T(t_0)S(t_0)x(t_0)] + \frac{1}{2}\, \text{trace} \int_{t_0}^{T} SGQ'G^T\, dt, \qquad (6.2\text{-}15)$$

and it is achieved for the optimal control

$$u^*(t) = -R^{-1}B^T S(t)x(t), \qquad (6.2\text{-}16)$$

where $S(t)$ satisfies (6.2-12). The boundary condition for the Riccati equation is $S(T)$, the final state weighting matrix in $J(t_0)$, since then $j^*(T) = j(T)$.

We have just discovered that the solution to the stochastic LQG control problem with complete state information is *identical to the deterministic LQ regulator* (Bryson and Ho 1975, Kwakernaak and Sivan 1972, Lewis 1986), except that the a term is added to $j^*(t_0)$ to account for the cost increase due to the process noise with covariance Q'. This is the same result obtained by dynamic programming in Example 6.1-3.

For computational purposes, note that the trace and integration operators commute.

In terms of the Kalman gain

$$K(t) = R^{-1}B^T S(t) \qquad (6.2\text{-}17)$$

the optimal control can be written as

$$u(t) = -K(t)x(t), \qquad (6.2\text{-}18)$$

which is a linear time-varying state variable feedback.

It would be useful to know the answer to another question. Suppose that the control input is selected as the suboptimal feedback (6.2-18), where $K(t)$ is any given state feedback gain, not necessarily the optimal gain (6.2-17). Then what is the resulting value of the expected cost $j(t_0)$? The motivation for our interest is that it is often convenient to use a suboptimal feedback of simplified form; for example, the constant steady-state feedback

$$K_\infty = R^{-1}B^T S(\infty), \tag{6.2-19}$$

where $S(\infty)$ is the limiting Riccati solution. We might want to determine the resulting $j(t_0)$ to see if the closed-loop plant behavior using this simplified control law is acceptable.

To answer this question, use (6.2-18) in (6.2-11) to obtain

$$j(t_0) = \frac{1}{2} E[x^T(t_0) S(t_0) x(t_0)]$$

$$= +\frac{1}{2} E \int_{t_0}^{T} x^T[\dot{S} + (A - BK)^T S + S(A - BK) + Q + K^T RK]x \, dt$$

$$+ \frac{1}{2} \text{trace} \int_{t_0}^{T} SGQ'G^T \, dt. \tag{6.2-20}$$

Now select $S(t)$ to satisfy the Lyapunov equation

$$-\dot{S} = (A - BK)^T S + S(A - BK) + Q + K^T RK \tag{6.2-21}$$

for $t \geq T$, which is expressed in terms of the known plant matrix of the closed-loop system

$$\dot{x} = (A - BK)x + Gw. \tag{6.2-22}$$

Then the expected cost resulting from *any* feedback control of the form $u = -Kx$ is equal to

$$j(t_0) = \frac{1}{2} E[x^T(t_0) S(t_0) x(t_0)] + \frac{1}{2} \text{trace} \int_{t_0}^{T} SGQ'G^T \, dt, \tag{6.2-23}$$

where $S(t)$ satisfies (6.2-21). The boundary condition for the Lyapunov equation is $S(T)$.

If $K(t)$ is selected as the optimal feedback gain (6.2-17), then (6.2-21) becomes the Joseph stabilized Riccati equation.

To determine how successful the control is at regulating the state in the presence of process noise, we can examine the mean-square value of the state

$$X(t) = E[x(t)x^T(t)]. \tag{6.2-24}$$

Since (6.2-22) is a stochastic system (Example 3.1-1), we have

$$\dot{X} = (A - BK)X + X(A - BK)^T + GQ'G^T, \tag{6.2-25}$$

for $t \geq t_0$, where $X(t_0) = P_0 + \bar{x}_0 \bar{x}_0^T$ is given. If $X(t)$ is small, then the regulating action of the control is satisfactory.

It is worth comparing the two Lyapunov equations (6.2-21) and (6.2-25).

In terms of the mean-square value of the initial state, the optimal expected cost can be written as

$$j(t_0) = \frac{1}{2} \operatorname{trace} \left[S(t_0) X(t_0) + \int_{t_0}^{T} SGQ'G^T \, dt \right]. \tag{6.2-26}$$

According to (6.2-18), the mean-square value of the control is given by

$$E[u(t)u^T(t)] = K(t)X(t)K^T(t). \tag{6.2-27}$$

It is useful in predicting the control effort required by the LQG regulator.

An illustration of the software simulation of the LQG regulator is provided by Example 6.1-4.

Incomplete State Information and the Separation Principle

Now let us assume that the entire state is not available. Instead, there are available measurements $z(t)$ given by (6.2-4). Then the optimal estimate of the state for *any* input $u(t)$ is given by the Kalman filter equations from Table 3.1-1:

$$\dot{P} = AP + PA^T + GQ'G^T - PH^T(R')^{-1}HP, \quad t \geq t_0, \quad P(t_0) = P_0 \tag{6.2-28}$$

$$L = PH^T(R')^{-1} \tag{6.2-29}$$

$$\dot{\hat{x}} = A\hat{x} + Bu + L(z - H\hat{x}), \qquad \hat{x}(0) = \bar{x}_0, \tag{6.2-30}$$

where $P(t)$ is the error covariance, $L(t)$ is the Kalman filter gain, and $\hat{x}(t)$ is the state estimate. Recall that $\hat{x}(t)$ and $P(t)$ are also the conditional mean and covariance of $x(t)$ given the control input and measurements prior to time t (Section 1.1).

The admissible controls are now functions of the *estimate* $\hat{x}(t)$, not the unknown state $x(t)$. To find the optimal control under these circumstances, we return to (6.2-14).

Only the second term there depends on the input $u(t)$. Defining the set of data available at time t as

$$Z_t = \{z(\tau) \mid t_0 \leq \tau \leq t\}, \tag{6.2-31}$$

we can write this term as

$$\frac{1}{2} E \int_{t_0}^{T} \| R^{-1}B^T Sx + u \|_R^2 \, dt = \frac{1}{2} E_{Z_t} \int_{t_0}^{T} E[(u + Kx)^T R(u + Kx)/Z_t] \, dt \tag{6.2-32}$$

with the Kalman control gain defined in (6.2-17) and E_{Z_t} denoting expectation over Z_t.

To proceed, we require a preliminary result (Åström 1970). Suppose that an arbitrary random variable x has mean $\bar{x}$ and covariance P_x. Then we can write, for a symmetric deterministic $R \geq 0$,

$$E(x^T R x) = E(x - \bar{x})^T R (x - \bar{x}) + \bar{x}^T R \bar{x}$$

$$= \bar{x}^T R \bar{x} + \text{trace } E[R(x - \bar{x})(x - \bar{x})^T],$$

or

$$E(x^T R x) = \bar{x}^T R \bar{x} + \text{trace}(RP_x). \tag{6.2-33}$$

This expression also holds if $\bar{x}$ and P_x are a conditional mean and covariance and E is a conditional expectation.

Using this identity, we can write (6.2-32) as

$$\frac{1}{2} E \int_{t_0}^{T} \| R^{-1} B^T S x + u \|_R^2 \, dt = \frac{1}{2} E_{Z_t} \int_{t_0}^{T} (u + K\hat{x})^T R (u + K\hat{x}) \, dt$$

$$+ \frac{1}{2} \text{trace} \int_{t_0}^{T} K^T R K P \, dt. \tag{6.2-34}$$

Since minimization with respect to $u(t)$ and the operator E_{Z_t} commute, the optimal control is

$$u(t) = - K(t)\hat{x}(t). \tag{6.2-35}$$

Using the minimum value of (6.2-34) in (6.2-14) yields the optimal expected cost

$$j^*(t_0) = \frac{1}{2} E[x^T(t_0) S(t_0) x(t_0)] + \frac{1}{2} \text{trace} \int_{t_0}^{T} SGQ'G^T \, dt$$

$$+ \frac{1}{2} \text{trace} \int_{t_0}^{T} K^T R K P \, dt. \tag{6.2-36}$$

(For computational purposes, trace and integration may be interchanged.)

This is quite an important result known as the *separation principle*. (It is also called the *certainty equivalence principle*, however, in adaptive control theory this term means something a little bit different.) It can be stated as follows. The solution to the *stochastic* LQG control problem with incomplete state information is to use the optimal *deterministic* LQ regulator, except that the control is a feedback (6.2-35) of the optimal state *estimate* which is constructed using the Kalman filter in Table 3.1-1. The resulting optimal expected cost is given by (6.2-36), where the first term comes from the deterministic LQ regulator, the second term represents the increase in cost due to the disturbing influence on $x(t)$ of the process noise, and the last term is the increase in cost due to measurement uncertainties.

The importance of the separation principle is that the LQG regulator design procedure can be accomplished in two separate stages: the Kalman filter design and the control feedback design. This means that all results derived separately for the deterministic optimal control problem and the optimal estimation problem are still valid. For instance, if the process and measurement noises are correlated, we need only use the modified filter gain in Section 3.6.

If the noise statistics are not Gaussian, then the separation principle yields the best *linear* regulator in terms of the best *linear* estimate of the state.

The form of the LQG regulator is shown in Fig. 6.2-1, where a reference or command input $r(t)$ has been included and $s(t)$ is an intermediate signal to be used later. The main regulator equations are

$$\dot{\hat{x}} = (A - LH)\hat{x} + Bu + Lz \qquad (6.2\text{-}37)$$

$$u = -K\hat{x} + r, \qquad (6.2\text{-}38)$$

which describe a *dynamic output feedback*. That is, the plant output $z(t)$ and the control input $u(t)$ are processed by a dynamical system (6.2-37), whose output $K(t)\hat{x}(t)$ is returned as the control. The function of the added dynamics in such a scheme is to reconstruct information about the state that is not directly measurable at the output. In our present application, the state of the dynamic output feedback system is actually an optimal estimate for the state of the plant. This output feedback should be contrasted with the state feedback we used in the case of complete state information.

Since we are considering a stochastic plant, we would like some feel for how small the regulator keeps the state. Thus, it would be useful to determine the mean-square value of the state $X(t) = E[x(t)x^T(t)]$ under the influence of our LQG regulator. See Bryson and Ho (1975).

The dynamics of the estimation error

$$\bar{x}(t) = x(t) - \hat{x}(t) \qquad (6.2\text{-}39)$$

are given by (Section 3.4)

$$\dot{\bar{x}} = (A - LH)\bar{x} + Gw - Lv. \qquad (6.2\text{-}40)$$

Note that $\bar{x}(t)$ is independent of the control, which is one of the reasons the

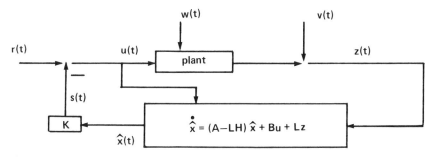

FIGURE 6.2-1   LQG regulator with incomplete state information.

separation principle works. [The other is our ability to express (6.2-34) in terms of $\hat{x}(t)$.] Since (6.2-40) is a stochastic system, we can use Example 3.1-1 to see that the error covariance satisfies

$$\dot{P} = (A - LH)P + P(A - LH)^T + GQ'G^T + LR'L^T, \qquad t \geq t_0$$
$$(6.2\text{-}41)$$

which is a Joseph stabilized version of (6.2-28).

Now we turn to (6.2-30). Using the control (6.2-35) and introducing the residual

$$\tilde{z}(t) = H\tilde{x}(t) + v(t), \qquad\qquad (6.2\text{-}42)$$

we can write the estimate dynamics as

$$\dot{\hat{x}} = (A - BK)\hat{x} + L\tilde{z}. \qquad\qquad (6.2\text{-}43)$$

Since $\tilde{z}(t)$ is a white noise process with mean zero and covariance R' which is independent of $\hat{x}$ (Section 3.4), the mean-square value of the *estimate*

$$\hat{X}(t) \triangleq E[\hat{x}(t)\hat{x}^T(t)] \qquad\qquad (6.2\text{-}44)$$

propagates according to

$$\dot{\hat{X}} = (A - BK)\hat{X} + \hat{X}(A - BK)^T + LR'L^T. \qquad (6.2\text{-}45)$$

What we are interested in determining here is $X(t)$. But

$$X(t) = E[xx^T] = E[(\hat{x} + \tilde{x})(\hat{x} + \tilde{x})^T]$$

so that

$$X(t) = \hat{X}(t) + P(t), \qquad\qquad (6.2\text{-}46)$$

where we have used the fact that $E(\tilde{x}) = 0$. Therefore, the mean-square value of the state $X(t)$ is the sum of the mean-square value of the estimate $\hat{X}(t)$ and the error covariance $P(t)$, which are found using (6.2-45) and (6.2-41).

Since the initial mean $\bar{x}_0$ and covariance P_0 of the state are known, we have

$$X(t_0) = P_0 + \bar{x}_0\bar{x}_0^T. \qquad\qquad (6.2\text{-}47)$$

Therefore, by (6.2-46),

$$\hat{X}(t_0) = X(t_0) - P_0 = x_0 x_0^T \qquad\qquad (6.2\text{-}48)$$

is the initial condition for (6.2-45).

In terms of the mean-square value of the initial state, the optimal expected cost can be written as

$$j(t_0) = \frac{1}{2} S(t_0) X(t_0) + \frac{1}{2} \operatorname{trace} \int_{t_0}^{T} SGQ'G^T \, dt + \frac{1}{2} \operatorname{trace} \int_{t_0}^{T} K^T RKP \, dt.$$
$$(6.2\text{-}49)$$

The mean-square value of the control is useful to know so that we can predict the control effort required. According to (6.2-35), it is given by

$$E[u(t)u^T(t)] = K(t)\hat{X}(t)K^T(t). \tag{6.2-50}$$

The software simulation of the LQG regulator with incomplete state information is a straightforward extension of the procedure in Example 6.1-4. Briefly, we could use program CTLQR in Appendix B. The backward dynamics (6.2-12) are implemented in subroutine FRIC, with the control gain computation (6.2-17) in subroutine FBGAIN. The forward dynamics (6.2-1) and (6.2-28)–(6.2-30) are implemented in subroutine F, with the control update (6.2-35) in subroutine CONUP. Equations (6.2-45), (6.2-46), and (6.2-50) can also be placed into subroutine F.

6.3 DISCRETE-TIME LINEAR QUADRATIC GAUSSIAN PROBLEM

Our concern here is the discrete linear stochastic plant

$$x_{k+1} = Ax_k + Bu_k + Gw_k, \qquad k \geqslant i, \tag{6.3-1}$$

with white process noise $w_k \sim N(0, Q')$ and initial state $x_i \sim N(\bar{x}_i, P_i)$. The performance index over the time interval of interest $[i, N]$ is

$$J_i(x_i) = \frac{1}{2} x_N^T S_N x_N + \frac{1}{2} \sum_{k=i}^{N-1} (x_x^T Q x_k + u_k^T R u_k), \tag{6.3-2}$$

with $S_N \geq 0$, $Q \geq 0$, $R > 0$. If the plant and weighting matrices are time varying, subscripts k can be added to our results.

We want to find the control u_k^* on the time interval $[i, N]$ which minimizes the expected cost

$$j_i = E[J_i(x_i)] \tag{6.3-3}$$

with x_N free and final time N fixed. This is the *discrete linear quadratic Gaussian (LQG) problem*.

The results for complete and incomplete state information are presented in turn. They represent the optimal controller if the noise is Gaussian, and for arbitrary noise statistics they give the best linear control law.

We do not derive the results. Derivations would be along the lines of the previous section, though somewhat more involved due to the more complicated form of the curvature matrix $(B^T S_{k+1} B + R)$ and residual covariance $(HP_k H^T + R')$ for discrete-time systems.

Complete State Information

If the state x_k is completely measurable at time k, then the LQG regulator is the same as the deterministic LQ regulator (Bryson and Ho 1975,

Kwakernaak and Sivan 1972, Lewis 1986). Thus the optimal control is

$$u_k = - K_k x_k \tag{6.3-4}$$

where the Kalman control gain is

$$K_k = (B^T S_{k+1} B + R)^{-1} B^T S_{k+1} A \tag{6.3-5}$$

and S_k satisfies the Riccati equation

$$S_k = A^T [S_{k+1} - S_{k+1} B (B^T S_{k+1} B + R)^{-1} B^T S_{k+1}] A + Q, \qquad k < N, \tag{6.3-6}$$

with boundary condition of S_N.

The optimal expected cost is

$$j_i^* = \frac{1}{2} \text{trace} \left[S_i X_i + \sum_{k=i}^{N-1} S_{k+1} G Q' G^T \right], \tag{6.3-7}$$

where the mean-square state

$$X_k \triangleq (x_k x_k^T), \tag{6.3-8}$$

has initial value of

$$X_i = P_i + \bar{x}_i \bar{x}_i^T. \tag{6.3-9}$$

The second term in j_i^* is the cost increase due to the disturbing effect of the process noise.

A derivation of these results could be carried out using the discrete stochastic optimality principle (6.1-15). Alternatively, we could use an approach like that in Section 6.2.

To see how effectively this LQG controller regulates the state, we can examine the mean-square value of the state. To find an equation for X_k, write the closed-loop system

$$x_{k+1} = (A - BK)x_k + G w_k. \tag{6.3-10}$$

Now note that x_k and w_k are orthogonal, postmultiply this equation by its transpose, and take expectations to get

$$X_{k+1} = (A - BK) X_k (A - BK)^T + G Q' G^T \tag{6.3-11}$$

for $k > i$. The initial condition is (6.3-9).

Using (6.3-4), the mean-square value of the optimal control is

$$E(u_k u_k^T) = K_k X_k K_k^T. \tag{6.3-12}$$

Incomplete State Information

If the state is not completely measurable at time k, but if noisy measurements are available of the form

$$z_k = H x_k + v_k \tag{6.3-13}$$

with white measurement noise $v_k \sim N(0, R')$, then a dynamic output feed-back must be used to control the plant.

Carrying out a derivation like the one in Section 6.2 shows that the LQG regulator with incomplete state information is composed of a state estimator like the one in Table 2.3-1, followed by a deterministic LQ regulator which feeds back the estimate $\hat{x}_k$ and not x_k. Thus the separation principle also holds for discrete systems. We gather the equations into one place here for convenience.

If $\hat{x}_k$, P_k are the a posteriori estimate and error covariance, and $\hat{x}_k^-$, P_k^- are the a priori quantities, then the optimal stochastic LQG regulator is given by:

Dynamic Output Feedback:

$$u_k^* = -K_k \hat{x}_k \tag{6.3-14}$$

$$\hat{x}_{k+1}^- = A\hat{x}_k + Bu_k, \qquad \hat{x}_i = \bar{x}_i \tag{6.3-15}$$

$$\hat{x}_k = \hat{x}_k^- + L_k(z_k - H\hat{x}_k^-) \tag{6.3-16}$$

Computation of Gains:

$$K_k = (B^T S_{k+1} B + R)^{-1} B^T S_{k+1} A \tag{6.3-17}$$

$$L_k = P_k^- H^T (HP_k^- H^T + R')^{-1} = P_k H^T (R')^{-1} \tag{6.3-18}$$

For $k < N$:

$$S_k = A^T S_{k+1} A - K_k^T (B^T S_{k+1} B + R) K_k + Q; \qquad S_N \text{ given} \tag{6.3-19}$$

For $k \geq i$:

$$P_{k+1}^- = AP_k A^T + GQ'G^T, \qquad P_i \text{ given} \tag{6.3-20}$$

$$P_k = P_k^- - L_k(HP_k^- H^T + R')L_k^T. \tag{6.3-21}$$

The control gain is K_k and the filter gain is L_k. (We have written the Riccati equations in this form because it is convenient later.)

The optimal expected cost using this LQG regulator is

$$J_i^* = \frac{1}{2} \text{trace} \left[S_i X_i + \sum_{k=i}^{N-1} S_{k+1} GQ'G^T + \sum_{k=i}^{N-1} S_{k+1} BK_k P_k A^T \right], \tag{6.3-22}$$

where the third term is the cost increase due to measurement uncertainties.

If the plant, covariance, and cost-weighting matrices depend on time, subscripts k should be appended everywhere.

To determine the success of the LQG regulator at keeping the state small in the presence of process noise and to see how much control effort will be required, we can proceed as follows (Bryson and Ho, 1975).

Define the a priori and a posteriori estimation errors as

$$\tilde{x}_k^- = x_k - \hat{x}_k^- \tag{6.3-23}$$

$$\tilde{x}_k = x_k - \hat{x}_k. \tag{6.3-24}$$

Then the mean-square value of the state is

$$X_k = P_k^- + \hat{X}_k^-. \tag{6.3-25}$$

[We used (6.3-23), where the estimate $\hat{x}_k^-$ and the estimation error $\tilde{x}_k^-$ are orthogonal.] P_k^- is given by (6.3-20) but the mean-square value of the a priori estimate

$$\hat{X}_k^- \triangleq E[\hat{x}_k^-(\hat{x}_k^-)^T] \tag{6.3-26}$$

remains to be determined.

To find $\hat{X}_k^-$, use (6.3-13)–(6.3-16) to write the dynamics of the a priori estimate (cf. Table 2.3-3) as

$$\hat{x}_{k+1}^- = (A - BK_k)\hat{x}_k = (A - BK_k)[\hat{x}_k^- + L_k(z_k - H\hat{x}_k^-)]$$
$$= (A - BK_k)[\hat{x}_k^- + L_k H\tilde{x}_k^- + L_k v_k]. \tag{6.3-27}$$

Now note that $\hat{x}_k^-$, $\tilde{x}_k^-$, and v_k are pairwise orthogonal, multiply (6.3-27) on the right by its transpose, and take expectations to obtain

$$\hat{X}_{k+1}^- = (A - BK_k)(\hat{X}_k^- + L_k HP_k^- H^T L_k^T + L_k R' L_k^T)(A - BK_k)^T. \tag{6.3-28}$$

Using (6.3-21), there results the forward recursion

$$\hat{X}_{k+1}^- = (A - BK_k)(\hat{X}_k^- + P_k^- - P_k)(A - BK_k)^T \tag{6.3-29}$$

for the mean-square value of the a priori estimate.

We know the initial mean and covariance of the state, $\bar{x}_i$ and P_i. From (6.3-9) and (6.3-25)

$$\hat{X}_i^- = X_i - P_i^- = P_i - P_i^- + \bar{x}_i\bar{x}_i^T. \tag{6.3-30}$$

Using this in (6.3-29) yields

$$\hat{X}_{i+1}^- = (A - BK_i)\bar{x}_i\bar{x}_i^T(A - BK_i)^T, \tag{6.3-31}$$

which is used to start (6.3-29).

Using $\hat{X}_k^-$ in (6.3-25) we can determine X_k, which allows us to tell if the regulating action of the controller will be satisfactory.

To find the mean-square value of the control input, write

$$u_k = -K_k\hat{x}_k = -K_k(\hat{x}_k^- + L_k H\tilde{x}_k^- + L_k v_k), \tag{6.3-32}$$

so that

$$E(u_k u_k^T) = K_k(\hat{X}_k^- + P_k^- - P_k)K_k^T. \tag{6.3-33}$$

It is of some interest to compare the differencing between a priori and a posteriori error covariances in these equations with the Rauch–Tung–Striebel smoother in Section 2.8. For additional insight on the LQG regulator, see Åström (1970), Bryson and Ho (1975), and Kwakernaak and Sivan (1972).

PROBLEMS

Section 6.1

6.1-1: For the system in Example 6.1-2, select the control sequence to maximize the probability that $x_3 = 0$ for any initial state x_0.

6.1-2: Effect of Control Weighting on Field of Extremals. Redo Example 6.1-2 if

$$J_0 = (x_N - 1)^2 + \frac{1}{2} \sum_{k=0}^{N-1} r u_k^2$$

for $r = \frac{1}{4}$, 4, and 8 and compare the resulting fields of extremals.

6.1-3: Stochastic Bilinear System. Consider the scalar bilinear system

$$x_{k+1} = x_k u_k + u_k^2 + w_k$$

with cost index

$$J_0 = x_N^2 + \sum_{k=0}^{N-1} x_k u_k$$

where $N = 2$. The control is constrained to take on values of $u_k = \pm 1$, and the state x_k to take on values of -1, 0, 1, or 2. The process noise w_k has PDF

$$f_{w_k}(w) = \begin{cases} 1 & \text{with probability } \frac{1}{2} \\ 0 & \text{with probability } \frac{1}{4} \\ -1 & \text{with probability } \frac{1}{4} \end{cases}$$

a. Find an optimal state feedback control law minimizing $E(J_0)$.

b. Suppose the control of part a is used. Find the conditional PDF of x_2 given that $x_0 = 0$.

c. Suppose x_0 takes on each of the allowed state values with probability $\frac{1}{4}$. Find the expected cost using the control in part a.

d. Now find a control law to maximize the probability that x_2 is either 1 or 2.

6.1-4: Consider Newton's system

$$\ddot{x} = u + w$$

with process noise $w(t) \sim N(0, q')$ and $x(0) \sim N(\bar{x}_0, p_0)$. Select

$$J(0) = \frac{1}{2} x^2(T) + \frac{1}{2} \int_{t_0}^{T} ru^2 \, dt$$

where $T = 2$.

a. Write the HJB equation, eliminating $u(t)$.

b. Assume that

$$\bar{J}^*(x, t) = \frac{1}{2} s_1(t) x^2(t) + s_2 x(t) \dot{x}(t) + \frac{1}{2} s_3(t) \dot{x}^2(t) + \frac{1}{2} \int_{t}^{T} q' s_3(\tau) \, d\tau$$

for some functions s_1, s_2, s_3. Use the HJB equation to find coupled scalar equations for these s_i. Find boundary conditions for the equations. Express the optimal control as a linear state feedback in terms of the $s_i(t)$.

Section 6.2

6.2-1:   The plant is

$$\dot{x} = u + w$$

$$z = x + v$$

with $w \sim (0, q')$, $v \sim (0, r')$, and performance index

$$J = \frac{1}{2} s(T) x^2 + \frac{1}{2} \int_{0}^{T} (qx^2 + ru^2) \, dt.$$

Determine analytic expressions for the optimal feedback gain and optimal cost for the cases of:

a. Complete state information.

b. Incomplete state information. For this case, determine also the transfer function at steady state from the output $z(t)$ to the control $u(t)$. Draw a block diagram of the optimal steady-state regulator.

c. Write subroutines to simulate the controller in part b using CTLQR.

6.2-2:   Polynomial Formulation of Optimal Regulator. Examine Fig. 6.2-1.

a. Find transfer functions from $z(t)$ to intermediate signal $s(t)$ and from $u(t)$ to $s(t)$.

b. Show that the regulator can be put into the form in Fig. P6.2-1, where $T(s)$ is the Kalman filter characteristic polynomial [i.e., find $R(s)$, $S(s)$, and $T(s)$]. This is called a *two-degree of freedom* regulator.

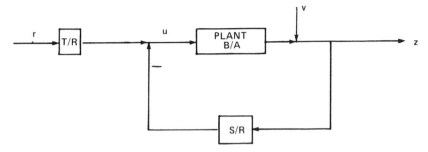

FIGURE P6.2-1 Two-degree of freedom regulator.

6.2-3: Scalar Plant. To the plant in Example 6.1-4 add measurements of

$$z = hx + v$$

with $v(t) \sim N(0, r')$ white and uncorrelated with $w(t)$ and $x(0)$.

a. Write down the complete set of equations for the LQG regulator.
b. Write subroutines to implement the regulator using CTLQR in Appendix B.
c. Find analytic solutions for $S(t)$ and the error covariance $P(t)$.

6.2-4: Suboptimal Scalar LQG Regulator. Let

$$\dot{x} = ax + bu + gw$$

$$z = hx + v$$

with $w(t) \sim N(0, q')$ and $v(t) \sim N(0, r')$ white and uncorrelated with each other and with $x(0) \sim N(\bar{x}_0, p_0)$. Suppose

$$J(0) = \frac{1}{2} \int_0^T (qx^2 + ru^2) \, dt.$$

Let $x \in R$

a. Find steady-state feedback gain K_∞.
b. Find steady-state Kalman gain L_∞. Find transfer function from z to $\hat{x}$ of the Wiener filter

$$\dot{\hat{x}} = (a - L_\infty h)\hat{x} + bu + L_\infty z.$$

Now we will combine the results of parts a and b to get a suboptimal regulator. Accordingly, suppose that gains K_∞, L_∞ are used for all time.

c. Find transfer function of the steady-state regulator (6.2-37), (6.2-38) from z to u. Draw a block diagram of the closed-loop system.
d. Find the associated suboptimal error covariance $p(t)$.
e. Find the resulting suboptimal mean-square state $X(t)$.

6.2-5: Steady-State LQG Regulator for Newton's System. Let

$$\ddot{y} = u + w$$

$$z = y + v$$

with $w(t) \sim N(0, \sigma^2)$, $v(t) \sim N(0, \rho^2)$ white and uncorrelated. Select the *infinite horizon* cost

$$J = \frac{1}{2} \int_0^{\infty} [q(y^2 + \dot{y}^2) + ru^2] \, dt.$$

a. Determine the optimal feedback gain K if the state is $x = [y \ \dot{y}]^T$.

b. Sketch the poles of the closed-loop plant $(A - BK)$ as q/r varies from 0 to ∞. Find the closed-loop damping ratio and natural frequency.

c. In terms of the signal-to-noise ratio $\Lambda = \sigma/\rho$ determine the steady-state Kalman gain.

d. Find the transfer function $H(s) = \hat{X}(s)/Z(s)$ of the Wiener filter.

e. Now suppose the Wiener filter is used to estimate the state for all t. Find the transfer function from the data z to the control u for the resulting suboptimal regulator.

f. The *observer polynomial* is $|sI - A + LH|$. Plot the observer poles. Find the observer damping ratio and natural frequency (in terms of Λ).

g. Find the suboptimal error covariance and mean-square state analytically.

h. Write subroutines to implement the *optimal* time-varying regulator using CTLQR.

6.2-6: The plant is described by

$$\dot{x} = u + v$$

and the state is measured according to

$$z = x + v$$

where noise $v(t)$ has spectral density

$$\Phi_z(\omega) = \frac{2\sigma^2}{\omega^2 + 1}.$$

It is desired to find a control $u(t)$ to minimize

$$J = \frac{1}{2} x^2(T) + \frac{1}{2} \int_0^T (qx^2 + u^2) \, dt.$$

a. Find the optimal steady-state feedback and observer gains, and the steady-state regulator transfer function from $z(t)$ to $u(t)$.

b. Plot root loci of steady-state closed-loop plant and Wiener filter poles versus σ^2 and/or q.

c. If the steady-state regulator is used for all t, find the resulting mean-square state $X(t)$.

Section 6.3

6.3-1: Derive the discrete LQG regulator equations in Section 6.3 for the case of:

a. Complete state information.

b. Incomplete state information.

6.3-2: Feedback of a Priori Estimate. Derive the discrete LQG regulator equations if a feedback of the form

$$u_k = - K_k \hat{x}_k^-$$

is used, where $\hat{x}_k^-$ is a one-step-ahead prediction of x_k. This feedback might be useful in a sampled system where there is a delay in implementing the control input.

6.3-3: Scalar Discrete LQG Regulator. Let

$$x_{k+1} = x_k + u_k$$

$$z_k = x_k + v_k$$

with $v_k \sim N(0, r')$ white and uncorrelated with $x_0 \sim (\bar{x}_0, p_0)$. Select

$$J = \frac{1}{2} x_N^2 + \frac{1}{2} \sum_{k=0}^{N-1} r u_k^2 .$$

a. Write the LQG regulator equations.

b. Solve for the observer and feedback gains.

6.3-4: Steady-State LQG Regulator. Let

$$x_{k+1} = ax_k + bu_k + w_k$$

$$z_k = hx_k + v_k$$

with $w_k \sim N(0, q')$ and $v_k \sim N(0, r')$ white and uncorrelated. Select the infinite horizon cost

$$J = \frac{1}{2} \sum_{k=0}^{\infty} (qx_k^2 + ru_k^2).$$

For simplicity, let $x_k \in R$.

a. Find the optimal feedback gain K_∞.

b. Find the steady-state Kalman gain L_∞. Find the transfer function from z_k to $\hat{x}_k$ of the Wiener filter.

c. Suppose that the Wiener filter is used to estimate x_k for all k. Find the transfer function from z_k to u_k of the resulting suboptimal regulator.

7

STOCHASTIC CONTROL
FOR POLYNOMIAL
SYSTEMS

In this chapter we discuss the optimal control of stochastic polynomial systems. Only discrete systems are covered, so that the results will be useful in digital control schemes. In the problems it is shown that the results also apply to continuous systems. We treat single-input single-output systems, but the extension to multivariable systems is not too difficult (Borison 1979, Koivo 1980).

7.1 POLYNOMIAL REPRESENTATION OF STOCHASTIC SYSTEMS

Here, discrete linear stochastic systems will be represented as

$$A(z^{-1})y_k = z^{-d}B(z^{-1})u_k + C(z^{-1})w_k, \qquad (7.1\text{-}1)$$

where $u_k \in R$, $y_k \in R$ are the control input and measured output, and $w_k \in R$ is a zero-mean white noise process with variance of q. The polynomials

$$A(z^{-1}) = 1 + a_1 z^{-1} + \cdots + a_n z^{-n} \qquad (7.1\text{-}2)$$

$$B(z^{-1}) = b_0 + b_1 z^{-1} + \cdots + b_m z^{-m}, \qquad b_0 \neq 0 \qquad (7.1\text{-}3)$$

$$C(z^{-1}) = 1 + c_1 z^{-1} + \cdots + c_n z^{-n} \qquad (7.1\text{-}4)$$

are expressed in terms of the delay operator z^{-1} (i.e., $z^{-1}y_k = y_{k-1}$). It is assumed that $C(z^{-1})$ is stable. The control delay is d.

Let us motivate the formulation (7.1-1) using two points of view.

Suppose y_k is a stochastic process with known spectral density $\Phi_y(e^{j\omega})$. Then (Section 1.5) we can perform the factorization

$$\Phi_y(e^{j\omega}) = H(e^{j\omega})qH(e^{-j\omega}), \tag{7.1-5}$$

where $H(e^{j\omega})$ is stable, with all zeros inside or on the unit circle. We shall assume $H(e^{j\omega})$ is of minimum phase, with relative degree of zero and monic numerator and denominator. Represent $H(z)$ as

$$H(z) = \frac{z^n + c_1 z^{n-1} + \cdots + c_n}{z^n + a_1 z^{n-1} + \cdots + a_n}, \tag{7.1-6}$$

(i.e., q has been selected to make $c_0 = 1$). Now it is seen that (7.1-1) with $u_k = 0$ represents the process y_k.

Next, consider the single-input single-output state representation

$$x_{k+1} = Fx_k + Gu_k + Je_k \tag{7.1-7a}$$

$$y_k = Hx_k + v_k, \tag{7.1-7b}$$

where F, G, J, H are matrices, process noise e_k is $(0, Q')$, measurement noise v_k is $(0, R')$, and the noises e_k and v_k are independent. In terms of transfer functions we have

$$Y(z) = H(zI - F)^{-1}GU(z) + H(zI - F)^{-1}JE(z) + V(z). \tag{7.1-8}$$

Represent the coefficient of $E(z)$ by $H_E(z)$ so that the noise term in (7.1-8) is $H_E(z)E(z) + V(z)$. Now perform a spectral factorization to determine $H(z)$ (minimum phase) and q in

$$H(e^{j\omega})qH(e^{-j\omega}) = H_E(e^{j\omega})Q'H_E(e^{-j\omega}) + R'. \tag{7.1-9}$$

Defining

$$H(z) = \frac{C_1(z)}{A_1(z)} \tag{7.1-10}$$

with both $C_1(z)$ and $A_1(z)$ monic, and

$$A_2(z) = |zI - F|, \tag{7.1-11}$$

$$B_2(z) = H[\text{adj}(zI - F)]G, \tag{7.1-12}$$

(7.1-8) can be written

$$Y(z) = \frac{B_2(z)}{A_2(z)} U(z) + \frac{C_1(z)}{A_1(z)} W(z), \tag{7.1-13}$$

where $w_k \sim (0, q)$ is white noise whose effect is equivalent to the combined effects of e_k and v_k.

If $n_1 = \deg[A_1(z)] = \deg[C_1(z)]$, $n_2 = \deg[A_2(z)]$, and $m_2 = \deg[B_2(z)]$, then it is evident that (7.1-1) represents (7.1-7) if

$$A(z^{-1}) = z^{-(n_1+n_2)}A_1(z)A_2(z) \tag{7.1-14}$$

$$B(z^{-1}) = z^{-(n_1+m_2)} A_1(z) B_2(z) \tag{7.1-15}$$

$$C(z^{-1}) = z^{-(n_1+n_2)} C_1(z) A_2(z), \tag{7.1-16}$$

and the control delay is

$$d = n_2 - m_2. \tag{7.1-17}$$

7.2 OPTIMAL PREDICTION

The treatment of the control problem is deferred to the next section; at this point we merely want to find an optimal d-step-ahead prediction for the output of (7.1-1) given any control sequence. That is, it is desired to predict y_{k+d} in terms of quantities that occur at times k or before. We shall follow Åström (1970).

In this section we shall assume $A(z^{-1})$ is stable.

To solve this problem, first perform a long division of $A(z^{-1})$ into $C(z^{-1})$ to define quotient $F(z^{-1})$ and remainder $z^{-d}G(z^{-1})$:

$$
A(z^{-1}) \overline{)\, C(z^{-1})} \\
\begin{array}{c} F(z^{-1}) \\ \vdots \\ \overline{z^{-d}G(z^{-1}).} \end{array}
\tag{7.2-1}
$$

The division is carried out d steps, that is, until z^{-d} can be factored out of the remainder to yield $G(z^{-1})$. Then

$$AF + z^{-d}G = C. \tag{7.2-2}$$

This is a *Diophantine equation* for $F(z^{-1})$ and $G(z^{-1})$. The solution to (7.2-2) is not unique, but the particular solution $F(z^{-1})$, $G(z^{-1})$ determined by long division has the form

$$F(z^{-1}) = 1 + f_1 z^{-1} + \cdots + f_{d-1} z^{-(d-1)} \tag{7.2-3}$$

$$G(z^{-1}) = g_0 + g_1 z^{-1} + \cdots + g_{n-1} z^{-(n-1)}. \tag{7.2-4}$$

The degree $(d-1)$ of $F(z^{-1})$ will be particularly important.

Multiply (7.1-1) by $F(z^{-1})$ to get

$$AFy_{k+d} = BFu_k + CFw_{k+d}; \tag{7.2-5}$$

then the Diophantine equation gives

$$Cy_{k+d} = z^{-d}Gy_{k+d} + BFu_k + CFw_{k+d},$$

or

$$y_{k+d} = \frac{G}{C} y_k + \frac{BF}{C} u_k + Fw_{k+d}. \tag{7.2-6}$$

Denote the optimal prediction of y_{k+d} in terms of $y_k, y_{k-1}, \ldots, u_k$, $u_{k-1}, \ldots$ as $\hat{y}_{k+d/k}$. The d-step prediction error is

$$\tilde{y}_{k+d/k} \triangleq y_{k+d} - \hat{y}_{k+d/k}. \qquad (7.2\text{-}7)$$

Let us determine $\hat{y}_{k+d/k}$ to minimize the mean-square error,

$$j_k = E(\tilde{y}_{k+d/k}^2) = E[(y_{k+d} - \hat{y}_{k+d/k})^2]. \qquad (7.2\text{-}8)$$

Using (7.2-6), this becomes

$$j_k = E\left[\left(\frac{G}{C} y_k + \frac{BF}{C} u_k - \hat{y}_{k+d/k}\right) + Fw_{k+d}\right]^2. \qquad (7.2\text{-}9)$$

The term in parentheses depends only on the input and output at times k and before, since $\hat{y}_{k+d/k}$ is only allowed to be expressed in terms of these values. Therefore, the term in parentheses depends only on the noise values $w_k, w_{k-1}, \ldots$. The remaining term looks like

$$F(z^{-1})w_{k+d} = w_{k+d} + f_1 w_{k+d-1} + \cdots + f_{d-1} w_{k+1}, \qquad (7.2\text{-}10)$$

since the degree of $F(z^{-1})$ is $(d-1)$, so it only depends on noise values subsequent to and including time $(k+1)$. Since the noise sequence is zero-mean and white, we conclude that the term in parentheses is orthogonal to $F(z^{-1})w_{k+d}$. Therefore, since y_k, u_k, and $\hat{y}_{k+d/k}$ are deterministic, the performance index separates into a deterministic part and a stochastic part:

$$j_k = \left(\frac{G}{C} y_k + \frac{BF}{C} u_k - \hat{y}_{k+d/k}\right)^2 + E(Fw_{k+d})^2. \qquad (7.2\text{-}11)$$

It is now evident that the minimum value of j_k occurs if we define the d-step prediction of the output as

$$\hat{y}_{k+d/k} = \frac{G}{C} y_k + \frac{BF}{C} u_k. \qquad (7.2\text{-}12)$$

Then the prediction error is

$$\tilde{y}_{k+d/k} = Fw_{k+d}. \qquad (7.2\text{-}13)$$

Thus the optimal prediction is the output of a linear deterministic system whose inputs are u_k and y_k, and the error is a moving average of the noise sequence. See Fig. 7.2-1. Since we assumed $C(z^{-1})$ is stable, the predictor is a stable system.

The output can be decomposed into two parts as

$$y_{k+d} = \hat{y}_{k+d/k} + \tilde{y}_{k+d/k}. \qquad (7.2\text{-}14)$$

These parts are orthogonal since the prediction depends only on values of the noise at times k and before, and the error depends only on noise values subsequent to time k. This decomposition is written explicitly as (7.2-6). Equation (7.2-6) is called the *predictive formulation* of the plant (7.1-1). We

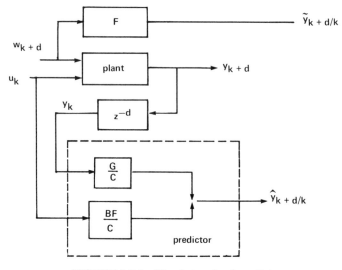

FIGURE 7.2-1   The d-step ahead predictor.

shall see that the solution to stochastic control problems depends on using this representation for the plant dynamics. In the present application, the predictive formulation allowed us to decompose the performance index J into two parts so we could solve the optimal prediction problem.

The minimum mean-square error is given by

$$j_k^* = E[F(z^{-1})w_{k+d}]^2$$
$$= E(w_{k+d} + f_1 w_{k+d-1} + \cdots + f_{d-1}w_{k+1})^2$$
$$= E(w_{k+d}^2) + f_1^2 E(w_{k+d-1}^2) + \cdots + f_{d-1}^2 E(w_{k+1}^2),$$

or

$$j_k^* = q(1 + f_1^2 + \cdots + f_{d-1}^2). \tag{7.2-15}$$

The cross-terms of the square drop out due to the whiteness of the noise. Since $E(\tilde{y}_{k+d/k}) = 0$, (7.2-15) is the minimum error *variance*, and hence (7.2-12) is a *minimum variance* estimate.

The results we have just derived are the equivalent for polynomial systems of the discrete Kalman filter in Chapter 2. The Kalman filter requires the solution of a Riccati equation, while the polynomial system predictor requires the solution of a Diophantine equation.

Example 7.2-1:   Minimum Variance Predictor

If the discretized plant dynamics are described by the difference equation

$$y_k + a_1 y_{k-1} + a_2 y_{k-2} = b_0 u_{k-d} + w_k + c_1 w_{k-1} \tag{1}$$

with white noise $w_k \sim (0, q)$, then (7.1-1) becomes

$$(1 + a_1 z^{-1} + a_2 z^{-2}) y_k = z^{-d} b_0 u_k + (1 + c_1 z^{-1}) w_k. \tag{2}$$

a. 1-Step Predictor

Let the control delay be

$$d = 1. \tag{3}$$

Then, dividing $A(z^{-1})$ into $C(z^{-1})$ yields

$$
\begin{array}{r}
1 \\
1 + a_1 z^{-1} + a_2 z^{-2} \overline{\smash{\big)}\ 1 + c_1 z^{-1}} \\
\underline{1 + a_1 z^{-1} + a_2 z^{-2}} \\
z^{-1}[(c_1 - a_1) - a_2 z^{-1}]
\end{array} \tag{4}
$$

so that

$$F(z^{-1}) = 1 \tag{5}$$

$$G(z^{-1}) = (c_1 - a_1) - a_2 z^{-1}. \tag{6}$$

Note that, for $d = 1$, there always results $F(z^{-1}) = 1$ and

$$G(z^{-1}) = [C(z^{-1}) - A(z^{-1})]z. \tag{7}$$

The optimal 1-step prediction satisfies (7.2-12), or

$$C(z^{-1}) \hat{y}_{k+1/k} = z[C(z^{-1}) - A(z^{-1})] y_k + B(z^{-1}) u_k, \tag{8}$$

so that $\hat{y}_{k+1/k}$ is given by the difference equation

$$\hat{y}_{k+1/k} = -c_1 \hat{y}_{k/k-1} + (c_1 - a_1) y_k - a_2 y_{k-1} + b_0 u_k, \qquad k > 0. \tag{9}$$

The initial conditions can be taken as

$$\hat{y}_{0/-1} = 0, \qquad y_{-1} = 0. \tag{10}$$

Since $C(z^{-1})$ is stable, the prediction error will eventually reach its optimal value regardless of the initial conditions.

The minimum error variance (7.2-15) is

$$j_k^* = q. \tag{11}$$

b. 2-Step Predictor

Let

$$d = 2 \tag{12}$$

so that we now need to predict ahead 2 steps.

Carrying the long division of $A(z^{-1})$ and $C(z^{-1})$ out 2 steps yields

$$
\begin{array}{r}
1 + (c_1 - a_1) z^{-1} \\
1 + a_1 z^{-1} + a_2 z^{-2} \overline{\smash{\big)}\ 1 + c_1 z^{-1}} \\
\underline{1 + a_1 z^{-1} + a_2 z^{-2}} \\
(c_1 - a_1) z^{-1} - a_2 z^{-2} \\
\underline{(c_1 - a_1) z^{-1} + a_1(c_1 - a_1) z^{-2} + a_2(c_1 - a_1) z^{-3}} \\
z^{-2}\{[-a_2 - a_1(c_1 - a_1)] - a_2(c_1 - a_1) z^{-1}\}
\end{array} \tag{13}
$$

so that

$$F(z^{-1}) = 1 + (c_1 - a_1)z^{-1} \tag{14}$$

$$G(z^{-1}) = -[a_2 + a_1(c_1 - a_1)] - a_2(c_1 - a_1)z^{-1}. \tag{15}$$

The predictor recursion (7.2-12) is now

$$\hat{y}_{k+2/k} = -c_1 \hat{y}_{k/k-1} - [a_2 + a_1(c_1 - a_1)]y_k$$
$$- a_2(c_1 - a_1)y_{k-1} + b_0 u_k + b_0(c_1 - a_1)u_{k-1} \tag{16}$$

and the optimal 2-step error variance is

$$y_k^* = q + (c_1 - a_1)^2 q. \tag{17}$$

Note that as we attempt to predict further into the future, the error variance increases. Note also that $G(z^{-1})$ generally has degree of $n-1$, while the degree $d-1$ of $F(z^{-1})$ increases with the delay. Thus, the d-step-ahead predictor requires $n-1$ previous values of the output and $d-1$ previous values of the input. ■

7.3 MINIMUM VARIANCE CONTROL

Now let us determine one type of optimal controller for the polynomial system

$$A(z^{-1})y_k = z^{-d}B(z^{-1})u_k + C(z^{-1})w_k, \tag{7.3-1}$$

with white noise $w_k \sim N(0, q)$, and polynomials given by (7.1-2)–(7.1-4). Our result will be the following: the control u_k will be selected to zero the d-step prediction $\hat{y}_{k+d/k}$ of the output.

To derive this result, we shall use the predictive formulation (7.2-6) of the plant, again following Åström (1970).

Select as a performance index the mean-square value of the output

$$j_k = E(y_{k+d}^2), \tag{7.3-2}$$

and use predictive formulation (7.2-6) to say that

$$j_k = E\left[\left(\frac{G}{C}y_k + \frac{BF}{C}u_k\right) + Fw_{k+d}\right]^2. \tag{7.3-3}$$

We want to select u_k to minimize j_k. Admissible controls u_k depend on previous values of the control, and on $y_k, y_{k-1}, \ldots$.

Since the term in parentheses in (7.3-3) depends only on $w_k, w_{k-1}, \ldots$, and $F(z^{-1})w_{k+d}$ depends only on $w_{k+1}, w_{k+2}, \ldots, w_{k+d}$, we have the decomposition

$$j_k = E\left(\frac{G}{C}y_k + \frac{BF}{C}u_k\right)^2 + E(Fw_{k+d})^2. \tag{7.3-4}$$

Thus, it is clear that the minimum value of j_k is attained if the control is selected to satisfy

$$u_k = -\frac{G}{BF}\, y_k.$$ (7.3-5)

Then the optimal cost is

$$j_k^* = E[F(z^{-1})w_{k+d}]^2$$
$$= q(1 + f_1^2 + \cdots + f_{d-1}^2),$$ (7.3-6)

exactly as (7.2-15).

This control scheme is shown in fig. 7.3-1, where a reference signal s_k has been added. Since the feedback is a function of z^{-1}, it contains some dynamics. Therefore, it is a dynamic output feedback, exactly like the LQG regulator in Chapter 6 which contained a state estimator and a deterministic LQ regulator. It can also be described as a *one-degree of freedom regulator*, since there is no feedforward term so that only the plant poles, and not the zeros, can be moved.

By examining (7.2-12), it is seen that the optimal control u_k given by (7.3-5) simply zeros the d-step prediction $\hat{y}_{k+d/k}$ of the output. The resulting mean-square value (7.3-6) of the output is exactly equal to the prediction error variance! It is therefore apparent that the dynamic output feedback (7.3-5) implicitly contains an estimator for the output. The upshot of all this is that the control law we have just derived could be interpreted as a *separation principle* for polynomial systems (Åström 1970). That is, to derive the minimum variance control, we could first find the optimal prediction of the output and *then* select the control input to zero the prediction.

We must now discuss the closed-loop behavior under the influence of our control law. Using (7.3-5) in (7.3-1), the transfer characteristics from the noise w_k to the output y_k are seen to be described by

$$\left(A + z^{-d}\frac{BG}{BF}\right)y_k = Cw_k$$ (7.3-7)

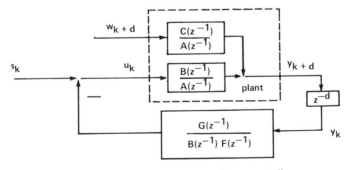

FIGURE 7.3-1 Minimum variance controller.

or

$$y_k = \frac{BFC}{B(AF + z^{-d}G)} w_k.$$ (7.3-8)

According to (7.2-2),

$$y_k = \frac{BFC}{BC} w_k.$$ (7.3-9)

$C(z^{-1})$ is assumed stable. If $B(z^{-1})$ is stable, that is, if the plant $B(z^{-1})/A(z^{-1})$ is minimum phase, then we can cancel to obtain

$$y_k = F(z^{-1})w_k,$$ (7.3-10)

which is the *regulation error*. [This agrees with the result obtained by using (7.3-5) in (7.2-6).] Therefore, the minimum variance regulation error is simply equal to the prediction error (7.2-13).

According to (7.3-9), if w_k has a mean of zero, then so does y_k in the closed-loop system (as long as reference input s_k is zero). (See Section 1.5.) Thus, j_k in (7.3-2) is the *variance* of the output, which justifies calling our control law a *minimum variance* controller.

Now let us turn to the closed-loop transfer relation between the reference input s_k and y_{k+d}. From Fig. 7.3-1,

$$y_{k+d} = \frac{B/A}{1 + GBz^{-d}/BAF} s_k,$$

or

$$y_{k+d} = \frac{BF}{B(AF + z^{-d}G)} s_k.$$ (7.3-11)

Taking into account the Diophantine equation gives

$$y_{k+d} = \frac{BF}{BC} s_k.$$ (7.3-12)

The closed-loop characteristic polynomial is therefore

$$\Delta^{cl}(z) = B(z^{-1})C(z^{-1})z^{m+n}.$$ (7.3-13)

(multiplication by z^{m+n} removes terms in z^{-1}). If the plant $B(z^{-1})/A(z^{-1})$ is minimum phase, then the closed-loop system is stable [since $C(z^{-1})$ is stable by assumption]. Then we can cancel to obtain

$$y_{k+d} = \frac{F(z^{-1})}{C(z^{-1})} s_k,$$ (7.3-14)

so that the closed-loop poles are the zeros of the noise polynomial $C(z^{-1})$.

Example 7.3-1:   Minimum Variance Controller

We can easily find the minimum variance controllers for the two cases in example 7.2-1, for the work has already been done.

a.   1-Step Controller

If the control delay is $d = 1$, then the 1-step predicted output can be set to zero using (7.3-5), which becomes

$$B(z^{-1})u_k = -z[C(z^{-1}) - A(z^{-1})]y_k, \qquad (1)$$

or, in our case

$$u_k = \frac{-(c_1 - a_1)}{b_0} y_k + \frac{a_2}{b_0} y_{k-1}. \qquad (2)$$

With this control, the regulation error is given by (7.3-10), or

$$y_k = w_k, \qquad (3)$$

and the regulation error variance is

$$j_k^* = q. \qquad (4)$$

b.   2-Step Controller

If $d = 2$, then the 2-step predicted output can be zeroed using the control defined by the recursion

$$b_0[1 + (c_1 - a_1)z^{-1}]u_k = \{[a_2 + a_1(c_1 - a_1)] + a_2(c_1 - a_1)z^{-1}\}y_k \qquad (5)$$

or

$$u_k = -(c_1 - a_1)u_{k-1} + \frac{a_2 + a_1(c_1 - a_1)}{b_0} y_k + \frac{a_2(c_1 - a_1)}{b_0} y_{k-1}. \qquad (6)$$

The regulation error is now

$$y_k = w_k + (c_1 - a_1)w_{k-1}, \qquad (7)$$

with a variance of

$$j_k^* = q + q(c_1 - a_1)^2. \qquad (8)$$

Note that the regulation error variance increases with the control delay.   ∎

7.4 POLYNOMIAL LINEAR QUADRATIC GAUSSIAN REGULATOR

The minimum variance controller was found in the previous section by minimizing the cost index (7.3-2). In this section, it is desired to find the control minimizing a more general cost index.

The linear plant is described by

$$Ay_k = z^{-d}Bu_k + Cw_k, \qquad (7.4-1)$$

with white noise $w_k \sim N(0, q)$, and

$$A(z^{-1}) = 1 + a_1 z^{-1} + \cdots + a_n z^{-n}$$

$$B(z^{-1}) = b_0 + b_1 z^{-1} + \cdots + b_m z^{-m}, \qquad b_0 \neq 0$$

$$C(z^{-1}) = 1 + c_1 z^{-1} + \cdots + c_n z^{-n}. \qquad (7.4\text{-}2)$$

Let us select the quadratic performance index

$$J_k = (Py_{k+d} - Qs_k)^2 + (Ru_k)^2, \qquad (7.4\text{-}3)$$

with weighting polynomials

$$P(z^{-1}) = 1 + p_1 z^{-1} + \cdots + p_{n_P} z^{-n_P}$$

$$Q(z^{-1}) = q_0 + q_1 z^{-1} + \cdots + q_{n_Q} z^{-n_Q}$$

$$R(z^{-1}) = r_0 + r_1 z^{-1} + \cdots + r_{n_R} z^{-n_R}. \qquad (7.4\text{-}4)$$

Signal s_k is a given reference signal.

We have chosen this cost index for two reasons: it results in a very nice solution, and it can be selected to specify a wide range of designs that result in desirable closed-loop behavior for different applications. Let us discuss some examples.

To keep the output small without using too much control energy, we might simply select $P = 1$, $Q = 0$, $R = r_0$, so that

$$J_k = y_{k+d}^2 + r_0^2 u_k^2. \qquad (7.4\text{-}5)$$

(In this case the reference input is not needed.) If we are more concerned about keeping small the *changes* in the control input, we could select $P = 1$, $Q = 0$, $R = r_0 (1 - z^{-1})$ so that

$$J_k = y_{k+d}^2 + r_0^2 (u_k - u_{k-1})^2. \qquad (7.4\text{-}6)$$

By choosing $P = Q = 1$, $R = r_0$, we get the cost index

$$J_k = (y_{k+d} - s_k)^2 + r_0^2 u_k^2, \qquad (7.4\text{-}7)$$

which is a polynomial tracker; the controlled output will follow a (delayed) reference signal s_k.

We can also design a regulator that makes the plant Bz^{-d}/A behave like any desired model, for if R is chosen as zero, then the optimal controller will attempt to ensure that

$$Py_{k+d} = Qs_k. \qquad (7.4\text{-}8)$$

This results in *model-following behavior*, where the plant output tracks the output of a desired model $Q(z^{-1})z^{-d}/P(z^{-1})$ which has input s_k. To see this, write (7.4-8) as

$$y_k = \frac{Q(z^{-1})z^{-d}}{P(z^{-1})} s_k. \qquad (7.4\text{-}9)$$

Therefore, $Q(z^{-1})$, $P(z^{-1})$ in (7.4-3) need only be selected to be the specified model numerator and denominator dynamics.

Finally, if $P = 1$, $Q = 0$, $R = 0$, then $E(J_k)$ reduces to (7.3-2).

The *polynomial linear quadratic Gaussian (LQG) problem* is to determine a control sequence that minimizes the expected cost

$$j_k = E(J_k). \tag{7.4-10}$$

Admissible controls u_k depend on u_{k-1}, u_{k-2}, ... , y_k, y_{k-1}, ... To solve this problem, we shall follow Clarke and Gawthrop (1975).

We cannot immediately differentiate the expected cost j_k to determine the optimal control u_k, since y_{k+d} depends on u_k. This dependence must first be made explicit. To do this, we shall find an optimal j-step-ahead predictor for values y_{k+j} of the output for $j \leq d$.

According to (7.2-14), y_{k+d} can be decomposed into two orthogonal parts as

$$y_{k+d} = \hat{y}_{k+d/k} + \tilde{y}_{k+d/k}, \tag{7.4-11}$$

where

$$\hat{y}_{k+d/k} = \frac{G_d}{C} y_k + \frac{BF_d}{C} u_k \tag{7.4-12}$$

is the optimal prediction of y_{k+d} given the values y_k, y_{k-1}, ... , u_k, u_{k-1}, ... , and

$$\tilde{y}_{k+d/k} = F_d w_{k+d} \tag{7.4-13}$$

is the prediction error. The polynomials $F_d(z^{-1})$ and $G_d(z^{-1})$ are defined by the Diophantine equation

$$AF_d + z^{-d}G_d = C, \tag{7.4-14}$$

which amounts to a d-step long division of $A(z^{-1})$ into $C(z^{-1})$.

We would like to find an optimal prediction for each y_{k+j}, $j \leq d$, in terms of quantities that occur at times k and before, since the values of the output appearing in (7.4-3) are y_{k+d}, y_{k+d-1}, ... , y_{k+d-n_p}, with n_p the degree of $P(z^{-1})$. If $\hat{y}_{k+j/k}$ denotes the optimal prediction of y_{k+j} in terms of quantities that occur at times k and before, and $\tilde{y}_{k+j/k}$ is the associated prediction error, then

$$y_{k+j} = \hat{y}_{k+j/k} + \tilde{y}_{k+j/k}, \qquad j \leq d, \tag{7.4-15}$$

with $\hat{y}_{k+j/k}$ and $\tilde{y}_{k+j/k}$ orthogonal.

For $j \leq 0$, it is clear that

$$\hat{y}_{k+j/k} = y_{k+j}, \qquad j \leq 0, \tag{7.4-16}$$

since then y_{k+j} itself is part of the known data. For $j = d$, we know that $\hat{y}_{k+j/k}$ and $\tilde{y}_{k+j/k}$ are given by (7.4-12), (7.4-13). However, it is *not* true that we can generalize (7.4-12) to say, for $j < d$,

$$\hat{y}_{k+j/k} = \frac{G_d}{C} y_{k-d+j} + \frac{BF_d}{C} u_{k-d+j}. \qquad (7.4\text{-}17)$$

The reason is that this estimate disregards the available data $y_{k-d+j+1}, \ldots, y_k$, so it is not optimal.

To construct a minimum variance estimate for y_{k+j}, $0 < j \le d$, in terms of y_k, $y_{k-1}, \ldots, u_k$, $u_{k-1}, \ldots$, we proceed as follows. Perform a j-step long division of $A(z^{-1})$ into $C(z^{-1})$ to define $F_j(z^{-1})$ and $G_j(z^{-1})$ by

$$AF_j + z^{-j}G_j = C. \qquad (7.4\text{-}18)$$

Then $F_j(z^{-1})$ has the form

$$F_j(z^{-1}) = 1 + f_1 z^{-1} + \cdots + f_{j-1} z^{-(j-1)}. \qquad (7.4\text{-}19)$$

The polynomials $F_j(z^{-1})$ and $G_j(z^{-1})$ must be determined for all j such that $d - n_p \le j \le d$. If we perform a single d-step long division of $A(z^{-1})$ into $C(z^{-1})$, then $F_j(z^{-1})$ is the quotient and $z^{-j}G_j(z^{-1})$ is the remainder at the jth step. See Example 7.4-1.

Multiply the plant (7.4-1) by F_j to get

$$AF_j y_{k+j} = BF_j u_{k-d+j} + F_j C w_{k+j}, \qquad (7.4\text{-}20)$$

and use the Diophantine equation (7.4-18) to see that

$$C y_{k+j} = z^{-j} G_j y_{k+j} + BF_j u_{k-d+j} + F_j C w_{k+j},$$

or

$$y_{k+j} = \left(\frac{G_j}{C} y_k + \frac{BF_j}{C} u_{k-d+j} \right) + F_j w_{k+j} \qquad (7.4\text{-}21)$$

Since the term in parentheses depends only on w_k, $w_{k-1}, \ldots$, and $F_j(z^{-1}) w_{k+j}$ depends only on w_{k+1}, $w_{k+2}, \ldots$, these two terms are orthogonal. Therefore an argument like the one in Section 7.2 shows that (7.4-15) holds, where the optimal j-step prediction is

$$\hat{y}_{k+j/k} = \frac{G_j}{C} y_k + \frac{BF_j}{C} u_{k-d+j}, \qquad 0 < j \le d, \qquad (7.4\text{-}22)$$

and the prediction error is

$$\tilde{y}_{k+j/k} = F_j w_{k+j} \qquad (7.4\text{-}23)$$

The error variance is equal to

$$E(\tilde{y}_{k+j/k}^2) = q(1 + f_1^2 + \cdots + f_{j-1}^2) \qquad (7.4\text{-}24)$$

where the f_i are the coefficients of $F_j(z^{-1})$.

Note that the prediction (7.4-22) uses *all* the available outputs, and that it reduces to (7.4-12) if $j = d$.

At this point we have succeeded in deriving a predictive formulation (7.4-15) for y_{k+j} for all $j \leq d$. The estimate and the error in (7.4-15) are orthogonal, and are given for $0 \leq j \leq d$ by (7.4-22) and (7.4-23). We may include (7.4-16) in (7.4-22) by defining

$$F_j = 0, \qquad G_j = z^j C, \qquad \text{for} \quad j \leq 0, \qquad (7.4\text{-}25)$$

for then (7.4-15), (7.4-22), and (7.4-23) hold for all $j \leq d$.

What is now required is to use the orthogonal decomposition (7.4-15) for y_{k+j} in the cost index (7.4-10). Then

$$j_k = E\{[P(\hat{y}_{k+d/k} + \tilde{y}_{k+d/k}) - Qs_k]^2 + (Ru_k)^2\}. \qquad (7.4\text{-}26)$$

Since each error $\tilde{y}_{k+j/k}$ depends only on w_{k+1} and subsequent noise values, $P(z^{-1})\tilde{y}_{k+d/k}$ is orthogonal to all the other terms. Furthermore, these other terms are all deterministic. Therefore,

$$j_k = (P\hat{y}_{k+d/k} - Qs_k)^2 + (Ru_k)^2 + E(P\tilde{y}_{k+d/k})^2. \qquad (7.4\text{-}27)$$

The importance of the predictive formulation (7.4-15), which is explicitly stated in (7.4-21), is now clear; it has allowed us to decompose the cost index into a stochastic portion independent of u_k, and a deterministic portion that can be minimized by differentiation.

Note that $\hat{y}_{k+d/k}$ depends on u_k, and that $\hat{y}_{k+j/k}$ for $j < d$ does not depend on u_k, but only on earlier values of the input. Note further that

$$\frac{\partial \hat{y}_{k+d/k}}{\partial u_k} = \frac{B(0)F_d(0)}{C(0)} = b_0. \qquad (7.4\text{-}28)$$

Furthermore,

$$\frac{\partial}{\partial u_k}(Ru_k)^2 = \frac{\partial}{\partial u_k}(r_0 u_k + r_1 u_{k-1} + \cdots + r_{n_R} u_{k-n_R})^2$$

$$= 2r_0 Ru_k. \qquad (7.4\text{-}29)$$

Therefore, differentiating j_k yields

$$\frac{\partial j_k}{\partial u_k} = 2b_0(P\hat{y}_{k+d/k} - Qs_k) + 2r_0 Ru_k = 0. \qquad (7.4\text{-}30)$$

The optimal LQG control law is therefore defined by the recursion

$$P\hat{y}_{k+d/k} + \frac{r_0}{b_0} Ru_k - Qs_k = 0 \qquad (7.4\text{-}31)$$

for u_k in terms of $u_{k-1}, u_{k-2}, \ldots, y_k, y_{k-1}, \ldots, s_k, s_{k-1}, \ldots$. The polynomial LQG control problem is now solved.

We can find a more explicit recursion for u_k as follows. Use the definition of $P(z^{-1})$ to write (7.4-31) as

$$\sum_{j=0}^{n_p} p_j \hat{y}_{k+d-j/k} + \frac{r_0}{b_0} Ru_k - Qs_k = 0,$$

with $p_0 = 1$. Now, by (7.4-22), this is

$$\sum_{j=0}^{n_p} p_j G_{d-j} y_k + \sum_{j=0}^{n_p} p_j B F_{d-j} u_{k-j} + \frac{r_0}{b_0} CR u_k - CQ s_k = 0. \qquad (7.4\text{-}32)$$

Defining new (composite) polynomials

$$G'(z^{-1}) = \sum_{j=0}^{n_p} p_j G_{d-j}(z^{-1}) \qquad (7.4\text{-}33)$$

$$F'(z^{-1}) = \sum_{j=0}^{n_p} p_j B(z^{-1}) F_{d-j}(z^{-1}) z^{-j} + \frac{r_0}{b_0} C(z^{-1}) R(z^{-1}), \qquad (7.4\text{-}34)$$

this becomes

$$F' u_k = -G' y_k + CQ s_k. \qquad (7.4\text{-}35)$$

It is now clear that what we have on our hands is a two-degree of freedom regulator (problems for Chapter 6). See Fig. 7.4-1. Since there are feedback and feedforward components, the plant poles *and* zeros can now be moved [e.g., see (7.4-9)].

If we select the weighting polynomials $P(z^{-1}) = 1$, $Q(z^{-1}) = 0$, and $R(z^{-1}) = 0$, then

$$j_k = E(y_{k+d}^2). \qquad (7.4\text{-}36)$$

In this case, $G' = G_d$, $F' = BF_d$, and the LQG control law (7.4-35) reduces to

$$BF_d u_k = -G_d y_k. \qquad (7.4\text{-}37)$$

This is exactly the minimum variance controller (7.3-5).

The polynomial LQG controller depends on the solution to the Diophantine equation(s) (7.4-18). Its dynamics contain an implicit estimator for the portion of the internal state not appearing directly in the output. It should be

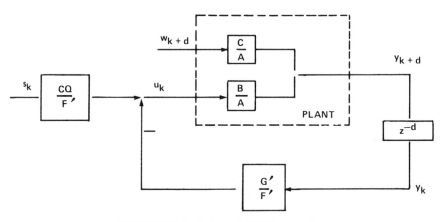

FIGURE 7.4-1 Polynomial LQG controller.

compared to the state variable LQG controller in Section 6.3, whose structure is considerably more complex!

According to Fig. 7.4-1, the closed-loop transfer function from s_k to y_{k+d} is given by

$$H^{cl}(z) = \frac{CQ}{F'} \cdot \frac{BF'}{AF' + BG'z^{-d}}. \tag{7.4-38}$$

Thus, the closed-loop characteristic polynomial is

$$\Delta^{cl}(z) = (AF' + BG'z^{-d})z^h, \tag{7.4-39}$$

where h is the highest power of z^{-1} in $H^{cl}(z)$ after canceling $F'(z^{-1})$.

To show how simple the design of a polynomial LQG controller really is, let us consider an example.

Example 7.4-1:   Polynomial LQG Controller

For purposes of illustration, we shall consider a discrete plant with simple and unrealistic numbers.

Suppose that the discretized plant has a transfer function description of

$$Y(z) = \frac{z - 0.5}{z^3 + 1.5z^2 - z} U(z) + \frac{z^3 + 0.25z^2}{z^3 + 1.5z^2 - z} W(z), \tag{1}$$

with $w_k \sim N(0, q)$ a white noise process. Then the polynomial description is

$$(1 + 1.5z^{-1} - z^{-2})y_k = z^{-2}(1 - 0.5z^{-1})u_k + (1 + 0.25z^{-1})w_k, \tag{2}$$

which is

$$Ay_k = z^{-d}Bu_k + Cw_k \tag{3}$$

with a control delay of

$$d = 2. \tag{4}$$

The plant is unstable with poles at

$$z = 0.5, -2. \tag{5}$$

Let the performance index be

$$J_k = [(y_{k+2} - y_{k+1}) - (s_k - s_{k-1})]^2 + [u_k - u_{k-1}]^2 \tag{6}$$

or

$$J_k = [(1 - z^{-1})y_{k+2} - (1 - z^{-1})s_k]^2 + [(1 - z^{-1})y_k]^2, \tag{7}$$

which is

$$J_k = (Py_{k+2} - Qs_k)^2 + (Ru_k)^2, \tag{8}$$

where s_k is a given reference input. This cost function will try to keep small the changes in input u_k. It will also try to make changes in the output follow changes in the (delayed) reference input. Thus, we are not concerned about matching the DC level in y_k to the DC level in the reference signal.

a. LQG Tracker

We call this a tracker since changes in the output with follow changes in s_k.
Carrying out a long division of $A(z^{-1})$ into $C(z^{-1})$ to $d = 2$ steps yields

$$
\begin{array}{r}
1 - 1.25z^{-1} \\
1 + 1.5z^{-1} - z^{-2} \overline{\big)\; 1 + 0.25z^{-1}} \\
\underline{1 + 1.5z^{-1} - z^{-2}} \\
-1.25z^{-1} + z^{-2} \\
\underline{-1.25z^{-1} - 1.875z^{-2} + 1.25z^{-3}} \\
2.875z^{-2} - 1.25z^{-3},
\end{array}
$$

whence we can immediately read off

$$F_1(z^{-1}) = 1, \qquad G_1(z^{-1}) = -1.25 + z^{-1}$$

$$F_2(z^{-1}) = 1 - 1.25z^{-1}, \qquad G_2(z^{-1}) = 2.875 - 1.25z^{-1}. \tag{9}$$

The composite polynomials in (7.4-33) and (7.4-34) are thus

$$
\begin{aligned}
G'(z^{-1}) &= p_0 G_2 + p_1 G_1 \\
&= (2.875 - 1.25z^{-1}) - (-1.25 + z^{-1}) \\
&= 4.125 - 2.25z^{-1} \tag{10}
\end{aligned}
$$

$$
\begin{aligned}
F'(z^{-1}) &= B(p_0 F_2 + p_1 F_1 z^{-1}) + \frac{r_0}{b_0} CR \\
&= (1 - 0.5z^{-1})(1 - 1.25z^{-1} - z^{-1}) + (1 + 0.25z^{-1})(1 - z^{-1}) \\
&= 2 - 3.5z^{-1} + 0.875z^{-2}. \tag{11}
\end{aligned}
$$

Also,

$$CQ = (1 + 0.25z^{-1})(1 - z^{-1}) = 1 - 0.75z^{-1} - 0.25z^{-2}. \tag{12}$$

The LQG regulator is given by (7.4-35):

$$
\begin{aligned}
(2 - 3.5z^{-1} + 0.875z^{-2})u_k &= -(4.125 - 2.25z^{-1})y_k \\
&\quad + (1 - 0.75z^{-1} - 0.25z^{-2})s_k, \tag{13}
\end{aligned}
$$

so that the optimal control can be determined by the recursion

$$
\begin{aligned}
u_k &= \tfrac{1}{2}(3.5u_{k-1} - 0.875u_{k-2} - 4.125y_k \\
&\quad + 2.25y_{k-1} + s_k - 0.75s_{k-1} - 0.25s_{k-2}). \tag{14}
\end{aligned}
$$

b. Closed-Loop System

According to (7.4-39) the closed-loop characteristic polynomial is

$$\Delta^{cl}(z) = (AF' + BG'z^{-d})z^h, \tag{15}$$

where h is the power of z required to clear negative powers of z. Performing the required operations,

$$\Delta^{cl}(z) = (2 - 0.5z^{-1} - 2.25z^{-2} + 0.5z^{-3} + 0.25z^{-4})z^h, \tag{16}$$

and it is now evident that $h = 4$, so that

$$\Delta^{cl}(z) = 2z^4 - 0.5z^3 - 2.25z^2 + 0.5z + 0.25. \tag{17}$$

The closed-loop poles are at

$$z = -0.25,\ 0.5,\ -1,\ 1, \tag{18}$$

and the closed-loop system is marginally stable even though the original plant was unstable.

PROBLEMS

Section 7.2

7.2-1: A process y_k is defined by

$$x_{k+1} = \tfrac{1}{2}x_k + u_{k-1} + e_k$$

$$y_k = x_k + v_k$$

with measurement noise $v_k \sim N(0, 1)$ white and process noise e_k having a spectral density of

$$\Phi_E(\omega) = \frac{(1-a^2)\sigma^2}{1+a^2 - 2a\cos\omega}$$

where $a = \tfrac{3}{4}$. Fine the optimal d-step predictors for x_k and for y_k as a function of σ^2. Find both prediction error variances.

7.2-2: Prediction of Sampled Signals. A Markov process $x(t)$ with spectral density

$$\phi_X(\omega) = \frac{2}{\omega^2 + 1}$$

is measured at discrete intervals kT according to

$$y_k = x(kT) + v_k$$

with $v_k \sim N(0, 1)$ white. Find the optimal 1-step and 2-step predictors for y_k as a function of the sampling period T. Find the prediction error variances.

Section 7.3

7.3-1:

a. For the plant in Problem 7.2-1, find the control input minimizing the variance of y_k. Compute the closed-loop transfer function and poles, and find the regulation error variance.

b. Repeat for the control input minimizing the variance of x_k, assuming that x_k is exactly measurable.

Section 7.4

7.4-1: For the plant in Problem 7.2-1, find the control minimizing the expected value of:

a. $J_k = (y_{k+d} - y_{k+d-1})^2$

(this control will keep small the *changes* in the output).

b. $J_k = y_{k+d}^2 + ru_k^2$

(a weighted-input minimum variance control).

c. $J_k = y_{k+d}^2 + r(u_k - u_{k-1})^2$.

7.4-2: Model Following Controller

a. Find a control to make the plant

$$y_k = \frac{z - \frac{1}{4}}{(z - \frac{1}{2})^2} u_k$$

behave like the model

$$y_k = \frac{(z - \frac{1}{2})}{(z - \frac{1}{4})^2} u_k$$

b. Find the first five terms of the impulse response of:

   i. The original plant.
   ii. The model.
   iii. The plant plus controller.

7.4-3: Model Following with Control Weighting. Solve Problem 7.4-2, including a term like ru_k^2 in the performance index. As r becomes small the control gives better model following behavior. Plot the closed-loop poles as a function of the control weighting r.

7.4-4: A system

$$x_{k+1} = 1.1x_k + u_{k-1}$$

is driven by a noisy actuator described by

$$u_{k+1} = 0.9u_k + v_k + e_k$$

where v_k is the input command and e_k is noise with spectral density

$$\Phi_E(\omega) = \frac{\frac{3}{4}}{\frac{5}{4} - \cos \omega}.$$

Find the regulator to generate input v_k from measurements described by

$$y_k = x_k + n_k,$$

where noise n_k has

$$\Phi_N(\omega) = \frac{\frac{15}{16}}{\frac{17}{16} - \frac{1}{8}\cos\omega},$$

if we want to minimize

$$J_k = (y_{k+d} - s_k)^2 + \frac{u_k^2}{10},$$

where d is the control delay and s_k is a reference input. Sketch your regulator.

7.4-5: Regulator Error. Find the regulation error for the optimal controller (7.4-35). Find the regulation error variance if $w_k \sim (0, q)$ is white.

7.4-6: Tracker for Newton's System. A plant

$$\dot{d} = v$$

$$\dot{v} = u + w$$

with $w(t) \sim N(0, \sigma^2)$ a white process noise, has discrete position measurements

$$y_k = d(kT) + n_k$$

with white measurement noise $n_k \sim N(0, 1)$. The cost index is

$$J_k = (y_{k+d} - s_k)^2 + (ru_k)^2$$

with s_k a reference input, and d the control delay.

a. Let $T = 0.2\,\text{msec}$, $r = 0.1$. Find the optimal regulator. Sketch the regulation error variance as a function of signal-to-noise ratio σ^2.
b. Let $T = 0.2\,\text{msec}$, $\sigma^2 = 0.1$. Find the optimal regulator. Sketch the regulation error variance as a function of control weighting r.
c. Let $\sigma^2 = 0.1$, $r = 0.1$. Find the optimal regulator. Sketch the regulation error variance as a function of sampling period T.

7.4-7: Polynomial Control for Continuous Systems. To explore the possibility of extending our results to the design of continuous controllers, consider the following. It is desired to compute a controller for the system of Problem 7.4-6 of the form

$$u = -H_1(s)y + H_2(s)z, \tag{1}$$

where $H_1(s)$ and $H_2(s)$ are feedback and feedforward regulators respectively and $z(t)$ is a reference input. We want to use the cost index

$$J = [y(t) - z(t)]^2 + [ru(t)]^2. \tag{2}$$

Assume position measurements of the form $y = d + n$, with noise $n \sim N(0, 1)$.

a. Carry out the design if there is no process noise so that $\sigma^2 = 0$. That is, find H_1 and H_2.

b. Repeat if $\sigma^2 \neq 0$.

c. Repeat if velocity measurements are taken so that $y = v + n$.

APPENDICES

APPENDIX A

REVIEW OF MATRIX ALGEBRA

We present here a brief review of some concepts which are assumed as background for the text. Good references include Gantmacher (1977) and Brogan (1974).

A.1 BASIC DEFINITIONS AND FACTS

The *determinant* of a square $n \times n$ matrix is symbolized as $|A|$. If A and B are both square, then

$$|A| = |A^T| \tag{A.1-1}$$

$$|AB| = |A| \cdot |B|, \tag{A.1-2}$$

where superscript "T" represents transpose. If $A \in C^{m \times n}$ and $B \in C^{n \times m}$ (where n can equal m), then

$$|I_m + AB| = |I_n + BA|, \tag{A.1-3}$$

with I_n the $n \times n$ identity matrix. (C represents the complex numbers.)

For any matrices A and B,

$$(AB)^T = B^T A^T \tag{A.1-4}$$

and if A and B are nonsingular, then

$$(AB)^{-1} = B^{-1} A^{-1}. \tag{A.1-5}$$

The *Kronecker product* of two matrices $A = [a_{ij}] \in C^{m \times n}$, $B = [b_{ij}] \in C^{p \times q}$

345

is

$$A \otimes B = [a_{ij}B] \in C^{mp \times nq}. \tag{A.1-6}$$

(It is sometimes defined as $A \otimes B = [Ab_{ij}]$.) If $A = [a_1 \quad a_2 \quad \dots \quad a_n]$ where a_i are the columns of A, the *stacking operator* is defined by

$$s(A) = \begin{bmatrix} a_1 \\ a_2 \\ \vdots \\ a_n \end{bmatrix} \tag{A.1-7}$$

It converts $A \in C^{m \times n}$ into a vector $s(A) \in C^{mn}$. An identity that is often useful is

$$s(ABD) = (D^T \otimes A)s(B). \tag{A.1-8}$$

If $A \in C^{m \times m}$ and $B \in C^{p \times p}$, then

$$|A \otimes B| = |A|^p \cdot |B|^m. \tag{A.1-9}$$

See Brewer (1978) for other results.

If λ_i is an eigenvalue for A with eigenvector v_i, then $1/\lambda_i$ is an eigenvalue for A^{-1} with the same eigenvector, for

$$Av_i = \lambda_i v_i \tag{A.1-10}$$

implies that

$$\lambda_i^{-1} v_i = A^{-1} v_i. \tag{A.1-11}$$

If λ_i is an eigenvalue of A with eigenvector v_i, and μ_j is an eigenvalue of B with eigenvector w_j, then $\lambda_i \mu_j$ is an eigenvalue of $A \otimes B$ with eigenvector $v_i \otimes w_j$ (Brewer 1978).

A.2   PARTITIONED MATRICES

If

$$D = \begin{bmatrix} A_{11} & 0 & 0 \\ 0 & A_{22} & 0 \\ 0 & 0 & A_{33} \end{bmatrix} \tag{A.2-1}$$

where A_{ij} are matrices, then we write $D = \text{diag}(A_{11}, A_{22}, A_{33})$ and call D *block diagonal*. If the A_{ii} are square, then

$$|D| = |A_{11}| \cdot |A_{22}| \cdot |A_{33}| \tag{A.2-2}$$

and if $|D| \neq 0$, then

$$D^{-1} = \text{diag}(A_{11}^{-1}, A_{22}^{-1}, A_{33}^{-1}). \tag{A.2-3}$$

If

$$D = \begin{bmatrix} A_{11} & A_{12} & A_{13} \\ 0 & A_{22} & A_{23} \\ 0 & 0 & A_{33} \end{bmatrix}, \qquad (A.2\text{-}4)$$

where A_{ij} are matrices, then D is *upper block triangular* and (A.2-2) still holds. *Lower block triangular* matrices have the form of the transpose of (A.2-4).

If

$$A = \begin{bmatrix} A_{11} & A_{12} \\ A_{21} & A_{22} \end{bmatrix} \qquad (A.2\text{-}5)$$

we define the *Schur complement of A_{22}* as

$$D_{22} = A_{22} - A_{21} A_{11}^{-1} A_{12} \qquad (A.2\text{-}6)$$

and the *Schur complement of A_{11}* as

$$D_{11} = A_{11} - A_{12} A_{22}^{-1} A_{21}. \qquad (A.2\text{-}7)$$

The inverse of A can be written as

$$A^{-1} = \begin{bmatrix} A_{11}^{-1} + A_{11}^{-1} A_{12} D_{22}^{-1} A_{21} A_{11}^{-1} & -A_{11}^{-1} A_{12} D_{22}^{-1} \\ -D_{22}^{-1} A_{21} A_{11}^{-1} & D_{22}^{-1} \end{bmatrix}, \qquad (A.2\text{-}8)$$

$$A^{-1} = \begin{bmatrix} D_{11}^{-1} & -D_{11}^{-1} A_{12} A_{22}^{-1} \\ -A_{22}^{-1} A_{21} D_{11}^{-1} & A_{22}^{-1} + A_{22}^{-1} A_{21} D_{11}^{-1} A_{12} A_{22}^{-1} \end{bmatrix} \qquad (A.2\text{-}9)$$

or

$$A^{-1} = \begin{bmatrix} D_{11}^{-1} & -A_{11}^{-1} A_{12} D_{22}^{-1} \\ -A_{22}^{-1} A_{21} D_{11}^{-1} & D_{22}^{-1} \end{bmatrix} \qquad (A.2\text{-}10)$$

depending of course on whether $|A_{11}| \neq 0$, $|A_{22}| \neq 0$, or both. These can be verified by checking that $AA^{-1} = A^{-1}A = I$. By comparing these various forms, we obtain the *well-known matrix inversion lemma*

$$(A_{11}^{-1} + A_{12} A_{22} A_{21})^{-1} = A_{11} - A_{11} A_{12} (A_{21} A_{11} A_{12} + A_{22}^{-1})^{-1} A_{21} A_{11}. \qquad (A.2\text{-}11)$$

The Schur complement arises naturally in the solution of linear simultaneous equations, for if

$$\begin{bmatrix} A_{11} & A_{12} \\ A_{21} & A_{22} \end{bmatrix} \begin{bmatrix} X \\ Y \end{bmatrix} = \begin{bmatrix} 0 \\ Z \end{bmatrix} \qquad (A.2\text{-}12)$$

then from the first equation

$$X = -A_{11}^{-1} A_{12} Y,$$

and using this in the second equation yields

$$(A_{22} - A_{21} A_{11}^{-1} A_{12}) Y = Z. \tag{A.2-13}$$

If A is given by (A.2-5), then

$$|A| = |A_{11}| \cdot |A_{22} - A_{21} A_{11}^{-1} A_{12}| = |A_{22}| \cdot |A_{11} - A_{12} A_{22}^{-1} A_{21}|. \tag{A.2-14}$$

Therefore, the determinant of A is the product of the determinant of A_{11} (or A_{22}) and the determinant of the Schur complement of A_{22} (or of A_{11}).

A.3 QUADRATIC FORMS AND DEFINITENESS

If $x \in C^n$ is a vector, then the square of the Euclidean norm is

$$\|x\|^2 = x^T x. \tag{A.3-1}$$

If S is any nonsingular transformation, the vector Sx has a norm squared of $(Sx)^T Sx = x^T S^T Sx$. Letting $P = S^T S$, we write

$$\|x\|_P^2 = x^T Px \tag{A.3-2}$$

as the norm squared of Sx. We call $\|x\|_P$ the *norm of x with respect to P*.

We call

$$x^T Qx \tag{A.3-3}$$

a *quadratic form*. We shall assume Q is real.

Every real square matrix Q can be decomposed into a symmetric part Q_s (i.e., $Q_s^T = Q_s$) and an antisymmetric part Q_a (i.e., $Q_a^T = -Q_a$):

$$Q = Q_s + Q_a, \tag{A.3-4}$$

where

$$Q_s = \frac{Q + Q^T}{2} \tag{A.3-5}$$

$$Q_a = \frac{Q - Q^T}{2}. \tag{A.3-6}$$

If the quadratic form $x^T Ax$ has A antisymmetric, then it must be equal to zero since $x^T Ax$ is a scalar so that $x^T Ax = (x^T Ax)^T = -x^T A^T x$. For a general real square Q, then

$$x^T Qx = x^T(Q_s + Q_a)x = x^T Q_s x = x^T(Q + Q^T)x/2 \tag{A.3-7}$$

We can therefore assume without loss of generality that Q in (A.3-3) is symmetric. Let us do so.

We say Q is:

Positive definite $(Q > 0)$ if $x^T Qx > 0$ for all nonzero x.

Positive semidefinite $(Q \geq 0)$ if $x^T Q x \geq 0$ for all nonzero x.
Negative semidefinite $(Q \leq 0)$ if $x^T Q x \leq 0$ for all nonzero x.
Negative definite $(Q < 0)$ if $x^T Q x < 0$ for all nonzero x.
Indefinite if $x^T Q x > 0$ for some x, $x^T Q x < 0$ for other x.

We can test for definiteness independently of the vectors x. If λ_i are the eigenvalues of Q, then

$$Q > 0 \text{ if all } \lambda_i > 0$$

$$Q \geq 0 \text{ if all } \lambda_i \geq 0$$

$$Q \leq 0 \text{ if all } \lambda_i \leq 0$$

$$Q < 0 \text{ if all } \lambda_i < 0. \qquad \text{(A.3-8)}$$

Another test is provided as follows. Let $Q = [q_{ij}] \in R^{n \times n}$. The *leading minors* of Q are

$$m_1 = q_{11}, \qquad m_2 = \begin{vmatrix} q_{11} & q_{12} \\ q_{21} & q_{22} \end{vmatrix}, \qquad m_3 = \begin{vmatrix} q_{11} & q_{12} & q_{13} \\ q_{21} & q_{22} & q_{23} \\ q_{31} & q_{32} & q_{33} \end{vmatrix}, \ldots, m_n = |Q|.$$

$$\text{(A.3-9)}$$

In terms of the leading minors, we have

$$Q > 0 \text{ if } m_i > 0, \quad \text{all } i$$

$$Q \geq 0 \text{ if } m_i \geq 0, \quad \text{all } i$$

$$Q \leq 0 \text{ if } \begin{cases} m_i \leq 0, & \text{all odd } i \\ m_i \geq 0, & \text{all even } i \end{cases}$$

$$Q < 0 \text{ if } \begin{cases} m_i < 0, & \text{all odd } i \\ m_i > 0, & \text{all even } i \end{cases}. \qquad \text{(A.3-10)}$$

Any positive semidefinite matrix Q can be factored into *square roots* either as

$$Q = \sqrt{Q} \sqrt{Q}^T \qquad \text{(A.3-11)}$$

or as

$$Q = \sqrt{Q}^T \sqrt{Q}. \qquad \text{(A.3-12)}$$

The ("left" and "right") square roots in (A.3-11) and (A.3-12) are not in general the same. Indeed, Q may have several roots since each of these factorizations is not even unique. If $Q > 0$, then all square roots are nonsingular.

If $P > 0$, then (A.3-2) is a norm. If $P \geq 0$, it is called a *seminorm* since $x^T P x$ may be zero even if x is not.

A.4   MATRIX CALCULUS

Let $x = [x_1 \quad x_2 \quad \ldots \quad x_n]^T \in C^n$ be a vector, $s \in C$ be a scalar, and $f(x) \in C^m$ be an m vector function of x. The differential in x is

$$dx = \begin{bmatrix} dx_1 \\ dx_2 \\ \vdots \\ dx_m \end{bmatrix} \quad\quad\quad (A.4\text{-}1)$$

and the derivative of x with respect to s (which could be time) is

$$\frac{dx}{ds} = \begin{bmatrix} dx_1/ds \\ dx_2/ds \\ \vdots \\ dx_n/ds \end{bmatrix}. \quad\quad\quad (A.4\text{-}2)$$

If s is a function of x, then the *gradient* of s with respect to x is the *column* vector

$$s_x \triangleq \frac{\partial s}{\partial x} = \begin{bmatrix} \partial s/\partial x_1 \\ \partial s/\partial x_2 \\ \vdots \\ \partial s/\partial x_n \end{bmatrix}. \quad\quad\quad (A.4\text{-}3)$$

(The gradient is defined as a row vector in some references.) Then the total differential in s is

$$ds = \left(\frac{\partial s}{\partial x}\right)^T dx = \sum_{i=1}^{n} \frac{\partial s}{\partial x_i} dx_i. \quad\quad\quad (A.4\text{-}4)$$

If s is a function of two vectors x and y, then

$$ds = \left(\frac{\partial s}{\partial x}\right)^T dx + \left(\frac{\partial s}{\partial y}\right)^T dy. \quad\quad\quad (A.4\text{-}5)$$

The *Hessian* of s with respect to x is the second derivative

$$s_{xx} \triangleq \frac{\partial^2 s}{\partial x^2} = \left[\frac{\partial^2 s}{\partial x_i \, \partial x_j}\right], \quad\quad\quad (A.4\text{-}6)$$

which is a symmetric $n \times n$ matrix. In terms of the gradient and the Hessian the *Taylor series expansion* of $s(x)$ about x_0 is

$$s(x) = s(x_0) + \left(\frac{\partial s}{\partial x}\right)^T (x - x_0) + \frac{1}{2}(x - x_0)^T \frac{\partial^2 s}{\partial x^2}(x - x_0) + 0(2), \quad (A.4\text{-}7)$$

where $0(2)$ represents terms of order 2 and derivatives are evaluated at x_0.

The *Jacobian* of f with respect to x is the $m \times n$ matrix

$$f_x \triangleq \frac{\partial f}{\partial x} = \left[\frac{\partial f}{\partial x_1} \quad \frac{\partial f}{\partial x_2} \quad \cdots \quad \frac{\partial f}{\partial x_n} \right], \qquad (A.4\text{-}8)$$

so that the total differential of f is

$$df = \frac{\partial f}{\partial x} \, dx = \sum_{i=1}^{n} \frac{\partial f}{\partial x_i} \, dx_i. \qquad (A.4\text{-}9)$$

We shall use the shorthand notation

$$\frac{\partial f^T}{\partial x} \triangleq \left(\frac{\partial f}{\partial x} \right)^T \in C^{n \times m}. \qquad (A.4\text{-}10)$$

If y is a vector, and A, B, D, Q are matrices, all with dimensions so that the following expressions make sense, then we have the following results:

$$\frac{d}{dt}(A^{-1}) = -A^{-1}\dot{A}A^{-1}. \qquad (A.4\text{-}11)$$

Some useful gradients are:

$$\frac{\partial}{\partial x}(y^T x) = \frac{\partial}{\partial x}(x^T y) = y \qquad (A.4\text{-}12)$$

$$\frac{\partial}{\partial x}(y^T A x) = \frac{\partial}{\partial x}(x^T A y) = A y \qquad (A.4\text{-}13)$$

$$\frac{\partial}{\partial x}[y^T f(x)] = \frac{\partial}{\partial x}[f^T(x)y] = f_x^T y \qquad (A.4\text{-}14)$$

$$\frac{\partial}{\partial x}(x^T A x) = A x + A^T x, \qquad (A.4\text{-}15)$$

and if Q is symmetric, then

$$\frac{\partial}{\partial x}(x^T Q x) = 2 Q x \qquad (A.4\text{-}16)$$

$$\frac{\partial}{\partial x}(x-y)^T Q(x-y) = 2 Q(x-y). \qquad (A.4\text{-}17)$$

The chain rule for two vector functions becomes

$$\frac{\partial}{\partial x}(f^T y) = f_x^T y + y_x^T f. \qquad (A.4\text{-}18)$$

Some useful Hessians are

$$\frac{\partial^2 x^T A x}{\partial x^2} = A + A^T, \qquad (A.4\text{-}19)$$

and if Q is symmetric

$$\frac{\partial^2 x^T Q x}{\partial x^2} = 2Q \qquad\qquad \text{(A.4-20)}$$

$$\frac{\partial^2}{\partial x^2}(x - y)^T Q(x - y) = 2Q. \qquad\qquad \text{(A.4-21)}$$

Some useful Jacobians are

$$\frac{\partial}{\partial x}(Ax) = A \qquad\qquad \text{(A.4-22)}$$

[contrast this with (A.4-12)], and the chain rule (with s a scalar)

$$\frac{\partial}{\partial x}(sf) = \frac{\partial}{\partial x}(fs) = sf_x + fs_x^T \qquad\qquad \text{(A.4-23)}$$

[contrast this with (A.4-18)].

Some useful derivatives involving the trace and determinant are

$$\frac{\partial}{\partial A}\operatorname{trace}(A) = I \qquad\qquad \text{(A.4-24)}$$

$$\frac{\partial}{\partial A}\operatorname{trace}(BAD) = B^T D^T \qquad\qquad \text{(A.4-25)}$$

$$\frac{\partial}{\partial A}\operatorname{trace}(ABA^T) = 2AB \qquad\qquad \text{(A.4-26)}$$

$$\frac{\partial}{\partial A}|BAD| = |BAD|A^{-T}, \qquad\qquad \text{(A.4-27)}$$

where $A^{-T} \triangleq (A^{-1})^T$.

APPENDIX B

COMPUTER SOFTWARE

This appendix provides some FORTRAN 5 software for digital computer simulation of systems.

The only way to obtain a practical intuitive feel for the concepts in this book is to perform computer simulations. It is but a short step from simulation to the implementation of digital controllers for actual physical systems.

There are many software packages available for simulation of systems. However, the student will gain more by writing his own programs, or at least by using programs that are simple and which he can modify for his own use. It is hoped that the examples throughout the book show how easy it is to simulate optimal controllers and filters using simple routines. The driver programs in this appendix make it possible for the student to run simulations using his own subroutines.

The software presented here is suitable for instructional purposes, but is not necessarily the most efficient or numerically stable implementation. For serious applications, professionally coded algorithms should be employed; such algorithms can be obtained from:

Dr. G. J. Bierman
Factorized Estimation Applications, Inc.
7017 Deveron Ridge Road
Canoga Park, CA 91307
(818) 347-4871

See also Bierman (1977).

The author is indebted to Dr. B. L. Stevens at Lockheed-Georgia Advanced Research Institute, Lockheed-Georgia Co., Marietta, Georgia. The philosophy and the better ideas on the structure of these programs are

TABLE B.1-1   Sample Format of Program Output

Time	Function 1	Function 2	Function 3
0.0	10.000	5.000	0.000
0.02	9.998	5.010	−0.001
0.04	9.987	5.020	−0.009

his. He also pointed out the usefulness of Runge–Kutta integration as an alternative to the transition matrix approach for the simulation of continuous-time state variable systems.

B.1   PLOTTING ROUTINES

It is essential to have good plotting routines. Seeing the results of a simulation displayed as a table of numbers is one thing, but seeing graphs of the results is quite another. Without graphical displays, very little intuition can be gained from software simulation.

The output of the time response programs we are about to describe is in the format of Table B.1-1. The output file is headed by a line that contains two numbers: "number of graphs" (3 in Table B.1-1), and "number of points in each graph." For time plots, a graphics routine is required which plots columns $2, 3, \ldots$ versus column 1.

To obtain phase-plane plots, a routine is also required that plots any column versus any other column.

The graphs in this book were obtained using the plotting software available for the VERSATEC plotter on the CDC CYBER 855 at the Georgia Institute of Technology.

B.2   TIME RESPONSE OF SYSTEMS

Programs to find the time response of discrete nonlinear systems

$$x_{k+1} = f(x_k, u_k, k) \tag{B.2-1}$$

are very easy to write, since these systems are described by a simple recursive equation. Therefore, we shall not dwell on the topic.

The time response of the linear time-invariant continuous system $\dot{x} = Ax + Bu$ can be obtained by discretizing and then using the resulting recursion.

Figure B.2-1 shows a program, TRESP, which depends on an alternative approach to finding time responses for continuous systems, and which works

```
C
C      PROGRAM TO FIND TIME HISTORY
C      USES NONADAPTIVE STEP SIZE RUNGE-KUTTA

C      NEEDS SUBROUTINES:
C        F(TIME,X,XP) FOR CONTINUOUS DYNAMICS
C        D( ) FOR DISCRETE DYNAMICS


          PROGRAM TRESP
          PARAMETER (NDI=100)
          REAL IX(10),Y(0:1024,0:10),X(NDI),SX(NDI)
          CHARACTER *20 FILNAM,ANS
          COMMON/COMMAND/U(32)

10        WRITE(*,*)'NAME OF INITIAL CONDITION FILE ("N" FOR NONE)'
          READ(*,'(A)') FILNAM
          IF(FILNAM.NE.'N') THEN
                  OPEN(20,FILE=FILNAM)
                  REWIND 20
                  READ(20,*) NX,NU
                  READ(20,*) (SX(I), I= 1,NX)
                  IF(NU.GT.0) READ(20,*) (U(I), I= 1,NU)
                  REWIND 20
                  CLOSE(20)
          ELSE
                  WRITE(*,*)'HOW MANY STATES?'
                  READ(*,*) NX
                  WRITE(*,*)'ENTER INITIAL CONDITIONS'
                  READ(*,*) (SX(I), I= 1,NX)
          END IF

          WRITE(*,*)'DISCRETE FUNCTION REQUIRED? (Y OR N)'
          READ(*,'(A)') ANS
30        WRITE(*,*) 'HOW MANY STATES TO BE PLOTTED?'
          READ(*,*) MX
          IF(MX.EQ.0) GO TO 50
          WRITE(*,*)'WHICH ONES?'
          READ(*,*) (IX(I), I= 1,MX)

50        WRITE(*,*) 'RUN TIME?'
          READ(*,*) TR
          WRITE(*,*)'PRINTING TIME INTERVAL ON SCREEN?'
          READ(*,*) TPR
          WRITE(*,*)'PLOTTING TIME INTERVAL?'
          READ(*,*) TPL
          WRITE(*,*)'SAMPLE PERIOD?'
          READ(*,*) TS
          NPR= NINT(TR/TPR)
          NPL= NINT(TPR/TPL)
          NT= NINT(TPL/TS)

          TIME= 0.
          IT= 0
          IP= 0
          DO 60 I= 1,NX
60        X(I)= SX(I)
          Y(0,0)= TIME
          DO 70 I= 1,MX
70        Y(0,I)= X(IX(I))
```

FIGURE B.2-1

355

```
            DO 110 I= 1,NPR
            DO 100 J= 1,NPL
            DO 90 K= 1,NT
            IF(ANS.EQ.'Y') CALL D(IT,X)

            IF(IT.EQ.0) THEN
               WRITE(*,*)
               WRITE(*,80) (IX(IND), IND= 1,MX)
80             FORMAT(35X,'STATES'/'      TIME',10(I12))
               WRITE(*,'(11(1PE12.3))') (Y(0,IND), IND= 0,MX)
            END IF

            CALL RUNKUT(TIME,TS,X,NX)
90          IT= IT+1
            IP= IP+1
            Y(IP,0)= TIME
            DO 100 L= 1,MX
100         Y(IP,L)= X(IX(L))
110         WRITE(*,'(11(1PE12.3))') (Y(IP,L), L= 0,MX)

120         WRITE(*,130)
130         FORMAT(//2X,'ENTER 0 TO FILE ANSWERS'/8X,'1 TO QUIT',
     &      /8X,'2 TO RESTART',/8X,'3 TO PICK NEW STATES',
     &      /8X,'4 TO CHANGE TIME SCALE')
            READ(*,*) I
            GO TO (150,10,30,50) I

            WRITE(*,*)'OUTPUT FILE NAME?'
            READ(*,'(A)') FILNAM
            OPEN(20,FILE= FILNAM)
            REWIND 20
            WRITE(20,*) MX, NPR*NPL
            DO 140 I= 0,IP
140         WRITE(20,'(11(1PE14.6))') (Y(I,J), J= 0,MX)
            REWIND 20
            CLOSE (20)
            GO TO 120

150         STOP
            END




C

C    FOURTH-ORDER RUNGE-KUTTA INTEGRATION SUBROUTINE

C    REQUIRES SUBROUTINE F(TIME,X,XP) TO DESCRIBE PLANT DYNAMICS

            SUBROUTINE RUNKUT(TIME,TS,X,N)

C    TS     SAMPLE PERIOD
C    X      STATE VECTOR
C    N      NUMBER OF STATES
C    XP     DERIVATIVE OF STATE VECTOR

            PARAMETER (NDIM=32)
            REAL X(*), XP(NDIM), X1(NDIM), XP1(NDIM)
```

FIGURE B.2-1 (*continued*)

356

```
         CALL F(TIME,X,XP)
         DO 10 I= 1,N
10       X1(I)= X(I) + .5*TS*XP(I)

         TIME= TIME + .5*TS
         CALL F(TIME,X1,XP1)
         DO 20 I= 1,N
         XP(I)= XP(I) + 2.*XP1(I)
20       X1(I)= X(I) + .5*TS*XP1(I)

         CALL F(TIME,X1,XP1)
         DO 30 I= 1,N
         XP(I)= XP(I) + 2.*XP1(I)
30       X1(I)= X(I) + TS*XP1(I)

         TIME= TIME + .5*TS
         CALL F(TIME,X1,XP1)
         DO 40 I= 1,N
40       X(I)= X(I) + TS*( XP(I)+XP1(I) )/6.

         RETURN
         END
```

FIGURE B.2-1 Program to compute time response of nonlinear continuous system.

for the general nonlinear time-varying system

$$\dot{x} = f(x, u, t). \tag{B.2-2}$$

It does not require discretization. The approach uses a Runge–Kutta integrator.

The driver program TRESP requires only a subroutine F(TIME, X, XP) describing the dynamics (B.2-2), where X and XP are the state vector and its derivative. (See Fig. 3.2-6.) It also has a provision for computing a control input in subroutine D(IT, X), where IT is the iteration number. This control should be stored in the common block COMMAND for passing to F(TIME, X, XP). By using subroutine D(IT, X) to describe a controller, closed-loop systems with digital controls can be studied with TRESP. ·

For some systems, the simple Runge–Kutta integrator RUNKUT included in TRESP is not satisfactory, and a more sophisticated adaptive step size integrator may be required.

B.3 SIMULATION OF OPTIMAL CONTROLLERS

The simulation of discrete linear quadratic controllers is quite simple. See Lewis (1986).

In Fig. B.3-1, we show a driver program, CTLQR, which can be used to simulate continuous-time linear quadratic regulators. It is basically a modification of TRESP (Section B.2). For an illustration of its use, see Example 6.1-4.

CTLQR requires four user-supplied subroutines. FRIC provides the

```
C
'C      PROGRAM TO IMPLEMENT CONTINUOUS OPTIMAL CONTROL
'C      USES NONADAPTIVE STEP SIZE RUNGE-KUTTA


C      NEEDS SUBROUTINES:
C         FRIC(TIME,S,SP)     FOR RICCATI EQUATION DYNAMICS
C         FBGAIN(K,S,AK)       TO COMPUTE OPTIMAL FB GAINS
C         CONUP(K,X,AK)          TO COMPUTE OPTIMAL CONTROL
C         F(TIME,X,XP)       FOR CONTINUOUS PLANT DYNAMICS


            PROGRAM CTLQR
            PARAMETER (NX=30,NLEN=1000)
            DIMENSION IX(10),Y(0:NLEN,0:10),X(NX),SX(NX)
            DIMENSION S(NX*(NX+1)/2), AK(NX,NLEN)
            DIMENSION SS(NX,NLEN)
            CHARACTER *20 FILNAM,ANS
            EXTERNAL FRIC,F
            COMMON/COMMAND/U(NX)

10          WRITE(*,*)'NAME OF INITIAL CONDITION FILE ("N" FOR NONE)'
            READ(*,'(A)') FILNAM
            IF(FILNAM.NE.'N') THEN
                    OPEN(20,FILE=FILNAM)
                    REWIND 20
                    READ(20,*) N,M
                    READ(20,*) (SX(I), I= 1,N)
                    REWIND 20
                    CLOSE(20)
            ELSE
                    WRITE(*,*)'HOW MANY STATES, INPUTS?'
                    READ(*,*) N,M
                    WRITE(*,*)'ENTER INITIAL STATES'
                    READ(*,*) (SX(I), I= 1,N)
            END IF


30          WRITE(*,*) 'HOW MANY STATES TO BE PLOTTED?'
            READ(*,*) MX
            IF(MX.EQ.0) GO TO 50
            WRITE(*,*)'WHICH ONES?'
            READ(*,*) (IX(I), I= 1,MX)

50          WRITE(*,*) 'RUN TIME?'
            READ(*,*) TR
            WRITE(*,*)'PLOTTING TIME INTERVAL?'
            READ(*,*) TPL
            WRITE(*,*)'SAMPLE PERIOD?'
            READ(*,*) TS
            NPL= NINT(TR/TPL)
            NT= NINT(TPL/TS)
            NN= NPL*NT

            IP= 0
            DO 60 I= 1,N
60          X(I)= SX(I)

            WRITE(*,*)'OPEN-LOOP SYSTEM? (Y OR N)'
            READ(*,'(A)') ANS
```

FIGURE B.3-1

358

```
        IF(ANS.EQ.'Y') GO TO 90

        NS= N*(N+1)/2
        WRITE(*,*)'ENTER FINAL STATE WEIGHTS BY ROW'
        READ(*,*) (S(I), I= 1, NS)
        WRITE(*,*)'ENTER PLOT FILE BASE'
        READ(*,*) NFB

C    BACKWARD INTEGRATION FOR RICCATI EQUATION SOLUTION S

        DO 61 I= 1,NS
61      SS(I,NN+1)= S(I)
        DO 70 K= NN,1,-1
        CALL RUNKUT(FRIC,TIME,TS,S,NS)
        CALL FBGAIN(K,S,AK)
        DO 65 I= 1,3
65      SS(I,K)= S(I)
70      CONTINUE

C    WRITE TO FILES FOR PLOT

        REWIND (NFB+1)
        REWIND (NFB+2)
        REWIND (NFB+3)
        WRITE(NFB+1,*) NS, NN
        WRITE(NFB+2,*) N, NN-1
        WRITE(NFB+3,*) M, NN-1
        DO 75 K= 1,NN
        WRITE(NFB+1,*) K, (SS(I,K), I= 1, NS)
        WRITE(NFB+2,*) K, (AK(I,K), I= 1,N)
75      CONTINUE
        WRITE(NFB+1,*) NN+1, (SS(I,NN+1), I= 1,NS)

C    FORWARD INTEGRATION FOR SIMULATION RUN

90      TIME= 0.
        DO 100 I= 1,NPL

        Y(IP,0)= TIME
        DO 80 L= 1,MX
80      Y(IP,L)= X(IX(L))
        IP= IP+1

        DO 100 K= 1,NT
        CALL CONUP(K,X,AK)
        CALL RUNKUT(F,TIME,TS,X,N)
        IF(ANS.NE.'Y')WRITE(NFB+3,*) I*K, (U(INPR), INPR= 1,M)
100     CONTINUE

        Y(IP,0)= TIME
        DO 110 I= 1,MX
110     Y(IP,I)= X(IX(I))

120     WRITE(*,130)
130     FORMAT(//2X,'ENTER 0 TO FILE ANSWERS'/8X,'1 TO QUIT',
     &  /8X,'2 TO RESTART',/8X,'3 TO PICK NEW STATES',
     &  /8X,'4 TO CHANGE TIME SCALE')
        READ(*,*) I
        GO TO (150,10,30,50) I
```

FIGURE B.3-1 (*continued*)

```
        WRITE(*,*)'OUTPUT FILE NAME?'
        READ(*,'(A)') FILNAM
        OPEN(20,FILE= FILNAM)
        REWIND 20
        WRITE(20,*) MX, NPL
        DO 140 I= 0,IP
140     WRITE(20,'(11(1PE14.6))') (Y(I,J), J= 0,MX)
        REWIND 20
        CLOSE (20)
        GO TO 120

150     STOP
        END




C

C   FOURTH-ORDER RUNGE-KUTTA INTEGRATION SUBROUTINE

C   REQUIRES SUBROUTINE F(TIME,X,XP)

        SUBROUTINE RUNKUT(F,TIME,TS,X,N)

C   TS   SAMPLE PERIOD
C   X    STATE VECTOR
C   N    NUMBER OF STATES

C   XP   DERIVATIVE OF STATE VECTOR

        PARAMETER (NX=30)
        REAL X(*), XP(NX), X1(NX), XP1(NX)

        CALL F(TIME,X,XP)
        DO 10 I= 1,N
10      X1(I)= X(I) + .5*TS*XP(I)

        TIME= TIME + .5*TS
        CALL F(TIME,X1,XP1)
        DO 20 I= 1,N
        XP(I)= XP(I) + 2.*XP1(I)
20      X1(I)= X(I) + .5*TS*XP1(I)

        CALL F(TIME,X1,XP1)
        DO 30 I= 1,N
        XP(I)= XP(I) + 2.*XP1(I)
30      X1(I)= X(I) + TS*XP1(I)

        TIME= TIME + .5*TS
        CALL F(TIME,X1,XP1)
        DO 40 I= 1,N
40      X(I)= X(I) + TS*( XP(I)+XP1(I) )/6.

        RETURN
        END
```

FIGURE B.3-1   Program implementing continuous optimal controller.

continuous Riccati equation dynamics and FBGAIN computes and stores the Kalman gain. These subroutines are used in the backward integration phase. For the forward integration (simulation) phase, CONUP is needed to update the control input, and F is needed to provide the continuous plant dynamics.

B.4 KALMAN FILTER SIMULATION

The discrete Kalman filter is easily implemented, since all the equations are just forward recursions.

The continuous Kalman filter consists only of differential equations that must be integrated forward in time. Therefore, it can be implemented using TRESP by putting both the Riccati equation and filter dynamics in subroutine F.

In Sections 3.7 and 5.3, we discussed the implementation of a continuous Kalman filter with discrete measurements. The basic scheme is shown in Fig. 3.7-1. Program CDKAL in Fig. B.4-1 implements this *continuous-*

```
C
C    FILE CDKAL
C    PROGRAM TO IMPLEMENT CONTINUOUS-DISCRETE KALMAN FILTER
C    USES NONADAPTIVE STEP SIZE RUNGE-KUTTA


C    NEEDS SUBROUTINES:
C       F(TIME,X,XP)        FOR CONTINUOUS DYNAMICS
C       MEASUP(TIME,X,Z)    FOR DISCRETE MEASUREMENT UPDATE


         PROGRAM CDKAL
         PARAMETER (NDI=100)
         DIMENSION IX(10),Y(0:1024,0:10),X(NDI),SX(NDI)
         CHARACTER *20 FILNAM
         COMMON/U/U(32)

10       WRITE(*,*)'NAME OF INITIAL CONDITION FILE ("N" FOR NONE)'
         READ(*,'(A)') FILNAM
         IF(FILNAM.NE.'N') THEN
                 OPEN(20,FILE=FILNAM)
                 REWIND 20
                 READ(20,*) NX,NU
                 READ(20,*) (SX(I), I= 1,NX)
                 IF(NU.GT.0) READ(20,*) (U(I), I= 1,NU)
                 REWIND 20
                 CLOSE(20)
         ELSE
                 WRITE(*,*)'HOW MANY STATES?'
                 READ(*,*) NX
                 WRITE(*,*)'ENTER INITIAL STATES'
                 READ(*,*) (SX(I), I= 1,NX)
         END IF
```

FIGURE B.4-1

```
30          WRITE(*,*) 'HOW MANY STATES TO BE PLOTTED?'
            READ(*,*) MX
            IF(MX.EQ.0) GO TO 50
            WRITE(*,*)'WHICH ONES?'
            READ(*,*) (IX(I), I= 1,MX)

50          WRITE(*,*) 'RUN TIME?'
            READ(*,*) TR
            WRITE(*,*)'MEASUREMENT UPDATE INTERVAL?'
            READ(*,*) TMU
            WRITE(*,*)'PLOTTING TIME INTERVAL?'
            READ(*,*) TPL
            WRITE(*,*)'SAMPLE PERIOD?'
            READ(*,*) TS
            NMU= NINT(TR/TMU)
            NPL= NINT(TMU/TPL)
            NT= NINT(TPL/TS)

            TIME= 0.
            IP= 0
            DO 60 I= 1,NX
60          X(I)= SX(I)

            DO 100 I= 1,NMU
            DO 90 J= 1,NPL
            Y(IP,0)= TIME
            DO 80 L= 1,MX
80          Y(IP,L)= X(IX(L))
            IP= IP+1

            DO 90 K= 1,NT
90          CALL RUNKUT(TIME,TS,X,NX)
100         CALL MEASUP(TIME,X,Z)

            Y(IP,0)= TIME
            DO 110 I= 1,MX
110         Y(IP,I)= X(IX(I))

120         WRITE(*,130)
130         FORMAT(//2X,'ENTER 0 TO FILE ANSWERS'/8X,'1 TO QUIT',
     &      /8X,'2 TO RESTART',/8X,'3 TO PICK NEW STATES',
     &      /8X,'4 TO CHANGE TIME SCALE')
            READ(*,*) I
            GO TO (150,10,30,50) I

            WRITE(*,*)'OUTPUT FILE NAME?'
            READ(*,'(A)') FILNAM
            OPEN(20,FILE= FILNAM)
            REWIND 20
            WRITE(20,*) MX, NMU*NPL
            DO 140 I= 0,IP
140         WRITE(20,'(11(1PE14.6))') (Y(I,J), J= 0,MX)
            REWIND 20
            CLOSE (20)
            GO TO 120

150         STOP
            END
```

FIGURE B.4-1 Program implementing continuous-discrete Kalman filter.

discrete Kalman filter. (This program also requires subroutine RUNKUT which is in Fig. B.2-1.)

CDKAL requires two user-supplied subroutines. MEASUP provides the measurement update (i.e., Riccati equation), and F provides the continuous dynamics. These dynamics can be nonlinear, so that CDKAL can be used to implement the extended Kalman filter in Section 5.3. Samples of the subroutines MEASUP and F are given in Examples 3.7-1 and 5.3-1.

Subroutine MEASUP can be written for the particular problem, if it is not too complicated, or it can be based on one of the efficient measurement update routines in Bierman (1977).

REFERENCES

Anderson, B. D. O., "An Algebraic Solution to the Spectral Factorization Problem," *IEEE Trans. Automatic Control*, **AC-12**(4), 410–414 (August 1967).

Anderson, B. D. O., "Exponential Data Weighting in the Kalman-Bucy Filter," *Information Sci.*, **5**, 217–230 (1973).

Anderson, B. D. O. and J. B. Moore, *Optimal Filtering*, Englewood Cliffs, New Jersey: Prentice-Hall, 1979.

Åström, K. J., *Introduction to Stochastic Control Theory*, New York: Academic Press, 1970.

Åström, K. J., "Self-Tuning Regulators-Design Principles and Applications," *Applications of Adaptive Control*, K. S. Narendra and R. V. Monopoli, eds., New York: Academic Press, 1980, pp. 1–68.

Bartle, Robert G., *The Elements of Real Analysis*, New York: Wiley, Second Edition, 1976.

Bellman, R. E., *Dynamic Programming*, Princeton, New Jersey: Princeton University Press, 1957.

Bellman, R. E. and S. E. Dreyfus, *Applied Dynamic Programming*, Princeton, New Jersey: Princeton University Press, 1962.

Bellman, R. E. and R. E. Kalaba, *Dynamic Programming and Modern Control Theory*, New York: Academic Press, 1965.

Benedict, T. R. and G. W. Bordner, "Synthesis of an Optimal Set of Radar Track-While-Scan Smoothing Equations," *IRE Trans. Automatic Control*, 27–32 (July 1962).

Bierman, G. J., *Factorization Methods for Discrete Sequential Estimation*, New York: Academic Press, 1977.

Bierman, G. J., "Efficient Time Propagation of U-D Covariance Factors," *IEEE Trans. Automatic Control*, **AC-26**(4), 890–894 (August 1981).

Blakelock, J. H., *Automatic Control of Aircraft and Missiles*, New York: Wiley, 1965.

Bode, H. W. and C. E. Shannon, "A Simplified Derivation of Linear Least Square Smoothing and Prediction Theory," *Proc. IRE*, **38**, 417–425 (April 1950).

Borison, U., "Self-Tuning Regulators for a Class of Multivariable Systems," *Automatica* **15**, 209–215 (1979).

Brewer, J. W., "Kronecker Products and Matrix Calculus in System Theory," *IEEE Trans. Circuits Systems*, **CAS-25**(9), 772–781 (September 1978).

Brogan, W. L., *Modern Control Theory*, New York: Quantum, 1974.

Bryson, Jr., A. E. and Y. C. Ho, *Applied Optimal Control*, New York: Hemisphere, 1975.

Bryson, Jr., A. E. and D. E. Johansen, "Linear Filtering for Time-Varying Systems Using Measurements Containing Colored Noise," *IEEE Trans. Automatic Control*, **AC-10**(1), 4–10 (January 1965).

Bucy, R. S., "Optimal Filtering for Correlated Noise," *J. Math. Analysis Applications*, **2**, 1–8 (1967).

Businger, P. and G. H. Golub, "Linear Least Squares Solution by Householder Transformations," *Numer. Math.*, **7**, 269–276 (1965).

Casti, J., *Dynamical Systems and Their Applications: Linear Theory*, New York: Academic, 1977.

Clarke, D. W. and P. J. Gawthrop, "Self-Tuning Controller," *Proc. IEE*, **122**(9), (September 1975).

D'Appolito, J. A., "The Evaluation of Kalman Filter Designs for Multisensor Integrated Navigation Systems," The Analytic Sciences Corp., AFAL-TR-70-271, (AD 881286), January 1971.

D'Appolito, J. A. and C. E. Hutchinson, "Low Sensitivity Filters for State Estimation in the Presence of Large Parameter Uncertainties," *IEEE Trans. Automatic Control*, 310–312 (June 1969).

Darmon, C. A., "A Ship Tracking System Using a Kalman–Schmidt Filter," *Practical Aspects of Kalman Filtering Implementation*, AGARD-LS-82, Chapter 6, March 1976.

Davis, M. C., "Factoring the Spectral Matrix," *IEEE Trans. Automatic Control*, **AC-8**(4), 296–305 (October 1963).

Deyst, J. J., "A Derivation of the Optimum Continuous Linear Estimator for Systems with Correlated Measurement Noise," *AIAA J.* **7**(11), 2116–2119 (November 1969).

Dyer, P. and S. R. McReynolds, "Extension of Square Root Filtering to Include Process Noise," *J. Optimization Theory Applics.*, **3**(6), 444 (1969).

Elbert, T. F., *Estimation and Control of Systems*, New York: Van Nostrand, 1984.

Fel'dbaum, A. A., *Optimal Control Systems*, New York: Academic Press, 1965.

Franklin, G. F. and J. D. Powell, *Digital Control of Dynamic Systems*, Reading, Massachusetts: Addison-Wesley, 1980.

Fraser, D. C., "A New Technique for the Optimal Smoothing of Data," Ph.D. Thesis, Massachusetts Institute of Technology, January 1967.

Friedland, B., "Treatment of Bias in Recursive Filtering," *IEEE Trans. Automatic Control*, **AC-14**(4), 359–367 (August 1969).

Gantmacher, F. R., *The Theory of Matrices*, New York: Chelsea, 1977.

Gauss, K. G., *Theory of Motion of Heavenly Bodies*, New York: Dover, 1963.

Gawthrop, P. J., "Some Interpretations of the Self-Tuning Controller," *Proc. Inst. Elec. Eng.*, **124**(10), 889–894 (October 1977).

Gelb, A., ed., *Applied Optimal Estimation*, Cambridge, Massachusetts: MIT Press, 1974.

Huddle, J. R., "On Suboptimal Linear Filter Design," *IEEE Trans. Automatic Control*, 317–318 (June 1967).

Huddle, J. R. and D. A. Wismer, "Degradation of Linear Filter Performance Due to Modeling Error," *IEEE Trans. Automatic Control*, 421–423 (August 1968).

IMSL, *Library Contents Document*, Eighth Edition, IMSL, Houston, Texas, 1980.

Jazwinski, A. H., *Stochastic Processes and Filtering Theory*, New York: Academic Press, 1970.

Kailath, T., "A View of Three Decades of Linear Filtering Theory," *IEEE Trans. Information Theory*, **20**(2), 146–181 (March 1974).

Kailath, T., *Lectures on Wiener and Kalman Filtering*, CISM Courses and Lectures, No. 140, New York: Springer-Verlag, 1981.

Kailath, T., "An Innovations Approach to Least-Squares Estimation, Part I: Linear Filtering in Additive White Noise," *IEEE Trans. Automatic Control*, **AC-13**(6), 646–655 (December 1968).

Kailath, T., *Linear Systems*, Englewood Cliffs, New Jersey: Prentice-Hall, 1980.

Kalman, R. E., "A New Approach to Linear Filtering and Prediction Problems," *Trans. ASME J. Basic Eng.*, **82**, 34–35 (1960).

Kalman, R. E., "New Methods in Wiener Filtering Theory," *Proc. Symp. Eng. Appl. Random Function Theory Probability*, Wiley, New York, 1963.

Kalman, R. E. and R. S. Bucy, "New Results in Linear Filtering and Prediction Theory," *Trans. ASME J. Basic Eng.*, **83**, 95–108 (1961).

Kaminski, P. G., A. E. Bryson, and S. F. Schmidt, "Discrete Square Root Filtering: A Survey of Current Techniques," *IEEE Trans. Automatic Control*, **AC-16**(6), 727–736 (December 1971).

Koivo, H. N., "A Multivariable Self-Tuning Controller," *Automatica*, **16**, 351–366 (1980).

Kolmogorov, A. N., "Sur l'interpolation et Extrapolation des Suites Stationnaires," *C. R. Acad. Sci.* **208**, 2043 (1939).

Kolmogorov, A., "Interpolation and Extrapolation," *Bull. Acad. Sci.*, U.S.S.R., Ser. Math., **5**, 3–14 (1941).

Kolmogorov, A. N., "Interpolation and Extrapolation of Stationary Random Sequences," translated by W. Doyle and J. Selin, Rept. RM-3090-PR, RAND Corp., Santa Monica, CA, 1962.

Kortüm, W., "Design and Analysis of Low-Order Filters Applied to the Alignment of Inertial Platforms," *Practical Aspects of Kalman Filtering*, AGARD-LS-82, Chapter 7, 1976.

Kwakernaak, H. and R. Sivan, *Linear Optimal Control Systems*, Wiley-Interscience, New York, 1972.

Letov, A. M., "Analytical Controller Design, I, II," *Autom. Remote Control*, **21**, 303–306 (1960).

Lewis, F. L., *Optimal Control*, New York: Wiley, 1986.

Luenberger, D. G., *Introduction to Dynamic Systems*, New York: Wiley, 1979.

Luenberger, D. G., *Optimization by Vector Space Methods*, New York: Wiley, 1969.

Maybeck, P. S., *Stochastic Models, Estimation, and Control*, Vol. 1, Mathematics in Science and Engineering, Series Vol. 141-1, New York: Academic Press, 1979.

McReynolds, S. R., Ph.D. Thesis, Harvard University, Cambridge, Massachusetts, 1966.

Melsa, J. L. and D. L. Cohn, *Decision and Estimation Theory*, New York: McGraw-Hill, 1978.

Morf, M. and T. Kailath, "Square Root Algorithms for Least-Squares Estimation," *IEEE Trans. Automatic Control*, **AC-20**(4), 487–497 (August 1975).

Nash, R. A., J. A. D'Appolito, and K. J. Roy, "Error Analysis of Hybrid Inertial Navigation Systems," AIAA Guidance and Control Conference, Paper No. 72-848, Stanford, California, August 1972.

Parkus, H., *Optimal Filtering*, New York: Springer-Verlag, 1971.

Papoulis, A., *Probability, Random Variables, and Stochastic Processes*, New York: McGraw-Hill, 1965.

Potter, J. E., "Matrix Quadratic Solutions," *J. SIAM Appl. Math.*, **14**(3), 496–501 (May 1966).

Rauch, H. E., F. Tung, and C. T. Striebel, "Maximum Likelihood Estimates of Linear Dynamic Systems," *AIAA J.* **3**(8), 1445–1450 (August 1965).

Rhodes, I. B., "A Tutorial Introduction to Estimation and Filtering," *IEEE Trans. Automatic Control*, 688–706 (December 1971).

Schmidt, S. F., "Application of State-Space Methods to Navigation Problems," *Advan. Control Systems*, **3**, 293–340 (1966).

Schmidt, S. F., "Estimation of State with Acceptable Accuracy Constraints," TR 67-16, 1967, Analytical Mechanics Assoc., Palo Alto, California.

Schmidt, S. F., "Computational Techniques in Kalman Filtering," *Theory and Applications of Kalman Filtering*, Chapter 3, NATO Advisory Group for Aerospace Research and Development, AGARDograph 139, February 1970.

Schooler, S. F., "Optimal $\alpha-\beta$ Filters for Systems with Modeling Inaccuracies," *IEEE Trans. Aerospace Electronic Systems*, **AES-11**(6), 1300–1306 (November 1975).

Schultz, D. G. and J. L. Melsa, *State Functions and Linear Control Systems*, New York: McGraw-Hill, 1967.

Shaked, U., "A General Transfer Function Approach to Linear Stationary Filtering and Steady-State Optimal Control Problems," *Int. J. Control*, **24**(6), 741–770 (1976).

Simon, K. W. and A. R. Stubberud, "Reduced Order Kalman Filter," *Int. J. Control*, **10**(5), 501–509 (1969).

Sorenson, H. W., "Comparison of Kalman, Bayesian, and Maximum-Likelihood Estimation Techniques," AGARDograph 139, Chapter 6, 1970a.

Sorenson, H. W., "Least-Squares Estimation: From Gauss to Kalman," *IEEE Spectrum*, 63–68 (July 1970b).

Stear, E. B. and A. R. Stubberud, "Optimal Filtering for Gauss-Markov Noise," *Int. J. Control*, **8**(2), 123–130 (1968).

Strintzis, M. G., "A Solution to the Matrix Factorization Problem," *IEEE Trans. Inform. Theory*, **IT-18**(2), 225–232 (March 1972).

Swerling, P., "A Proposed Stagewise Differential Correction Procedure for Satellite Tracking and Prediction," Report P-1292, The Rand Corp., Santa Monica, California, January 1958.

Swerling, P., "Topics in Generalized Least Squares Signal Estimation," *J. SIAM Applied Mathematics*, **14**(5), 998–1031 (September 1966).

Swerling, Peter, "Modern State Estimation Methods from the Viewpoint of the Method of Least Squares," *IEEE Trans. Automatic Control*, **AC-16**(6), 707–719 (December 1971).

Van Trees, H. L., *Detection, Estimation, and Modulation Theory, Part 1*, New York: Wiley, 1968.

Wagner, W. E., "Re-Entry Filtering, Prediction, and Smoothing," *J. Spacecraft Rockets*, **3**, 1321–1327 (1966).

Wauer, J. C., "Practical Considerations in Implementing Kalman Filters," *Practical Aspects of Kalman Filtering Implementation*, AGARD-LS-82, Chapter 2, March 1976.

Weiss, I. M., "A Survey of Discrete Kalman–Bucy Filtering with Unknown Noise," AIAA Paper No. 70-955, AIAA Guidance, Control, and Flight Mechanics, Conference, August 1970.

Wiener, N., *The Extrapolation, Interpolation, and Smoothing of Stationary Time Series*, New York: Wiley, 1949.

Wiener, N. and L. Masani, "The Prediction Theory of Multivariable Stochastic Processes, Parts 1 and 2," *Acta Mathematics*, **98** (June 1958).

Wong, E. and J. B. Thomas, "On the Multidimensional Prediction and Filtering Problem and the Factorization of Spectral Matrices," *J. Franklin Inst.*, **272**(2), 87–99 (August 1961).

Youla, D. C. "On the Factorization of Rational Matrices," *IRE Trans. Inform. Theory*, **IT-17**(3), 172–189 (July 1961).

Yovits, M. C. and J. L. Jackson, "Linear Filter Optimization with Game Theory Considerations," *IRE National Convention Record*, Pt. 4, 193–199 (1955).

INDEX

A

Ackermann's formula, 58, 110, 177
Actual estimation error, 207, 214, 229, 236
Admissible:
  control, 305, 328
  cost, 286
Aircraft:
  angle of attack, 115
  cargo vibration isolation, 140
  gust noise, 119, 203
  lateral dynamics, 177
  longitudinal dynamics, 104, 119, 203
  sideslip angle, 177
Alpha-beta tracker, 88, 187
A posteriori:
  error covariance, 7, 106, 136
  error system, 106
  estimate, 6, 106
  mean, 7
A priori:
  error covariance, 5, 70, 206
  error system, 98, 123
  estimate, 4, 70
  mean, 5
Autocorrelation function, 34

B

Batch processing of data, 28
Bayes:
  cost, 4
  estimate, 4
  risk, 4
  rule, 6, 253
Bellman's principle of optimality, 283
  for continuous nonlinear stochastic system, 297
  for discrete nonlinear stochastic system, 287
Bilinear system, control of, 317
Block triangular matrices, 347
Butterworth spectrum, 266

C

Certainty equivalence principle, 310
Chain rule:
  for differentiation, 351
  for probability density functions, 251
Chang-Letov equation, 110, 176, 202
Chapman-Kolmogorov equation, 251, 292
Cholesky decomposition, 244

Circuit analog to maximum likelihood estimator, 23

Closed-loop:
  characteristic polynomial, 108, 174, 202, 330, 337
  poles, 105, 112, 169, 179, 199, 202, 320
  system, 70, 101, 106, 147, 155, 180, 199, 201
  system for control, 301, 314, 320, 329, 337

Colored noise, *see* Correlated noise

Complete state information for control, 305, 313

Computer simulation:
  of Gaussian noise, 304
  of Kalman filter, 157, 187, 197, 198, 265, 361
  of optimal controller, 302, 357
  preliminary analysis for, 71, 88, 91, 159, 191, 265

Conditional:
  covariance, 7, 250
  expected cost, 285, 295
  mean, 4, 7, 250
  mean-square error, 6

Consistent estimate, 25

Constant, estimation of, 75, 84, 151

Constraint:
  control, 288
  state, 288

Control:
  admissible, 305, 328, 333
  delay, 322
  minimum variance, 328
  model-following, 332
  suboptimal, 308

Controllability, *see* Reachability

Convergence, stochastic, 25

Correlated noise, 50
  measurement, 124, 183
  process, 118, 180
  process with measurement, 122, 181

Cost:
  admissible, 286
  Bayes, 4
  conditional, 285, 295
  expected, 284, 295, 305, 313, 333
  to go, 291

Covariance, 5
  a posteriori, 67, 136
  a priori, 67, 98, 206
  conditional, 7, 64
    Gaussian, 10
  error, 5, 36, 67, 146, 160, 250, 312
  propagation of, 62, 69, 146, 258
  weighted, 222

Curse of dimensionality, 295

Curvature matrix, 313

D

Damping ratio, 169

Definiteness of matrices, 348

Delay:
  control, 322
  operator, 322

Derived measurement, 125, 183, 233

Detectability, 58, 100, 172

Determinant of matrix, 345

Deyst filter, 149, 185

Diophantine equation, 324, 333

Discretization:
  of stochastic process, 81, 171
  of system, 80
  of time-varying system, 96

Divergence, 205, 214, 245

Dynamic:
  output feedback, 311, 315, 329
  programming, 283

E

Efficient estimate, 25

Error:
  conditional mean-square, 6
  correlation:
    actual, 207, 214
    predicted, 207
  covariance, 5, 7, 36, 67, 146, 160, 250, 312
  estimation, 3, 20, 36, 57, 97, 160, 170, 311
  mean-square, 3, 36, 325

modeling, 205
prediction, 325, 334
regulation, 330, 341
Error system, 57, 97, 123, 163, 170
  actual, 209, 214
  a posteriori, 106
  characteristic polynomial, 108, 174, 202
  poles, 105, 199, 202
Estimate, 3, 36
  a posteriori, 6, 67
  a priori, 4, 67
  conditional mean, 4, 7
  consistent, 25
  efficient, 24
  linear, 14
  from linear combination of data, 52
  for linear function of unknown, 50, 122
  quadratic, 18
  unbiased, 5, 98, 170
  update equation, 28, 69, 146
Estimation:
  Bayesian, 4
  error, 3, 20, 36, 57, 97, 160, 170, 259, 311
    actual, 207
    predicted, 207
  least-squares, 48
  linear mean-square, 14
  maximum a posteriori, 4
  maximum-likelihood, 19
  mean-square, 3, 250
  minimum-variance, 105, 334
  nonlinear examples, 8, 17, 20
  quadratic mean-square, 18
  recursive, 28
Euclidean norm, 3, 14, 348
Euler approximation, 83
Exponential data weighting, 222, 246
Extended Kalman filter, 263, 265
  computer simulation of, 269, 363

F

Feedback:
  of estimate, 310
  output, 311, 315, 329

state variable, 287, 292, 298, 300, 305
Fictitious process noise, 218
Field of extremals, 291, 317
Filter:
  optimal, 36, 54, 69
  performance analysis, 99, 171
  reduced-order, 213, 214, 233
  shaping, 113, 118, 160, 181
  steady-state, 37, 107, 114, 165, 174, 201
  whitening, 41, 109, 113, 175
Fisher information matrix, 24, 106
FM demodulation, 265
Fokker-Planck equation, 254
Free fall, tracking a body in, 278
Full rank stochastic process, 34

G

Gaussian:
  conditional covariance, 10
  conditional mean, 10
  conditional probability density function, 11
  measurements, linear, 11, 20, 250
  noise, computer simulation, 304
  optimal estimate, 16
  probability density function, 9
Gauss-Markov estimator, 21
Gradient vector, 350

H

Hamiltonian matrix, 201
Hamilton-Jacobi-Bellman equation, 298
Harmonic oscillator, 197, 198
  continuous Kalman filter, 157
  steady-state filter, 104, 159
  Wiener filter, 159, 202
Hessian matrix, 350
Hyperstate, 64, 98, 170, 249
  measurement update, 251, 254
  recursion for, 251
  time update, 251, 254

I

Inadmissible control, 289
Incomplete state information for control, 309, 314
Independent measurements, 22, 30
  optimal combination of, 52
Infinite horizon cost, 320, 321
Information:
  filter, 128, 136, 192
  matrix, 5, 26, 71, 106, 129, 192, 198, 225
Innovations:
  process, 41, 80, 98, 109, 171
  representation, 109, 175
Intermittent measurements, 138, 187, 264
Itô calculus, 253

J

Jacobian matrix, 261, 351
Joseph stabilized Riccati equation, 70, 98, 129, 308, 312

K

Kalman filter:
  a priori recursive formulation, 70
  computer simulation, 71, 91, 157, 187, 198, 361
  continuous, 147
  continuous-discrete, 186, 203
  derived from:
    hyperstate recursion, 252
    orthogonality principle, 79
    Wiener filter, 159
  discrete, 69
  extended, 263, 265
  information formulation, 128, 136
  innovations formulation, 109, 175
  linearized, 264
  maximum-likelihood, 71, 77
  multiple-sampling-rate, 94
  with no measurement noise, 185
  with no process noise, 71, 136, 198
  performance analysis, 99, 171, 212, 229, 236
  predictor-corrector formulation, 69
  reduced-order, 213, 214, 233
  with stability margin, 224, 246
  steady-state, 107, 114, 165, 174, 199, 204, 320
  suboptimal, 124, 213, 227, 235, 239, 247
  for weakly coupled systems, 227
Kalman gain:
  continuous dynamic, 146, 163, 182, 309
  for control, 301, 307, 314, 320
  discrete dynamic, 68, 123, 315
  static, 12
  steady-state, 101, 165, 172, 174, 218, 224, 239, 320
  suboptimal, 239, 308
Kalman-Schmidt filter, 237
Keplerian orbital elements, 269
Kinematic modeling, 234
Kirchhoff's Laws, 23
Kolmogorov equation, 254, 298
Kronecker product, 345

L

Law of large numbers, 77
Least-squares:
  estimation, 48
  normal equations, 49
Likelihood:
  equation, 19
  function, 19
Linear:
  estimate, 14, 259
  filtering, 36
  function of unknown, estimate for, 50, 122
  measurement model, 11, 20, 160, 250
  measurements, Wiener filter for, 46
  prediction, 36
  smoothing, 36
  system, effect on stochastic process, 34

Linear quadratic regulator:
  computer simulation of, 357
  continuous, 299, 305
  discrete, 313
  polynomial formulation, 318, 333
  steady-state, 308, 318, 319, 320, 321
  suboptimal, 308, 319
Line of best fit, recursive, 137
Lyapunov equation, 63, 98, 100, 143,
      149, 174, 186, 188, 198, 240, 258
  algebraic, 66, 135, 196, 301
  for control, 301, 308
  solution, 135, 196
  as vector equation, 135, 196

M

Maneuver noise, 86
Markov process:
  discrete, 62, 293
  Fokker-Planck equation for, 273
  prediction for, 339
  Wiener filter, 38, 43, 55
Markov property, 250
Matrix inversion lemma, 347
Maximum a posteriori estimation, 4
Maximum-likelihood:
  error covariance, 21
  estimation, 19
    recursive, 30, 71, 77, 198
Mean, 5
  conditional, 6, 64, 250
  propagation of, 62, 146, 257
  sample, 23, 28
Mean-square:
  error, 3, 36, 325
    conditional, 6
  error covariance, 5, 7
  estimation, 3, 250
    linear, 11
    recursive, 33
  state, 149, 300, 308, 311, 314, 316
  value of control, 301, 309, 313, 314,
      316
  value of estimate, 312, 316
Measurement, 5
  derived, 125, 183, 233

  intermittent, 138, 187, 264
  irregular, 95, 187, 264
  linear Gaussian, 11, 20, 250
  noise, 11, 38, 60
Measurement update, 67, 186, 209
  approximate, 262
  for hyperstate, 251, 254
  Joseph stabilized, 70
  linear, 259
  for nonlinear system, 258
Millman's theorem, 24, 53, 131, 193
Minimum:
  energy control problem, 285
  fuel control problem, 285
  phase system, 35, 118
  time control problem, 284
  variance control, 328
  variance estimation, 105, 326, 334
Minors, leading, 349
Model:
  definition, 206
  following control, 332
  inaccuracies, 205
  mismatch, 206
Moments, time update of, 274
Monitoring optimality of filter, 99, 171

N

Natural frequency, 169
Navigation of ship, example, 71
Negative definite matrix, 349
Newton's system, 197, 203
  continuous Kalman filter, 167
  control, 317
  discretization, 83
  inaccurate modeling of, 245
  polynomial tracker, 341
  smoothing for, 204
  steady-state control, 320
  steady-state filter, 199
Noise:
  additive, 11
  computer simulation, 304
  correlated, 118, 180. *See also* Corre-
      lated noise
  fictitious, 218

Noise (*Continued*)
  Gaussian, 11, 20
  maneuver, 86
  measurement, 11, 20, 38, 60
  multiplicative, 50
  nonwhite, *see* Correlated noise
  process, 38, 60
  white, 61
Nonlinear estimate, examples, 8, 17, 20
Nonlinear system:
  approximate updates, 261, 262
  computer simulation of, 354
  extended Kalman filter for, 263
  linear measurement update, 259
  measurement update, 258
  time update, 256
Norm, 348
Normal:
  equations, 19, 49
  random variable, *see* Gaussian

O

Observability, 58, 100
  gramian, 106, 173
  matrix, 58
  for time-varying systems, 106, 173
Observable canonical form, 180
Observer:
  deterministic, 56
  polynomial, 320
One-degree of freedom regulator, 329
Orbit estimation, 269
Orthogonality principle, 13, 16, 36,
    79, 99
Orthogonal transformation, 137
Output feedback, dynamic, 311, 315,
    329
Output injection, 58, 100

P

Parallel filtering, 88, 227
Partitioned matrix, inverse of, 347
Pendulum, EKF for, 275
Performance analysis for Kalman filter,
    99, 171, 212, 229, 236

Performance index, 284, 294, 295
  infinite horizon, 320, 321
  for polynomial systems, 328, 332
  quadratic, 285, 299, 305, 313, 332
Plant, definition, 205
Polynomial:
  cost weighting, 332
  system, 322
  tracker, 332, 338
Polynomial tracking filter, 88, 199
Positive definite matrix, 348
Predicted estimation error, 207
Prediction, 36, 54, 138, 187, 191, 264,
    324, 334
Predictive formulation of plant, 325,
    335
Predictor-corrector Kalman filter, 69
Prefiltering of data, 33, 244
Preliminary analysis for filter simplifi-
    cation, 71, 88, 91, 159, 191, 265
Prewhitening of data, 33, 41, 243
Principle of optimality, *see* Bellman's
    principle of optimality
Probability density function, 250
  chain rule for, 251
  conditional, 64, 250
  Gaussian, 9
  see also Hyperstate
Projection theorem, stochastic, 14
Propagation of mean and covariance,
    64, 69, 147, 149, 256

Q

Quadratic:
  estimation, 18
  form, 348
  performance index, *see* Performance
    index, quadratic

R

Radar tracking, 85, 187
  in x-y coordinates, 277
Random variables, sum of, 273
Rauch-Tung-Striebel smoother, 133, 195
Reachability, 100, 172

gramian, 106, 173
for time-varying systems, 106, 173
Recursive:
computation of sample mean, 28
estimation, 28
line of best fit, 137
maximum-likelihood estimate, 30
mean-square estimation, 33
processing of data, 28
Reduced-order filtering, 213, 214, 233
Reentry trajectory estimation, 276
Regulation error, 330, 341
Regulator:
linear quadratic, see Linear quadratic
regulator
one-degree of freedom, 329
two-degree of freedom, 318, 336
Representation theorem, 35
Residual, 15, 20, 32, 57, 312
covariance of, 110, 171, 174, 200, 312,
313
Kalman filter, 68, 80, 97, 109, 148,
170, 262
Riccati equation:
algebraic, 58, 99, 139, 172, 199, 201
analytic solution, 199
continuous, 146, 165
for control, 300, 307, 314
discrete, 71, 98
information formulation, 129, 136,
193
Joseph formulation, 70, 98, 129, 143,
308, 312
limiting behavior, 99, 141
periodic solutions to, 197
square-root formulations, 136
Rocket dynamics, EKF for, 276
Root locus, 46, 105, 111, 169, 176, 199,
320
Runge-Kutta integrator, 154, 185, 357

S

Sample:
mean, 23, 28, 77
variance, 29
Sampling rate, 80
multiple, 94

Satellite orbit estimation, 269
Scalar system, examples:
constant scalar unknown, 75, 84, 151
continuous Kalman filter, 152, 165
deterministic state observer, 58
discrete Kalman filter, 77
dynamic programming, 287
optimal control of, 302, 319, 321
propagation of mean and variance,
64, 149
smoothing for, 204
steady-state behavior, 64, 149, 321
suboptimal control of, 319
Wiener filter, 38, 43, 165, 319
Schur complement, 347
Seminorm, 349
Separation:
principle, 310, 315, 325
of variables, 151
Sequential, see Recursive
Shaping filter, 113, 118, 160, 181, 237
Signal-to-noise ratio, 39, 153, 159, 166,
168, 320
Smoothability, 133, 195
Smoothing, 36, 127, 191
Rauch-Tung-Striebel, 133, 195
steady-state, 37
Spectral:
densities, figures, 120, 121
density, 34, 66, 173
density matrix, 82, 149, 171, 185
factorization, 34, 107, 174
Square root:
formulation of Riccati equation, 136
of matrix, 349
Stability margin in Kalman filter, 224
Stabilizability, 100, 172
Stacking operator, 345
State-transition matrix, 96, 106, 163,
173
Statistical steady state, 66, 106, 173,
174
Steady state:
control, 301, 308, 318, 320
filter, 37, 107, 114, 165, 174, 199, 201,
204, 320
Kalman gain, 101, 165, 172, 174, 218,
224, 320
smoothing, 37

Steady state (*Continued*)
  statistical, 66, 106, 173, 174
Stochastic:
  convergence, 25
  observability, 106, 173
  process:
    effect of linear system on, 34
    full rank, 34
    reachability, 106, 173
    steady-state, 66, 106, 173, 174
    system, 61, 135, 173, 197, 240
Suboptimal filter, 124, 213, 227, 235,
    239, 247
Sum of random variables, 273
Symmetric part of matrix, 348

T

Target, 288, 294
Taylor series, 261, 297, 350
Time update, 67, 186, 208
  approximate, 261
  for hyperstate, 251, 254
  for nonlinear system, 256
  for scalar function, 257
Time-varying system:
  discretization of, 96
  filtering for, 69, 139, 147, 186, 202
  steady-state filter for, 106, 173
Total differential, 350
Tracker, for polynomial systems, 332,
    338
Transition probability, 292, 293
Truth model, 206
Two-degree of freedom regulator, 318,
    336

U

Unbiased estimate, 5, 52, 98, 170, 259
Unmodeled:
  dynamics, 86, 213, 218
  unstable states, 215
Update equation:
  error covariance, 31
  estimate, 28, 32
  measurement, 67, 251, 254, 258, 259
  time, 67, 149, 251, 254, 256

V

Variance, sample, 29

W

Weakly coupled systems, 227
Whitening:
  filter, 33, 41, 109, 113, 166, 175
  property, 41
White noise, 61
Wiener filter, 3, 33, 176, 199, 240, 319
  causal, 40
  derived from Kalman filter, 113
  for linear measurements, 46
  noncausal, 36
  with prediction, 54
Wiener-Hopf equation, 36, 159
Wiener process:
  filtering for, 156
  Fokker-Planck equation for, 254
  inaccurate modeling of, 210, 244
  suboptimal filter for, 240